THE STRUCTURE OF THE QUIET PHOTOSPHERE AND THE LOW CHROMOSPHERE

THE STRUCTURE OF THE QUIET PHOTOSPHERE AND THE LOW CHROMOSPHERE

PROCEEDINGS OF THE 'BILDERBERG' CONFERENCE
HELD NEAR ARNHEM, HOLLAND, APRIL 17–21, 1967

Edited by C. de Jager

D. REIDEL PUBLISHING COMPANY

DORDRECHT-HOLLAND

ISBN-13: 978-94-010-3472-2 e-ISBN-13: 978-94-010-3470-8
DOI: 10.1007/978-94-010-3470-8

Softcover reprint of the hardcover 1st edition 1968

PREFACE

From 17 to 21 April 1967 a Study Week was held in the hotel 'De Bilderberg' near Arnhem, Holland, with the purpose to establish a new, and if possible, generally acceptable working model for the quiet parts of the solar photosphere and low chromosphere.

The organizers of the conference hoped that even if this latter goal appeared too far to be reached, such a meeting would still be useful, if only for enumerating the crucial problems in solar photospheric research, and for defining future subjects of research. About twenty solar physicists from outside the Netherlands participated in the Study Week, while some others, though prevented from actively attending, submitted their comments before the meeting.

The two above-mentioned goals were reached: a working model *could* be established; yet it became clear that not everyone would agree about this model, and it became obvious too that future research is strongly needed, in particular in the field of line formation (coherence, or non-coherence; local thermal equilibrium), while also the motion field of the photosphere and chromosphere is insufficiently known, and its influence on the formation of spectral lines hardly understood.

After the Study Week the preliminary results were circulated and provisionally presented in a meeting of Commission 12 of the International Astronomical Union in Prague. This preprint circulation offered one the opportunity to modify the model in certain respects, while it also provoked some post-Bilderberg discussions, the results of which may also be found in the present proceedings of the Study Week, reprinted from *Solar Physics* **3** (1968), No. 1.

We gladly mention that the Study Week has been made possible by a grant of the Leids Kerkhoven-Bosscha Fonds. We remember with thankfulness the efficient help in the practical organization by Miss C. E. Boot and Mr. J. B. Vogel.

C. DE JAGER

First row (left to right): J.-C. Pecker; J.T. Jefferies; Carla E. Boot; Marijke Burger; R.W. Noyes; Edith A. Müller; R.M. Bonnet; Simone Dumont; D. L. Lambert; Yvette Cuny. *Second row (left to right)*: J. E. Blamont; J. Houtgast; R.G. Athay; Christiane Grevesse-Guillaume; N. Grevesse; O. Namba; C. de Jager. *Third row (left to right)*: J.P. Mutschlecner; O. Gingerich; G. Withbroe; M.G.J. Minnaert; T. de Groot; H. Holweger; A. Sauval; H. Vesters; P. Souffrin; H. Hubenet; M. Herse; R. J. Rutten; P. J. Léna; A. Skumanich; P. Delache; J. B. Vogel.

TABLE OF CONTENTS

THE BILDERBERG MODEL OF THE PHOTOSPHERE AND LOW CHROMOSPHERE

O. GINGERICH

Smithsonian Astrophysical Observatory, Cambridge, Mass., U.S.A.

and

C. DE JAGER

University Observatory 'Sonnenborgh' and Space Research Laboratory, Utrecht

(Received 25 September, 1967)

Abstract. From 17 through 21 April 1967, an international study week was held in the 'Bilderberg' near Arnhem, Netherlands, with the aim of obtaining an internationally acceptable model of the solar photosphere and low chromosphere. It was found that such a model, based on observed intensities and center-to-limb observations of the solar continuous spectrum, could indeed be established. This model, henceforth called the Bilderberg Continuum Atmosphere (BCA), is shown in Table I, which gives the temperature, gas and electron pressures, and other data as functions of the continuous optical depth at 5000 Å between $\tau_{5000} = 10^{-7}$ and 25. The model is characterized by a flat temperature minimum of 4600°K between $\tau_{5000} \approx 10^{-2}$ to 10^{-4}. The model is homogeneous, and in hydrostatic equilibrium. A hydrogen-helium ratio of 10 has been assumed.

Much divergence remains in the interpretation of line-profile observations with regard to the establishment of a photospheric model (Section 4). It proved to be as yet impossible to obtain reliable information on the variation with depth of the following functions: temperature fluctuations, turbulence velocities, convective velocities, and vibrational velocity amplitudes (Section 5). Provisionally, it is assumed that $v_{\text{macro}} = 2$ km/sec and $v_{\text{micro}} = 1$ km/sec, isotropic and independent of depth.

1. Introduction

In the course of the last twenty years, many models of the solar photosphere have been constructed. These have been used with more or less success in predicting certain aspects of the solar spectrum or for interpreting observations. Gradually it became obvious that a generally acceptable model of the solar photosphere and low chromosphere that could be used as a reference model by all the workers in the field might be useful.

It was with a view to this need that the so-called Utrecht Reference Model of the Photosphere and Low Chromosphere (URP; HEINTZE, HUBENET, and DE JAGER, 1964) was established. It fulfilled this requirement in the sense that it has been used by several solar physicists for the intercomparison of results of their computations based on their own photospheric model with the URP model; this proved to be valuable.

However, after three years of service, it became obvious that the URP model needed revision in several respects. To arrive at a model acceptable for many solar photospheric workers, and in particular to take account of the heterogeneous mass of observations of various kinds and their implications for our knowledge of solar structure, a study week was convened for the construction of a new and possibly internationally acceptable model of the photosphere and low chromosphere. This study week was organized by the Utrecht Observatory and held at the Hotel 'De Bilderberg' near Arnhem, Netherlands, from 17 through 21 April 1967.

2. Program of the Meeting

The general schedule of the meeting was as follows:

In a period of about five months before the meeting, many participants formulated their opinions about the solar model. These contributions, in the form of some 30 preprints, were distributed prior to the meeting.

In the first two days of the conference, the participants discussed their preprints for 10 to 15 minutes, and general discussions followed. On the third day, the results of the previous sessions were reviewed and a first attempt was made to synthesize the various contributions into a model of the photosphere.

On the fourth day, some general problems of methodology were discussed; in the afternoon, three working groups were convened to discuss:

(a) the establishment of a photospheric model based on continuum measurements;

(b) the establishment of a photospheric model based on observations of line profiles and equivalent widths;

(c) the variation with optical depth of the following functions: temperature fluctuations, turbulent velocities, convective velocities, and vibrational velocities.

On the fifth day, the three chairmen of these groups reported, and a model of the photosphere was established. This model appeared essentially to be based on the continuum observations and hence was named the Bilderberg Continuum Atmosphere (BCA).

The detailed program of the study week is given in Addendum I. The composition of the three working groups is shown in Addendum II.

3. The Bilderberg Continuum Atmosphere

In this and the two following sections, we briefly summarize some of the main conclusions of the three working groups. The establishment of a solar photospheric model based on continuum observations proved feasible, and, in fact, the participants were surprised to see how well the observations of the various parts of the solar spectrum, including the ultraviolet and infrared ranges, agreed in producing one uniform model of the solar photosphere. This model is outlined in detail here. A brief summary of the results from line-profile observations is given in Section 4; in Section 5, a summary of the meeting's opinions on the temperature and pressure variations and on the velocity fields in the photosphere will be given.

To determine empirically the run of temperature in the sun, we must examine the emergent intensities that arise from various depths. This is accomplished in three ways:

(a) by investigating absolute intensities over an extended spectral range, exploiting changes in opacity to cause different effective depths of formation;

(b) by measuring the intensities close to the extreme limb of the sun;

(c) by studying the profiles of strong lines where great differences in the absorption take place within a few angströms.

Although the first method permits the simplest interpretation (provided the theoretical absorption coefficients are linked without systematic error throughout the wavelength range), the ordinary visual continuum does not provide much leverage on the solar temperature distribution, because the opacity changes very little between 4000 Å and 10000 Å. However, in both the micron and millimeter regions, as well as in the ultraviolet, the absorption increases enormously and the emergent radiation originates in higher layers of the sun's atmosphere.

Until recently, there were too few observations to sketch a systematic picture of the region of the solar temperature minimum and the low chromosphere by this method, but in the past few years more observations from this extended wavelength range have become available, and their significance for a solar model has been more clearly appreciated. As a result, a number of new measurements were collected and reported for the first time at the Bilderberg Conference and incorporated in the BCA. This new empirical solar model defines the temperature throughout twice as many orders of magnitude in optical depth as the previous URP.

Although the observations considered fall into place with remarkable consistency, solar physicists should be aware that conflicting results from line-profile analyses were completely ignored in the formulation of this model. Furthermore, the agreement of the far-ultraviolet and infrared observations may point more toward the meagerness of the data than to the certainty of our deduced temperature distribution.

Before the particular observations and the temperature regions they define are discussed, some general remarks about the different spectral regions are in order.

In the far ultraviolet, calibrations rest primarily on the rocket results of DETWILER *et al.* (1961). The ultraviolet opacities, although now much better understood through the work of GINGERICH and RICH (1966, 1968), are still comparatively poorly known. Nevertheless, they point to the spectral region between 1525 Å and 1683 Å as arising from the solar temperature minimum, and since the minimum region is apparently rather broad, even large errors in the opacity will not appreciably affect the emergent intensity. Furthermore, because the Planck function is extremely sensitive to temperature in this wavelength region, an order-of-magnitude error in the absolute intensity calibration will make only a 400° difference in the emergent brightness temperature.

In the visible spectrum violetward of 4500 Å, marked discrepancies still exist between the observed intensities and the model-atmosphere predictions, as reported to the Bilderberg Conference by CARBON, GINGERICH, and KURUCZ (1968). In the absence of any truly convincing explanation for this disagreement, it is best to omit these wavelengths in constructing an empirical solar temperature distribution. This near-ultraviolet region is increasingly blanketed by absorption lines, but these seem to account only partially for the discrepancy.

The wavelength region extending from about 5000 Å onward into the infrared provides several distinct advantages for determining the solar temperature distribution. The comparative paucity of absorption lines eliminates the difficulty of establishing the true continuum. The opacity is well known, arising in the infrared regions from the free-free absorption of negative hydrogen ions. Finally, at least some of the non-LTE effects haunting the ultraviolet regions are absent here. Undoubtedly, future models will place increasing reliance on this spectral region as more observations become available.

The run of temperatures in the BCA can be roughly divided into three sections: the *photosphere*, below $\log \tau_{5000} = -1$, the *temperature minimum* from $\log \tau_{5000} = -2$ to -4, and the *low-chromosphere* temperature rise above $\log \tau_{5000} = -4$.

3.1. The Photosphere

A great wealth of limb-darkening data from the visible regions of the spectrum delineates the shape of the temperature distribution between optical depth 0.1 and unity. The models of Sauval (1968), Holweger (1967), Mutschlecner (1968), Lambert (1967), and Elste (as presented by Withbroe, 1968) differ from each other or from the URP by less than 200°. Much of the existing disagreement derives from the various values adopted for the absolute intensities of the continuum at various wavelengths. It is clear that not all the values in the literature can be taken with equal weight, and it should be noted that averages including the original, high results of Labs (such as those given in Allen's Astrophysical Quantities) are less trustworthy than the more recent values given by Labs and Neckel (1962).

Some concept of the internal accuracy of measurements can be formed by comparing the temperatures derived from different wavelengths. For example, the results of Sauval and Holweger show that either the opacities or their adopted continuum intensities are incorrect in parts of the near-infrared. Since any opacity errors would probably show their most conspicuous anomalies in the relatively transparent region of the H^- absorption minimum at 16400 Å, and since the results of Sauval and Holweger were quite smooth in this particular spectrum region, we conclude that their adopted intensities beyond 20000 Å contain errors. The work of Lambert, which was based primarily on the absolute intensities of Labs and Neckel and the relative intensities of Pierce scaled to the intensities of Labs and Neckel, did not show any such anomalies in the infrared. Because the model of Lambert not only appeared internally consistent but closely resembled the model of Mutschlecner and the middle column of Elste's three-column model, his model was adopted for the region $\log \tau_{5000} = -1$ to 0. At the same time, the value for $I_{5000}(0,0)$ of 4.05×10^{14} ergs cm^{-3} sec^{-1} ster^{-1} was adopted from the work of Labs and Neckel.

The temperature distribution below optical depth unity is rather poorly determined by the observations, since there is no particularly transparent region of the spectrum that permits unambiguous observations of these depths. Models constructed with the mixing-length theory of convection give some guidance to plotting the temperature in this vicinity, and since the URP already reflected these convective trends, the temperatures at greater depth were simply scaled from that model. In retrospect, it appears that the temperatures below $\tau_{5000} = 2$ were pressed too low. It is perhaps this decision that has produced the lower predicted $I_{5000}(0,0)$ of 3.91×10^{14} ergs cm^{-3} sec^{-1} ster^{-1} in the BCA rather than the desired value of 4.05×10^{14}.

3.2. The temperature minimum

The value of 4600° adopted for the temperature minimum in the BCA rests on two lines of evidence. First, this is close to the color temperature found by Tousey (1963) and his colleagues for 1600 Å; the reasons for believing that radiation from this wavelength region emerges from the temperature minimum have been given by Gingerich and Rich (1968). Second, the work of Rich (1966) on the carbon monoxide spectrum in the solar ultraviolet presents a consistent picture if the lowest temperatures lie in the vicinity of 4600°. Because the carbon monoxide is formed preferentially and with great sensitivity on the high-pressure side of the temperature minimum, it provides a delicate indicator. Should a minimum of 4400° have been adopted, the predicted carbon monoxide spectrum would have been much too strong.

With the temperature placed at 4600°, it was easier to draw a smooth transition both to the deeper photosphere layers and to the overlying chromosphere than if the minimum were 4400°.

Limb-darkening measurements both in the infrared and in the ultraviolet provide evidence for the broadness of the temperature minimum. The observations by Léna (1968) at 20.4 and 24.2 μ and by Noyes, Beckers, and Low (1968) at 22.5 and 24.3 μ show a center-to-limb intensity variation going from definite darkening at 20 μ to an essentially flat distribution at 24 μ. This is in distinct contradiction to the URP, which predicts definite limb brightening because its temperature-minimum region is too narrow. The BCA model, on the other hand, yields infrared predictions in satisfactory agreement with the observations. Unfortunately, these infrared measurements are as yet only relative, and consequently, they offer no information about the actual temperature in the minimum region.

In the ultraviolet, the spectral region between 1525 Å and 1683 Å shows neither darkening nor brightening toward the limb. As pointed out in the paper by Gingerich and Rich, the URP predicts considerable limb brightening throughout this region, whereas the BCA is much more satisfactory in this regard.

The transition between the photosphere and the temperature minimum (i.e., the region between -1 and -2 in log τ_{5000}) is comparatively poorly determined. The URP reaches a minimum value of 4500° at an optical depth of 0.025, whereas the models of Lambert and Mutschlecner show at the same point a temperature of about 4800°. The evidence for a comparatively sharp dip in this region comes largely from intensity measurements at the extreme limb made during eclipses (Heintze, 1965). Such measurements inevitably carry a rather low degree of confidence because of the difficulties in correcting for the geometry of the eclipse and because of the rough and spiculed nature of the sun's limb. It would be far preferable to determine the temperatures in this transition region from absolute intensities and limb-darkening measurements in some suitably opaque spectral region, for example, from 1900 Å to 2800 Å. The BCA satisfactorily represents the central intensities determined by Bonnet and

BLAMONT (1968), but it has not been possible to achieve a satisfactory interpretation of their limb-darkening observations. Consequently, in the BCA the temperature distribution in this transition region is simply a smooth curve connecting the points at $\log \tau_{5000} = -1$ and -2.

3.3. THE LOW CHROMOSPHERE

Absolute-intensity measurements from the millimeter region as compiled by NOYES, BECKERS, and LOW (1968) and from the far ultraviolet as interpreted by SAUVAL (1968) established the temperature rise from $\log \tau_{5000} = -5$ to -6. The results from these two extremely diverse spectral regions showed such remarkably good agreement as to border on the unbelievable. If the interpretation of these observations is correct, then inhomogeneities must have comparatively little influence on the model in this region – otherwise, the temperatures derived from the ultraviolet should have been higher than those derived from the infrared because of the non-linear increase in the Planck function for high frequencies.

Given the run of temperatures established for the vicinity of the low chromosphere, an interpolation between this region and that of the temperature minimum was comparatively unambiguous. For shallower layers, however, the millimeter-region observations implied a steeper rise in temperature. Although these temperatures are not well determined, we decided to specify them in the BCA as a warning that the temperature distribution between $\log \tau_{5000} = -5$ and -6 should not simply be extrapolated outward.

3.4. NOTES ON THE COMPUTATION OF THE MODEL

The model is homogeneous, and is in hydrostatic equilibrium. The abundances are those of GOLDBERG, MÜLLER, and ALLER (1960), except that the ratio of hydrogen to helium by number is 10. The electron contributions of six elements (Si, Mg, Fe, C, Na, and Al) have been considered explicitly; as can be seen in Table I, these elements contribute over 80% of the electrons throughout the intermediate regions of this model. It is necessary to perform detailed calculations of the ionization of these elements that contribute significantly to the electron pressure, because throughout the intermediate regions of the model the negative hydrogen ion is the predominant opacity source at 5000 Å. Our studies show that the six elements considered explicitly constitute a sufficiently extensive set.

Tables I to IV have been calculated on the Control Data Corporation 6400 computer at the Smithsonian Astrophysical Observatory with the program described by GINGERICH (1963). In Table I, the monochromatic opacity at 5000 Å is tabulated. The sources of opacity used in the calculation are neutral H, H^-, H_2^+, metals, and Rayleigh and electron scattering. The abundances of Si, Mg, and Al affect the opacity, and they play a very important role in the ultraviolet rather than at 5000 Å. The treatment of the opacities has been described by GINGERICH (1964) and GINGERICH and RICH (1966).

The optical depth/geometrical depth relation has been independently calculated by Linsky, with the conclusion that this integration is quite stable, even at shallow optical

TABLE I

BILDERBERG CONTINUUM ATMOSPHERE

T EFF = 5780, LOG G = 4.440, HELIUM/HYDROGEN BY NUMBER= .100, ABUNDANCES G-M-A

OPTICAL DEPTH	TEMP (K)	PRESSURE (CGS)	ELECTRON PRESSURE	OPACITY (PER GM)	HYDROGEN IONIZED	METAL TOTAL	ELECTRON CONTRIBUTORS SI	MG	FE	C	DENSITY (GM/CC)	GRADIENT RADIATIVE	ADIABATIC	DEPTH (CM)
.0000001	9500.0	1.922E-02	9.204E-03	2.87E-01	1.00E+00	.1	.0	.0	.0	.1	1.63E-14	-1.41E+14	1.90E+10	-2.21E+08
.0000001	8780.0	2.173E-02	1.035E-02	2.84E-01	1.00E+00	.1	.0	.0	.0	.1	2.01E-14	-2.25E+10	1.55E+10	-2.16E+08
.0000002	8220.0	2.489E-02	1.184E-02	2.84E-01	9.99E-01	.1	.0	.0	.0	.1	2.46E-14	-1.44E+10	1.20E+10	-2.10E+08
.0000002	7740.0	2.887E-02	1.371E-02	2.84E-01	9.94E-01	.1	.0	.0	.0	.1	3.03E-14	-1.01E+10	7.84E+09	-2.05E+08
.0000003	7300.0	3.392E-02	1.594E-02	2.79E-01	9.75E-01	.1	.0	.0	.0	.1	3.81E-14	-7.33E+09	3.52E+09	-2.00E+08
.0000003	6900.0	4.055E-02	1.819E-02	2.57E-01	8.95E-01	.1	.0	.0	.0	.1	5.02E-14	-5.21E+09	1.44E+09	-1.94E+08
.0000004	6550.0	5.044E-02	1.920E-02	1.96E-01	6.76E-01	.1	.0	.0	.0	.1	7.38E-14	-3.59E+09	9.83E+08	-1.88E+08
.0000005	6250.0	6.960E-02	1.801E-02	1.12E-01	3.84E-01	.2	.0	.0	.0	.1	1.28E-13	-2.38E+09	1.08E+09	-1.81E+08
.0000006	6010.0	1.167E-01	1.662E-02	5.38E-02	1.82E-01	.3	.0	.0	.0	.3	2.58E-13	-1.48E+09	1.35E+09	-1.72E+08
.0000008	5830.0	2.437E-01	1.705E-02	2.49E-02	8.22E-02	.7	.0	.0	.0	.6	6.02E-13	-8.67E+08	1.63E+09	-1.61E+08
.0000010	5700.0	5.807E-01	1.968E-02	1.22E-02	3.81E-02	1.3	.1	.1	.0	1.2	1.52E-12	-4.90E+08	1.82E+09	-1.48E+08
.0000013	5610.0	1.387E+00	2.444E-02	6.85E-03	1.93E-02	2.4	.2	.1	.0	2.0	3.76E-12	-3.43E+08	1.85E+09	-1.36E+08
.0000016	5500.0	3.153E+00	2.763E-02	4.02E-03	9.34E-03	4.1	.3	.3	.0	3.4	8.80E-12	-2.48E+08	1.94E+09	-1.26E+08
.0000020	5435.0	6.485E+00	3.332E-02	2.99E-03	5.35E-03	5.8	.6	.4	.1	4.7	1.84E-11	-1.42E+08	1.59E+09	-1.16E+08
.0000025	5370.0	1.176E+01	3.754E-02	2.49E-03	3.25E-03	7.7	.9	.7	.1	5.9	3.38E-11	-1.13E+08	1.24E+09	-1.09E+08
.0000032	5305.0	1.940E+01	4.018E-02	2.24E-03	2.06E-03	9.9	1.4	1.1	.2	7.1	5.65E-11	-8.97E+07	9.45E+08	-1.02E+08
.0000040	5240.0	2.987E+01	4.142E-02	2.09E-03	1.34E-03	12.3	2.1	1.6	.2	8.1	8.81E-11	-7.12E+07	7.10E+08	-9.71E+07
.0000050	5175.0	4.376E+01	4.154E-02	2.01E-03	8.88E-04	15.1	3.0	2.4	.4	9.0	1.31E-10	-5.46E+07	5.33E+08	-9.24E+07
.0000063	5115.0	6.177E+01	4.148E-02	1.97E-03	6.04E-04	18.4	4.3	3.4	.5	9.6	1.87E-10	-4.15E+07	3.98E+08	-8.83E+07
.0000079	5055.0	8.476E+01	4.087E-02	1.95E-03	4.13E-04	22.3	5.9	4.7	.7	10.1	2.59E-10	-3.17E+07	3.01E+08	-8.45E+07
.0000100	5000.0	1.138E+02	4.055E-02	1.96E-03	2.87E-04	26.8	8.0	6.4	.9	10.4	3.52E-10	-2.30E+07	2.27E+08	-8.10E+07
.0000126	4950.0	1.500E+02	4.066E-02	1.99E-03	2.03E-04	32.0	10.5	8.4	1.2	10.4	4.69E-10	-1.58E+07	1.71E+08	-7.78E+07
.0000158	4910.0	1.943E+02	4.190E-02	2.06E-03	1.49E-04	37.3	13.1	10.6	1.6	10.2	6.12E-10	-1.10E+07	1.28E+08	-7.48E+07
.0000200	4870.0	2.479E+02	4.316E-02	2.15E-03	1.09E-04	43.2	16.2	13.1	1.9	9.8	7.88E-10	-7.78E+06	9.66E+07	-7.20E+07
.0000251	4840.0	3.121E+02	4.578E-02	2.28E-03	8.26E-05	48.8	19.2	15.5	2.3	9.2	9.98E-10	-5.21E+06	7.24E+07	-6.94E+07
.0000316	4810.0	3.880E+02	4.865E-02	2.44E-03	6.25E-05	54.7	22.3	18.2	2.7	8.5	1.25E-09	-4.14E+06	5.46E+07	-6.69E+07
.0000398	4780.0	4.773E+02	5.191E-02	2.61E-03	4.69E-05	60.8	25.6	20.9	3.1	7.6	1.55E-09	-3.29E+06	4.14E+07	-6.46E+07
.0000501	4750.0	5.817E+02	5.572E-02	2.82E-03	3.49E-05	66.9	28.9	23.7	3.5	6.7	1.90E-09	-2.61E+06	3.14E+07	-6.24E+07
.0000631	4720.0	7.029E+02	6.022E-02	3.08E-03	2.58E-05	72.7	32.0	26.5	3.9	5.8	2.31E-09	-1.77E+06	2.38E+07	-6.03E+07
.0000794	4700.0	8.419E+02	6.660E-02	3.40E-03	2.00E-05	77.0	34.3	28.7	4.2	5.0	2.77E-09	-1.10E+06	1.79E+07	-5.83E+07
.0001000	4680.0	9.999E+02	7.382E-02	3.77E-03	1.55E-05	81.0	36.3	30.7	4.5	4.2	3.31E-09	-7.76E+05	1.36E+07	-5.64E+07
.0001259	4665.0	1.179E+03	8.249E-02	4.20E-03	1.23E-05	84.0	37.7	32.4	4.7	3.6	3.91E-09	-5.20E+05	1.03E+07	-5.46E+07
.0001585	4650.0	1.381E+03	9.215E-02	4.69E-03	9.81E-06	86.7	38.9	33.9	5.0	3.1	4.60E-09	-3.52E+05	7.87E+06	-5.29E+07
.0001995	4640.0	1.608E+03	1.035E-01	5.25E-03	8.07E-06	88.6	39.7	35.1	5.1	2.7	5.36E-09	-2.19E+05	6.03E+06	-5.12E+07
.0002512	4630.0	1.864E+03	1.160E-01	5.87E-03	6.66E-06	90.3	40.2	36.2	5.3	2.3	6.23E-09	-1.74E+05	4.65E+06	-4.96E+07
.0003162	4620.0	2.152E+03	1.298E-01	6.57E-03	5.50E-06	91.7	40.6	37.3	5.4	2.0	7.21E-09	-1.38E+05	3.59E+06	-4.81E+07
.0003981	4610.0	2.476E+03	1.450E-01	7.34E-03	4.54E-06	92.9	40.8	38.3	5.5	1.7	8.32E-09	-8.53E+04	2.79E+06	-4.65E+07
.0005012	4605.0	2.841E+03	1.625E-01	8.19E-03	3.90E-06	93.8	40.7	39.1	5.6	1.5	9.55E-09	-4.35E+04	2.18E+06	-4.51E+07
.0006310	4600.0	3.254E+03	1.818E-01	9.13E-03	3.35E-06	94.6	40.6	40.0	5.7	1.3	1.10E-08	-1.93E+04	1.70E+06	-4.36E+07
.0007943	4600.0	3.720E+03	2.039E-01	1.02E-02	2.98E-06	95.0	40.3	40.6	5.8	1.2	1.25E-08	0.	1.34E+06	-4.21E+07

Table I (continued)

OPTICAL DEPTH	TEMP (K)	PRESSURE (CGS)	ELECTRON PRESSURE	OPACITY (PER GM)	HYDROGEN IONIZED	METAL ELECTRON CONTRIBUTORS TOTAL	SI	MG	FE	C	DENSITY (GM/CC)	GRADIENT RADIATIVE	GRADIENT ADIABATIC	DEPTH (CM)
.0010000	4600.0	4.247E+03	2.283E-01	1.13E-02	2.66E-06	95.5	40.0	41.3	5.9	1.1	1.43E-08	0.	1.05E+06	-4.07E+07
.0012589	4600.0	4.844E+03	2.554E-01	1.26E-02	2.38E-06	95.9	39.6	42.0	6.0	1.0	1.63E-08	0.	8.31E+05	-3.93E+07
.0015849	4600.0	5.520E+03	2.853E-01	1.40E-02	2.13E-06	96.2	39.1	42.6	6.0	.9	1.86E-08	0.	6.56E+05	-3.79E+07
.0019953	4600.0	6.286E+03	3.184E-01	1.55E-02	1.91E-06	96.6	38.6	43.4	6.1	.8	2.12E-08	0.	5.19E+05	-3.65E+07
.0025119	4600.0	7.156E+03	3.550E-01	1.72E-02	1.71E-06	96.9	38.0	44.1	6.2	.8	2.41E-08	0.	4.11E+05	-3.51E+07
.0031623	4600.0	8.143E+03	3.954E-01	1.91E-02	1.54E-06	97.1	37.4	44.8	6.3	.7	2.74E-08	0.	3.26E+05	-3.37E+07
.0039811	4600.0	9.264E+03	4.402E-01	2.11E-02	1.38E-06	97.4	36.7	45.6	6.3	.7	3.12E-08	0.	2.59E+05	-3.23E+07
.0050119	4600.0	1.054E+04	4.896E-01	2.34E-02	1.24E-06	97.6	35.9	46.3	6.4	.6	3.55E-08	0.	2.05E+05	-3.09E+07
.0063096	4600.0	1.199E+04	5.441E-01	2.59E-02	1.12E-06	97.8	35.1	47.1	6.4	.6	4.04E-08	1.53E+03	1.63E+05	-2.95E+07
.0079433	4605.0	1.363E+04	6.074E-01	2.87E-02	1.04E-06	97.9	34.4	47.7	6.5	.5	4.59E-08	2.75E+03	1.30E+05	-2.81E+07
.0100000	4610.0	1.551E+04	6.777E-01	3.17E-02	9.72E-07	98.0	33.7	48.4	6.5	.5	5.21E-08	5.08E+03	1.03E+05	-2.67E+07
.0125893	4630.0	1.764E+04	7.673E-01	3.51E-02	1.01E-06	97.9	33.5	48.5	6.5	.5	5.90E-08	8.46E+03	8.22E+04	-2.53E+07
.0158489	4660.0	2.007E+04	8.772E-01	3.90E-02	1.11E-06	97.7	33.5	48.3	6.5	.5	6.67E-08	9.48E+03	6.56E+04	-2.39E+07
.0199526	4700.0	2.282E+04	1.012E+00	4.32E-02	1.32E-06	97.3	33.9	47.7	6.5	.6	7.52E-08	9.71E+03	5.25E+04	-2.25E+07
.0251189	4750.0	2.593E+04	1.179E+00	4.80E-02	1.65E-06	96.7	34.4	46.8	6.4	.7	8.46E-08	8.68E+03	4.20E+04	-2.11E+07
.0316228	4800.0	2.946E+04	1.373E+00	5.34E-02	2.06E-06	96.0	34.8	45.8	6.3	.8	9.51E-08	7.51E+03	3.36E+04	-1.97E+07
.0398107	4860.0	3.346E+04	1.614E+00	5.95E-02	2.71E-06	94.9	35.2	44.6	6.2	.9	1.07E-07	6.57E+03	2.69E+04	-1.83E+07
.0501187	4920.0	3.797E+04	1.897E+00	6.63E-02	3.53E-06	93.6	35.5	43.2	6.0	1.1	1.20E-07	5.61E+03	2.15E+04	-1.68E+07
.0630957	4990.0	4.307E+04	2.254E+00	7.42E-02	4.83E-06	91.6	35.6	41.5	5.8	1.3	1.34E-07	4.84E+03	1.72E+04	-1.53E+07
.0794328	5060.0	4.879E+04	2.681E+00	8.32E-02	6.51E-06	89.2	35.3	39.7	5.6	1.6	1.49E-07	4.09E+03	1.37E+04	-1.39E+07
.1000000	5140.0	5.518E+04	3.234E+00	9.40E-02	9.12E-06	85.9	34.6	37.5	5.3	1.9	1.66E-07	3.59E+03	1.09E+04	-1.24E+07
.1258925	5225.0	6.228E+04	3.940E+00	1.07E-01	1.28E-05	81.6	33.4	34.9	5.0	2.2	1.85E-07	3.10E+03	8.58E+03	-1.09E+07
.1584893	5320.0	7.007E+04	4.893E+00	1.24E-01	1.86E-05	75.9	31.4	31.7	4.5	2.7	2.04E-07	2.80E+03	6.71E+03	-9.48E+06
.1995262	5430.0	7.846E+04	6.263E+00	1.46E-01	2.78E-05	68.4	28.5	27.9	4.0	3.2	2.24E-07	2.50E+03	5.16E+03	-8.05E+06
.2511886	5550.0	8.731E+04	8.233E+00	1.77E-01	4.19E-05	59.7	24.9	23.7	3.4	3.6	2.44E-07	2.12E+03	3.91E+03	-6.68E+06
.3162278	5675.0	9.644E+04	1.102E+01	2.17E-01	6.19E-05	50.8	21.0	19.6	2.8	3.9	2.63E-07	1.79E+03	2.93E+03	-5.37E+06
.3981072	5810.0	1.057E+05	1.511E+01	2.73E-01	9.13E-05	42.0	17.2	15.7	2.3	4.1	2.82E-07	1.50E+03	2.17E+03	-4.14E+06
.5011872	5950.0	1.149E+05	2.096E+01	3.46E-01	1.32E-04	34.1	13.7	12.3	1.8	4.2	2.99E-07	1.26E+03	1.59E+03	-2.98E+06
.6309573	6100.0	1.240E+05	2.959E+01	4.46E-01	1.91E-04	27.1	10.6	9.4	1.4	4.1	3.15E-07	1.08E+03	1.16E+03	-1.91E+06
.7943282	6265.0	1.328E+05	4.278E+01	5.84E-01	2.80E-04	21.1	7.9	7.0	1.0	4.0	3.28E-07	9.30E+02	8.40E+02 C	-9.15E+05
1.000000	6440.0	1.412E+05	6.239E+01	7.72E-01	4.07E-04	16.3	5.8	5.1	.7	3.7	3.39E-07	7.73E+02	6.05E+02 C	0.
1.258925	6620.0	1.493E+05	9.065E+01	1.02E+00	5.85E-04	12.6	4.3	3.7	.5	3.4	3.49E-07	6.24E+02	4.38E+02 C	8.46E+05
1.584893	6800.0	1.570E+05	1.298E+02	1.33E+00	8.21E-04	9.9	3.2	2.7	.4	3.1	3.57E-07	5.02E+02	3.21E+02 C	1.64E+06
1.995262	6985.0	1.644E+05	1.846E+02	1.72E+00	1.14E-03	7.8	2.4	2.0	.3	2.8	3.64E-07	4.14E+02	2.35E+02 C	2.39E+06
2.511886	7180.0	1.716E+05	2.629E+02	2.25E+00	1.58E-03	6.2	1.7	1.5	.2	2.6	3.70E-07	3.42E+02	1.72E+02 C	3.10E+06
3.162278	7380.0	1.786E+05	3.712E+02	2.91E+00	2.18E-03	5.0	1.3	1.1	.2	2.3	3.74E-07	2.76E+02	1.26E+02 C	3.78E+06
3.981072	7580.0	1.854E+05	5.156E+02	3.73E+00	2.94E-03	4.1	1.0	.8	.1	2.1	3.78E-07	2.24E+02	9.29E+01 C	4.44E+06
5.011872	7790.0	1.921E+05	7.155E+02	4.80E+00	3.98E-03	3.4	.7	.6	.1	1.8	3.80E-07	1.44E+02	6.79E+01 C	5.08E+06
6.309573	7900.0	1.990E+05	8.516E+02	5.50E+00	4.59E-03	3.1	.6	.5	.1	1.7	3.88E-07	1.10E+02	5.63E+01 C	5.74E+06
7.943282	8120.0	2.062E+05	1.171E+03	7.07E+00	6.13E-03	2.5	.5	.4	.1	1.5	3.91E-07	9.65E+01	4.08E+01 C	6.41E+06
10.00000	8240.0	2.137E+05	1.396E+03	8.14E+00	7.07E-03	2.3	.4	.3	.1	1.4	3.99E-07	7.17E+01	3.35E+01 C	7.09E+06
12.58925	8460.0	2.214E+05	1.876E+03	1.04E+01	9.22E-03	1.9	.3	.3	.0	1.3	4.02E-07	7.62E+01	2.43E+01 C	7.80E+06
15.84893	8680.0	2.291E+05	2.486E+03	1.31E+01	1.19E-02	1.7	.3	.2	.0	1.1	4.04E-07	6.06E+01	1.77E+01 C	8.49E+06
19.95262	8900.0	2.368E+05	3.250E+03	1.65E+01	1.51E-02	1.4	.2	.2	.0	1.0	4.06E-07	4.81E+01	1.30E+01 C	9.17E+06
25.11886	9120.0	2.445E+05	4.197E+03	2.07E+01	1.90E-02	1.2	.2	.1	.0	.9	4.08E-07	8.80E-01	9.58E+00	9.86E+06

TABLE II

BILDERBERG CONTINUUM ATMOSPHERE

T EFF = 5780, LOG G = 4.440, HELIUM/HYDROGEN BY NUMBER= .100, ABUNDANCES G-M-A

OPTICAL DEPTH	TEMP (K)	RAYLEIGH /TOTAL	ELECTRON /TOTAL	OPACITY PER NEUTRAL H (*E26) A(H-)	A(H)	A(H2+)	NEUTRAL H PER GM	MEAN WEIGHT	FRACTIONAL IONIZATION OF METALS SI	MG	FE	C		
.0000001	9500.0	.0000	.9981	6.188E-03	2.552E+03	5.902E-04	2.178E+19	.671	1.000	1.000	1.000	1.000	1.000	1.000
.0000001	8780.0	.0000	.9970	9.015E-03	7.397E+02	8.233E-04	1.164E+20	.674	1.000	1.000	1.000	1.000	1.000	1.000
.0000002	8220.0	.0000	.9954	1.288E-02	2.435E+02	1.127E-03	5.337E+20	.675	1.000	1.000	1.000	1.000	1.000	1.000
.0000002	7740.0	.0000	.9931	1.835E-02	8.280E+01	1.542E-03	2.350E+21	.676	1.000	1.000	1.000	1.000	1.000	1.000
.0000003	7300.0	.0000	.9896	2.623E-02	2.724E+01	2.122E-03	1.060E+22	.683	1.000	1.000	1.000	1.000	1.000	1.000
.0000003	6900.0	.0002	.9845	3.669E-02	8.778E+00	2.862E-03	4.473E+22	.710	1.000	1.000	1.000	.998	1.000	1.000
.0000004	6550.0	.0007	.9784	4.693E-02	2.914E+00	3.543E-03	1.378E+23	.798	1.000	1.000	1.000	.994	1.000	1.000
.0000005	6250.0	.0024	.9721	5.250E-02	1.027E+00	3.850E-03	2.621E+23	.955	1.000	1.000	1.000	.984	1.000	1.000
.0000006	6010.0	.0066	.9624	5.631E-02	4.143E-01	4.028E-03	3.477E+23	1.104	1.000	1.000	1.000	.963	1.000	1.000
.0000008	5830.0	.0161	.9411	6.498E-02	1.998E-01	4.552E-03	3.903E+23	1.198	1.000	1.000	1.000	.924	1.000	1.000
.0000010	5700.0	.0344	.8960	8.195E-02	1.147E-01	5.630E-03	4.091E+23	1.244	1.000	1.000	1.000	.856	1.000	1.000
.0000013	5610.0	.0625	.8158	1.084E-01	7.691E-02	7.307E-03	4.171E+23	1.265	.999	1.000	1.000	.762	1.000	1.000
.0000016	5500.0	.1076	.6858	1.326E-01	4.638E-02	8.692E-03	4.213E+23	1.276	.999	1.000	1.000	.628	1.000	1.000
.0000020	5435.0	.1450	.5376	1.678E-01	3.408E-02	1.073E-02	4.230E+23	1.281	.999	1.000	1.000	.506	1.000	1.000
.0000025	5370.0	.1744	.4001	1.985E-01	2.485E-02	1.235E-02	4.239E+23	1.284	.998	1.000	1.000	.397	1.000	1.000
.0000032	5305.0	.1948	.2892	2.232E-01	1.799E-02	1.348E-02	4.244E+23	1.285	.997	1.000	.999	.307	1.000	1.000
.0000040	5240.0	.2083	.2067	2.419E-01	1.291E-02	1.413E-02	4.247E+23	1.286	.996	1.000	.999	.235	1.000	1.000
.0000050	5175.0	.2172	.1475	2.554E-01	9.197E-03	1.435E-02	4.249E+23	1.286	.995	1.000	.999	.178	1.000	1.000
.0000063	5115.0	.2214	.1063	2.676E-01	6.671E-03	1.438E-02	4.250E+23	1.287	.994	.999	.999	.136	1.000	1.000
.0000079	5055.0	.2238	.0771	2.769E-01	4.803E-03	1.408E-02	4.251E+23	1.287	.993	.999	.998	.103	1.000	1.000
.0000100	5000.0	.2231	.0568	2.876E-01	3.530E-03	1.370E-02	4.251E+23	1.287	.991	.999	.998	.078	1.000	1.000
.0000126	4950.0	.2194	.0425	3.008E-01	2.652E-03	1.326E-02	4.252E+23	1.287	.988	.999	.998	.059	1.000	1.000
.0000158	4910.0	.2114	.0326	3.207E-01	2.101E-03	1.299E-02	4.252E+23	1.287	.986	.999	.997	.046	1.000	1.000
.0000200	4870.0	.2029	.0252	3.420E-01	1.658E-03	1.249E-02	4.252E+23	1.287	.983	.998	.996	.036	1.000	1.000
.0000251	4840.0	.1910	.0200	3.723E-01	1.385E-03	1.222E-02	4.252E+23	1.288	.979	.998	.996	.028	1.000	1.000
.0000316	4810.0	.1791	.0160	4.063E-01	1.154E-03	1.176E-02	4.252E+23	1.288	.975	.998	.995	.022	1.000	1.000
.0000398	4780.0	.1669	.0130	4.452E-01	9.598E-04	1.113E-02	4.252E+23	1.288	.969	.997	.994	.017	1.000	.999
.0000501	4750.0	.1545	.0106	4.909E-01	7.962E-04	1.035E-02	4.252E+23	1.288	.962	.997	.992	.013	1.000	.999
.0000631	4720.0	.1418	.0087	5.451E-01	6.589E-04	9.448E-03	4.252E+23	1.288	.953	.996	.990	.010	1.000	.999
.0000794	4700.0	.1283	.0072	6.140E-01	5.801E-04	8.926E-03	4.252E+23	1.288	.943	.995	.988	.008	1.000	.999
.0001000	4680.0	.1157	.0061	6.932E-01	5.101E-04	8.336E-03	4.252E+23	1.288	.932	.994	.986	.007	1.000	.999
.0001259	4665.0	.1038	.0052	7.855E-01	4.629E-04	7.926E-03	4.252E+23	1.288	.919	.992	.983	.005	1.000	.999
.0001585	4650.0	.0929	.0044	8.897E-01	4.198E-04	7.484E-03	4.252E+23	1.288	.904	.991	.980	.004	1.000	.998
.0001995	4640.0	.0831	.0038	1.008E+00	3.932E-04	7.237E-03	4.252E+23	1.288	.889	.989	.976	.004	1.000	.998
.0002512	4630.0	.0743	.0033	1.141E+00	3.681E-04	6.975E-03	4.252E+23	1.288	.872	.988	.972	.003	1.000	.998
.0003162	4620.0	.0664	.0029	1.289E+00	3.446E-04	6.705E-03	4.252E+23	1.288	.852	.985	.968	.003	1.000	.997
.0003981	4610.0	.0594	.0025	1.454E+00	3.225E-04	6.432E-03	4.252E+23	1.288	.831	.983	.962	.002	1.000	.997
.0005012	4605.0	.0533	.0022	1.637E+00	3.119E-04	6.357E-03	4.251E+23	1.288	.811	.981	.957	.002	1.000	.996
.0006310	4600.0	.0478	.0019	1.839E+00	3.017E-04	6.280E-03	4.251E+23	1.288	.789	.978	.951	.002	1.000	.996
.0007943	4600.0	.0429	.0017	2.063E+00	3.017E-04	6.401E-03	4.251E+23	1.288	.769	.975	.945	.001	1.000	.995

Table II (continued)

OPTICAL DEPTH	TEMP (K)	RAYLEIGH /TOTAL	ELECTRON /TOTAL	OPACITY PER NEUTRAL H (*E26) A(H-)	A(H)	A(H2+)	NEUTRAL H PER GM	MEAN WEIGHT	FRACTIONAL IONIZATION OF METALS SI	MG	FE	C		
.0010000	4600.0	.0385	.0015	2.310E+00	3.017E-04	6.525E-03	4.251E+23	1.288	.748	.972	.939	.001	1.000	.995
.0012589	4600.0	.0347	.0013	2.584E+00	3.017E-04	6.653E-03	4.251E+23	1.288	.726	.969	.932	.001	1.000	.994
.0015849	4600.0	.0312	.0012	2.887E+00	3.017E-04	6.786E-03	4.250E+23	1.288	.704	.966	.925	.001	1.000	.994
.0019953	4600.0	.0281	.0010	3.222E+00	3.017E-04	6.925E-03	4.250E+23	1.288	.680	.962	.917	.001	1.000	.993
.0025119	4600.0	.0253	.0009	3.592E+00	3.017E-04	7.070E-03	4.250E+23	1.288	.656	.958	.908	.001	1.000	.992
.0031623	4600.0	.0228	.0008	4.001E+00	3.017E-04	7.222E-03	4.249E+23	1.288	.632	.953	.899	.001	1.000	.991
.0039811	4600.0	.0206	.0007	4.454E+00	3.017E-04	7.381E-03	4.249E+23	1.288	.606	.948	.889	.001	1.000	.990
.0050119	4600.0	.0186	.0006	4.953E+00	3.017E-04	7.548E-03	4.248E+23	1.288	.581	.943	.878	.001	1.000	.989
.0063096	4600.0	.0168	.0005	5.506E+00	3.017E-04	7.725E-03	4.248E+23	1.288	.555	.937	.866	.001	1.000	.988
.0079433	4605.0	.0152	.0005	6.117E+00	3.119E-04	8.156E-03	4.247E+23	1.288	.534	.931	.856	.000	.999	.987
.0100000	4610.0	.0137	.0004	6.793E+00	3.225E-04	8.616E-03	4.246E+23	1.288	.513	.925	.845	.000	.999	.986
.0125893	4630.0	.0124	.0004	7.548E+00	3.681E-04	9.974E-03	4.246E+23	1.289	.506	.923	.841	.000	.999	.985
.0158489	4660.0	.0112	.0003	8.391E+00	4.481E-04	1.224E-02	4.245E+23	1.289	.510	.924	.843	.000	.999	.985
.0199526	4700.0	.0101	.0003	9.335E+00	5.801E-04	1.590E-02	4.245E+23	1.289	.523	.927	.849	.001	.999	.985
.0251189	4750.0	.0091	.0003	1.039E+01	7.962E-04	2.177E-02	4.245E+23	1.289	.544	.931	.858	.001	.999	.985
.0316228	4800.0	.0082	.0003	1.157E+01	1.086E-03	2.961E-02	4.245E+23	1.289	.564	.936	.867	.001	.999	.986
.0398107	4860.0	.0073	.0003	1.290E+01	1.562E-03	4.224E-02	4.245E+23	1.289	.591	.941	.879	.001	.999	.986
.0501187	4920.0	.0066	.0002	1.440E+01	2.228E-03	5.966E-02	4.245E+23	1.289	.617	.946	.889	.001	.999	.987
.0630957	4990.0	.0059	.0002	1.612E+01	3.335E-03	8.777E-02	4.246E+23	1.289	.648	.952	.900	.001	.999	.988
.0794328	5060.0	.0052	.0002	1.809E+01	4.938E-03	1.273E-01	4.246E+23	1.289	.675	.956	.910	.002	.999	.988
.1000000	5140.0	.0046	.0002	2.045E+01	7.633E-03	1.905E-01	4.246E+23	1.288	.706	.961	.920	.002	.999	.989
.1258925	5225.0	.0041	.0002	2.329E+01	1.195E-02	2.855E-01	4.247E+23	1.288	.735	.966	.930	.003	.999	.990
.1584893	5320.0	.0035	.0002	2.687E+01	1.939E-02	4.349E-01	4.247E+23	1.288	.763	.970	.938	.004	.999	.990
.1995262	5430.0	.0030	.0002	3.166E+01	3.327E-02	6.794E-01	4.248E+23	1.288	.791	.974	.947	.005	.999	.991
.2511886	5550.0	.0025	.0002	3.812E+01	5.851E-02	1.055E+00	4.248E+23	1.288	.816	.977	.953	.007	.999	.991
.3162278	5675.0	.0020	.0002	4.668E+01	1.027E-01	1.596E+00	4.248E+23	1.288	.836	.979	.959	.009	.999	.992
.3981072	5810.0	.0016	.0002	5.836E+01	1.837E-01	2.384E+00	4.249E+23	1.288	.853	.981	.963	.012	.999	.992
.5011872	5950.0	.0013	.0002	7.380E+01	3.265E-01	3.471E+00	4.249E+23	1.288	.867	.983	.967	.016	.999	.992
.6309573	6100.0	.0010	.0002	9.466E+01	5.873E-01	4.993E+00	4.249E+23	1.288	.879	.984	.970	.021	.999	.992
.7943282	6265.0	.0007	.0002	1.236E+02	1.085E+00	7.171E+00	4.249E+23	1.288	.890	.986	.972	.027	.999	.992
1.000000	6440.0	.0006	.0002	1.625E+02	2.010E+00	1.017E+01	4.249E+23	1.287	.900	.987	.974	.034	.999	.992
1.258925	6620.0	.0004	.0002	2.130E+02	3.666E+00	1.416E+01	4.249E+23	1.287	.908	.988	.976	.044	.999	.992
1.584893	6800.0	.0003	.0002	2.761E+02	6.479E+00	1.924E+01	4.248E+23	1.287	.915	.988	.978	.054	.999	.992
1.995262	6985.0	.0003	.0002	3.562E+02	1.129E+01	2.582E+01	4.247E+23	1.286	.921	.989	.979	.067	.999	.992
2.511886	7180.0	.0002	.0002	4.592E+02	1.965E+01	3.447E+01	4.245E+23	1.286	.927	.990	.981	.082	.999	.992
3.162278	7380.0	.0001	.0002	5.877E+02	3.367E+01	4.550E+01	4.243E+23	1.285	.932	.990	.982	.100	.999	.991
3.981072	7580.0	.0001	.0002	7.427E+02	5.609E+01	5.909E+01	4.239E+23	1.284	.937	.991	.983	.120	.999	.991
5.011872	7790.0	.0001	.0002	9.367E+02	9.322E+01	7.645E+01	4.235E+23	1.283	.941	.991	.984	.144	.999	.991
6.309573	7900.0	.0001	.0002	1.062E+03	1.203E+02	8.774E+01	4.233E+23	1.282	.943	.991	.985	.156	.999	.991
7.943282	8120.0	.0001	.0003	1.328E+03	1.965E+02	1.124E+02	4.226E+23	1.280	.947	.992	.986	.183	.999	.991
10.00000	8240.0	.0001	.0003	1.506E+03	2.540E+02	1.289E+02	4.222E+23	1.279	.948	.992	.986	.198	.999	.991
12.58925	8460.0	.0000	.0003	1.851E+03	3.991E+02	1.619E+02	4.213E+23	1.277	.952	.992	.987	.229	.999	.991
15.84893	8680.0	.0000	.0003	2.250E+03	6.129E+02	2.008E+02	4.202E+23	1.274	.955	.993	.987	.261	.999	.991
19.95262	8900.0	.0000	.0003	2.706E+03	9.218E+02	2.462E+02	4.188E+23	1.270	.957	.993	.988	.294	.999	.991
25.11886	9120.0	.0000	.0003	3.224E+03	1.360E+03	2.987E+02	4.172E+23	1.266	.960	.993	.989	.328	.999	.991

TABLE III

MONOCHROMATIC OPTICAL DEPTHS AT SELECTED WAVELENGTHS
BILDERBERG CONTINUUM ATMOSPHERE

TAU	1200.0	1400.0	1527.2-	1527.2+	1600.0	1683.3-	1683.3+	1800.0	2077.5-	2077.5+	2190.0	2400.0	2665.0
.000	.0072	.0127	.0126	.0014	.0014	.0015	.0003	.0002	.0001	.0001	.0001	.0001	.0000
.000	.0494	.1067	.1071	.0091	.0097	.0104	.0012	.0009	.0004	.0004	.0003	.0002	.0001
.000	.2772	.6540	.6604	.0472	.0524	.0578	.0036	.0028	.0013	.0010	.0008	.0008	.0003
.000	1.2464	3.0528	3.0897	.2034	.2297	.2558	.0095	.0081	.0037	.0022	.0020	.0024	.0007
.001	4.5366	11.2836	11.4305	.7255	.8256	.9213	.0245	.0226	.0104	.0048	.0047	.0074	.0017
.003	13.8686	34.7215	35.1898	2.2087	2.5226	2.8099	.0626	.0613	.0300	.0106	.0115	.0221	.0039
.010	38.3303	96.2748	97.6052	6.1269	7.0108	7.7719	.1594	.1630	.0880	.0246	.0297	.0663	.0096
.032	88.6403	222.7637	225.9443	14.5028	16.6082	18.3662	.3761	.3942	.2272	.0573	.0741	.1781	.0249
.100	159.9620	401.2370	407.2624	27.4815	31.4692	34.9118	.7637	.8052	.4634	.1279	.1668	.3954	.0679
.316	231.5441	578.7545	587.9423	42.2029	48.2953	53.7769	1.3394	1.4076	.8157	.2822	.3560	.7734	.1966
1.000	278.3424	692.9844	704.5271	53.5724	61.2450	68.3348	2.0833	2.1853	1.3796	.6639	.7930	1.4610	.6272
3.162	304.9034	756.1357	769.2516	61.6427	70.3747	78.5797	3.3484	3.5429	2.7303	1.8577	2.1207	3.2131	2.2975
10.000	324.3857	800.1754	814.7723	69.9479	79.6675	88.9549	6.7695	7.3469	7.3364	6.3062	7.0837	9.4242	9.4506

TAU	2885.0	3200.0	3400.0	3647.1-	3647.1+	3800.0	4000.0	4200.0	4400.0	4600.0	4800.0	5000.0	5500.0
.000	.0000	.0000	.0000	.0000	.0000	.0000	.0000	.0000	.0000	.0000	.0000	.0000	.0000
.000	.0001	.0001	.0001	.0001	.0000	.0000	.0000	.0000	.0000	.0000	.0000	.0000	.0000
.000	.0002	.0002	.0002	.0002	.0001	.0001	.0001	.0001	.0001	.0001	.0001	.0001	.0001
.000	.0006	.0005	.0005	.0005	.0004	.0004	.0003	.0003	.0003	.0003	.0003	.0003	.0003
.001	.0014	.0013	.0013	.0013	.0012	.0011	.0010	.0010	.0010	.0010	.0010	.0010	.0010
.003	.0035	.0034	.0035	.0037	.0037	.0032	.0029	.0029	.0030	.0030	.0031	.0032	.0033
.010	.0092	.0095	.0101	.0110	.0109	.0094	.0089	.0090	.0092	.0095	.0098	.0100	.0107
.032	.0252	.0273	.0293	.0324	.0321	.0283	.0272	.0279	.0288	.0299	.0310	.0316	.0340
.100	.0711	.0793	.0860	.0955	.0941	.0869	.0844	.0872	.0905	.0941	.0978	.1000	.1078
.316	.2114	.2399	.2612	.2903	.2761	.2677	.2633	.2735	.2847	.2964	.3085	.3162	.3416
1.000	.6899	.7966	.8727	.9746	.8181	.8231	.8233	.8588	.8961	.9344	.9731	1.0000	1.0824
3.162	2.6035	3.1045	3.4588	3.9333	2.4505	2.5203	2.5646	2.6872	2.8129	2.9396	3.0660	3.1623	3.4396
10.000	11.0536	13.6806	15.5541	18.0889	7.3545	7.6599	7.9132	8.3381	8.7705	9.2052	9.6392	10.0000	11.0077

TAU	6000.0	6500.0	8000.0	10000.0	12000.0	16500.0	20000.0	100000.0	200000.0	240000.0	500000.0	1000000.	2000000.
.000	.0000	.0000	.0000	.0000	.0000	.0000	.0000	.0001	.0003	.0004	.0018	.0073	.0291
.000	.0000	.0000	.0000	.0000	.0000	.0000	.0000	.0002	.0008	.0012	.0052	.0209	.0834
.000	.0001	.0001	.0001	.0001	.0001	.0000	.0000	.0006	.0025	.0036	.0156	.0625	.2500
.000	.0003	.0004	.0004	.0003	.0003	.0001	.0001	.0020	.0079	.0113	.0489	.1955	.7815
.001	.0011	.0011	.0012	.0011	.0009	.0002	.0003	.0063	.0250	.0360	.1557	.6220	2.4863
.003	.0035	.0037	.0039	.0037	.0029	.0006	.0009	.0203	.0804	.1157	.5007	2.0006	7.9974
.010	.0113	.0120	.0127	.0120	.0093	.0020	.0028	.0651	.2587	.3722	1.6109	6.4370	25.7325
.032	.0362	.0383	.0406	.0385	.0298	.0067	.0095	.2199	.8737	1.2569	5.4413	21.7448	86.9299
.100	.1151	.1217	.1300	.1232	.0963	.0249	.0354	.8177	3.2499	4.6752	20.2403	80.8860	323.3641
.316	.3652	.3858	.4143	.3936	.3119	.1009	.1431	3.2896	13.0755	18.8104	81.4437	325.4860	*01.2480
1.000	1.1576	1.2230	1.3240	1.2582	1.0257	.4435	.6282	14.4010	57.2650	82.3898	356.8274	*26.2483	*02.3653
3.162	3.6928	3.9170	4.3367	4.0690	3.5036	2.0857	2.9859	70.0238	278.6625	400.9870	*37.4246	*46.0469	*74.5215
10.000	11.9569	12.8453	14.9797	13.4869	12.7705	10.2595	15.0173	366.4012	*58.3532	*98.6211	*94.3947	*61.5826	*02.5756

TABLE IV

SPECIFIC INTENSITY AND PHI FOR LIMB DARKENING BILDERBERG CONTINUUM ATMOSPHERE

MU	1200.0	1200.0	1400.0	1400.0	1527.2-	1527.2-	1527.2+	1527.2+	1600.0	1600.0
1.000	1.8479E-12	3.8471E+08	5.1716E-11	7.9102E+09	2.4068E-10	3.0936E+10	1.5519E-10	1.9947E+10	3.4269E-10	4.0131E+10
.750	2.0108E-12	1.0881E+00	5.6275E-11	1.0882E+00	2.5982E-10	1.0796E+00	1.5897E-10	1.0244E+00	3.5121E-10	1.0248E+00
.500	2.2956E-12	1.2423E+00	6.3935E-11	1.2363E+00	2.9170E-10	1.2120E+00	1.6629E-10	1.0716E+00	3.6736E-10	1.0720E+00
.250	2.9765E-12	1.6107E+00	8.1096E-11	1.5681E+00	3.6193E-10	1.5038E+00	1.8565E-10	1.1963E+00	4.0922E-10	1.1941E+00
.100	4.4307E-12	2.3976E+00	1.1397E-10	2.2038E+00	4.9297E-10	2.0483E+00	2.2957E-10	1.4793E+00	5.0131E-10	1.4629E+00
.080	4.9106E-12	2.6574E+00	1.2417E-10	2.4010E+00	5.3289E-10	2.2142E+00	2.4429E-10	1.5741E+00	5.3166E-10	1.5514E+00
.060	5.6385E-12	3.0513E+00	1.3878E-10	2.6836E+00	5.8958E-10	2.4497E+00	2.6611E-10	1.7148E+00	5.7624E-10	1.6815E+00
.040	6.8926E-12	3.7299E+00	1.6294E-10	3.1506E+00	6.8187E-10	2.8331E+00	3.0324E-10	1.9540E+00	6.5132E-10	1.9006E+00
.020	9.7882E-12	5.2969E+00	2.1539E-10	4.1648E+00	8.7848E-10	3.6501E+00	3.8808E-10	2.5007E+00	8.2021E-10	2.3934E+00
.010	1.4006E-11	7.5793E+00	2.8476E-10	5.5061E+00	1.1326E-09	4.7058E+00	5.0942E-10	3.2826E+00	1.0543E-09	3.0764E+00

SPECIFIC INTENSITY AND PHI FOR LIMB DARKENING BILDERBERG CONTINUUM ATMOSPHERE

MU	1683.3-	1683.3-	1683.3+	1683.3+	1800.0	1800.0	2077.5-	2077.5-	2077.5+	2077.5+
1.000	7.7420E-10	8.1916E+10	9.1438E-08	9.6745E+12	1.4113E-07	1.3058E+13	8.5698E-07	5.9528E+13	1.7579E-06	1.2209E+14
.750	7.9353E-10	1.0250E+00	4.1070E-08	4.4915E-01	6.4896E-08	4.5983E-01	5.1225E-07	5.9774E-01	1.2525E-06	7.1253E-01
.500	8.2973E-10	1.0717E+00	1.1104E-08	1.2144E-01	1.9301E-08	1.3676E-01	2.2347E-07	2.6076E-01	7.4427E-07	4.2340E-01
.250	9.2188E-10	1.1908E+00	1.7175E-09	1.8783E-02	3.9490E-09	2.7982E-02	5.0918E-08	5.9416E-02	2.7105E-07	1.5419E-01
.100	1.1192E-09	1.4457E+00	8.3637E-10	9.1468E-03	2.2001E-09	1.5589E-02	1.7029E-08	1.9871E-02	6.5960E-08	3.7523E-02
.080	1.1832E-09	1.5282E+00	8.0996E-10	8.8579E-03	2.1376E-09	1.5146E-02	1.5420E-08	1.7993E-02	4.8244E-08	2.7445E-02
.060	1.2766E-09	1.6489E+00	8.0498E-10	8.8036E-03	2.1207E-09	1.5027E-02	1.4301E-08	1.6687E-02	3.3735E-08	1.9191E-02
.040	1.4324E-09	1.8502E+00	8.3189E-10	9.0978E-03	2.1682E-09	1.5363E-02	1.3719E-08	1.6008E-02	2.2700E-08	1.2913E-02
.020	1.7771E-09	2.2954E+00	9.4019E-10	1.0282E-02	2.3743E-09	1.6823E-02	1.3954E-08	1.6283E-02	1.5796E-08	8.9859E-03
.010	2.2435E-09	2.8978E+00	1.1382E-09	1.2447E-02	2.7552E-09	1.9523E-02	1.4974E-08	1.7474E-02	1.4821E-08	8.4313E-03

SPECIFIC INTENSITY AND PHI FOR LIMB DARKENING BILDERBERG CONTINUUM ATMOSPHERE

MU	2190.0	2190.0	2400.0	2400.0	2665.0	2665.0	2885.0	2885.0	3200.0	3200.0
1.000	2.0798E-06	1.3000E+14	1.8406E-06	9.5800E+13	6.4771E-06	2.7340E+14	8.5491E-06	3.0793E+14	1.1613E-05	3.3998E+14
.750	1.4676E-06	7.0564E-01	1.1918E-06	6.4751E-01	5.1416E-06	7.9381E-01	6.8908E-06	8.0603E-01	9.5383E-06	8.2137E-01
.500	8.5643E-07	4.1179E-01	6.1861E-07	3.3608E-01	3.6372E-06	5.6154E-01	5.0079E-06	5.8579E-01	7.1447E-06	6.1525E-01
.250	3.0497E-07	1.4663E-01	2.0943E-07	1.1378E-01	1.9401E-06	2.9953E-01	2.8194E-06	3.2979E-01	4.2730E-06	3.6796E-01
.100	7.8944E-08	3.7958E-02	8.7875E-08	4.7741E-02	8.3091E-07	1.2828E-01	1.3166E-06	1.5400E-01	2.1967E-06	1.8916E-01
.080	6.0401E-08	2.9042E-02	7.9128E-08	4.2989E-02	6.8210E-07	1.0531E-01	1.1055E-06	1.2931E-01	1.8917E-06	1.6290E-01
.060	4.5298E-08	2.1780E-02	7.2146E-08	3.9196E-02	5.3517E-07	8.2624E-02	8.9292E-07	1.0445E-01	1.5787E-06	1.3595E-01
.040	3.3899E-08	1.6300E-02	6.7333E-08	3.6581E-02	3.9191E-07	6.0506E-02	6.8013E-07	7.9556E-02	1.2571E-06	1.0826E-01
.020	2.6922E-08	1.2945E-02	6.5961E-08	3.5836E-02	2.5717E-07	3.9704E-02	4.7146E-07	5.5148E-02	9.2993E-07	8.0079E-02
.010	2.6380E-08	1.2684E-02	6.8307E-08	3.7110E-02	2.0005E-07	3.0885E-02	3.7776E-07	4.4187E-02	7.7761E-07	6.6962E-02

SPECIFIC INTENSITY AND PHI FOR LIMB DARKENING BILDERBERG CONTINUUM ATMOSPHERE

MU	3400.0	3400.0	3647.1-	3647.1-	3647.1+	3647.1+	3800.0	3800.0	4000.0	4000.0
1.000	1.3537E-05	3.5105E+14	1.5829E-05	3.5677E+14	1.8663E-05	4.2063E+14	2.0749E-05	4.3078E+14	2.3548E-05	4.4122E+14
.750	1.1234E-05	8.2987E-01	1.3282E-05	8.3906E-01	1.5269E-05	8.1817E-01	1.7178E-05	8.2789E-01	1.9718E-05	8.3736E-01
.500	8.5492E-06	6.3156E-01	1.0282E-05	6.4959E-01	1.1407E-05	6.1125E-01	1.3072E-05	6.3001E-01	1.5264E-05	6.4820E-01
.250	5.2778E-06	3.8989E-01	6.5689E-06	4.1499E-01	6.9416E-06	3.7195E-01	8.2337E-06	3.9682E-01	9.9020E-06	4.2050E-01
.100	2.8534E-06	2.1079E-01	3.7483E-06	2.3680E-01	3.8183E-06	2.0460E-01	4.7278E-06	2.2786E-01	5.8872E-06	2.5001E-01
.080	2.4896E-06	1.8391E-01	3.3164E-06	2.0951E-01	3.3623E-06	1.8016E-01	4.1981E-06	2.0233E-01	5.2649E-06	2.2358E-01
.060	2.1131E-06	1.5610E-01	2.8657E-06	1.8104E-01	2.8925E-06	1.5499E-01	3.6441E-06	1.7563E-01	4.6076E-06	1.9567E-01
.040	1.7220E-06	1.2721E-01	2.3932E-06	1.5119E-01	2.4059E-06	1.2892E-01	3.0575E-06	1.4736E-01	3.9030E-06	1.6574E-01
.020	1.3183E-06	9.7387E-02	1.9002E-06	1.2005E-01	1.9025E-06	1.0194E-01	2.4265E-06	1.1694E-01	3.1309E-06	1.3296E-01

SPECIFIC INTENSITY AND PHI FOR LIMB DARKENING BILDERBERG CONTINUUM ATMOSPHERE

MU	4200.0	4200.0	4400.0	4400.0	4600.0	4600.0	4800.0	4800.0	5000.0	5000.0
1.000	2.5676E-05	4.3636E+14	2.7623E-05	4.2774E+14	2.9396E-05	4.1648E+14	3.0997E-05	4.0333E+14	3.2615E-05	3.9111E+14
.750	2.1673E-05	8.4409E-01	2.3484E-05	8.5015E-01	2.5152E-05	8.5562E-01	2.6679E-05	8.6069E-01	2.8238E-05	8.6579E-01
.500	1.6995E-05	6.6190E-01	1.8628E-05	6.7435E-01	2.0158E-05	6.8575E-01	2.1584E-05	6.9631E-01	2.3050E-05	7.0671E-01
.250	1.1290E-05	4.3973E-01	1.2640E-05	4.5759E-01	1.3942E-05	4.7428E-01	1.5188E-05	4.8997E-01	1.6480E-05	5.0527E-01
.100	6.9317E-06	2.6997E-01	7.9853E-06	2.8908E-01	9.0359E-06	3.0739E-01	1.0073E-05	3.2497E-01	1.1155E-05	3.4200E-01
.080	6.2449E-06	2.4322E-01	7.2416E-06	2.6216E-01	8.2431E-06	2.8042E-01	9.2385E-06	2.9804E-01	1.0277E-05	3.1508E-01
.060	5.5148E-06	2.1479E-01	6.4467E-06	2.3338E-01	7.3918E-06	2.5145E-01	8.3386E-06	2.6901E-01	9.3264E-06	2.8595E-01
.040	4.7259E-06	1.8406E-01	5.5824E-06	2.0209E-01	6.4608E-06	2.1979E-01	7.3499E-06	2.3711E-01	8.2773E-06	2.5379E-01
.020	3.8523E-06	1.5004E-01	4.6175E-06	1.6716E-01	5.4153E-06	1.8422E-01	6.2347E-06	2.0114E-01	7.0875E-06	2.1730E-01
.010	3.3974E-06	1.3232E-01	4.1132E-06	1.4891E-01	4.8685E-06	1.6562E-01	5.6525E-06	1.8235E-01	6.4648E-06	1.9821E-01

SPECIFIC INTENSITY AND PHI FOR LIMB DARKENING BILDERBERG CONTINUUM ATMOSPHERE

MU	5500.0	5500.0	6000.0	6000.0	6500.0	6500.0	8000.0	8000.0	10000.0	10000.0
1.000	3.5683E-05	3.5363E+14	3.7906E-05	3.1566E+14	3.9434E-05	2.7981E+14	4.1375E-05	1.9381E+14	4.0506E-05	1.2143E+14
.750	3.1291E-05	8.7693E-01	3.3594E-05	8.8623E-01	3.5259E-05	8.9411E-01	3.7737E-05	9.1209E-01	3.7504E-05	9.2588E-01
.500	2.6030E-05	7.2949E-01	2.8382E-05	7.4873E-01	3.0176E-05	7.6521E-01	3.3233E-05	8.0321E-01	3.3732E-05	8.3276E-01
.250	1.9255E-05	5.3961E-01	2.1580E-05	5.6931E-01	2.3470E-05	5.9518E-01	2.7124E-05	6.5556E-01	2.8498E-05	7.0355E-01
.100	1.3621E-05	3.8172E-01	1.5812E-05	4.1714E-01	1.7692E-05	4.4864E-01	2.1650E-05	5.2327E-01	2.3645E-05	5.8374E-01
.080	1.2674E-05	3.5519E-01	1.4829E-05	3.9119E-01	1.6695E-05	4.2337E-01	2.0680E-05	4.9981E-01	2.2763E-05	5.6197E-01
.060	1.1642E-05	3.2626E-01	1.3751E-05	3.6276E-01	1.5600E-05	3.9559E-01	1.9603E-05	4.7378E-01	2.1774E-05	5.3756E-01
.040	1.0493E-05	2.9405E-01	1.2544E-05	3.3092E-01	1.4367E-05	3.6432E-01	1.8376E-05	4.4414E-01	2.0634E-05	5.0940E-01
.020	9.1794E-06	2.5725E-01	1.1160E-05	2.9442E-01	1.2952E-05	3.2844E-01	1.6956E-05	4.0980E-01	1.9284E-05	4.7607E-01
.010	8.4972E-06	2.3813E-01	1.0450E-05	2.7568E-01	1.2235E-05	3.1026E-01	1.6241E-05	3.9252E-01	1.8578E-05	4.5865E-01

SPECIFIC INTENSITY AND PHI FOR LIMB DARKENING BILDERBERG CONTINUUM ATMOSPHERE

MU	12000.0	12000.0	16500.0	16500.0	20000.0	20000.0	100000.0	100000.0	200000.0	200000.0
1.000	3.8340E-05	7.9820E+13	3.2499E-05	3.5786E+13	2.4142E-05	1.8094E+13	1.2185E-06	3.6528E+10	3.0622E-07	2.2950E+09
.750	3.5887E-05	9.3600E-01	3.1098E-05	9.5691E-01	2.3183E-05	9.6028E-01	1.1969E-06	9.8229E-01	3.0314E-07	9.8993E-01
.500	3.2735E-05	8.5381E-01	2.9199E-05	8.9848E-01	2.1895E-05	9.0690E-01	1.1699E-06	9.6016E-01	2.9959E-07	9.7836E-01
.250	2.8210E-05	7.3578E-01	2.6219E-05	8.0676E-01	1.9884E-05	8.2363E-01	1.1327E-06	9.2966E-01	2.9570E-07	9.6565E-01
.100	2.3819E-05	6.2125E-01	2.2864E-05	7.0354E-01	1.7648E-05	7.3102E-01	1.1004E-06	9.0308E-01	2.9408E-07	9.6034E-01
.080	2.2995E-05	5.9975E-01	2.2154E-05	6.8170E-01	1.7179E-05	7.1158E-01	1.0954E-06	8.9899E-01	2.9408E-07	9.6036E-01
.060	2.2057E-05	5.7530E-01	2.1301E-05	6.5543E-01	1.6616E-05	6.8827E-01	1.0905E-06	8.9501E-01	2.9422E-07	9.6082E-01
.040	2.0955E-05	5.4656E-01	2.0212E-05	6.2194E-01	1.5904E-05	6.5876E-01	1.0864E-06	8.9163E-01	2.9462E-07	9.6211E-01
.020	1.9593E-05	5.1103E-01	1.8654E-05	5.7400E-01	1.4897E-05	6.1706E-01	1.0848E-06	8.9030E-01	2.9582E-07	9.6605E-01
.010	1.8804E-05	4.9046E-01	1.7455E-05	5.3710E-01	1.4138E-05	5.8561E-01	1.0870E-06	8.9209E-01	2.9792E-07	9.7289E-01

SPECIFIC INTENSITY AND PHI FOR LIMB DARKENING BILDERBERG CONTINUUM ATMOSPHERE

MU	240000.0	240000.0	500000.0	500000.0	1000100.0	1000100.0	2000200.0	2000200.0	4000400.0	4000400.0
1.000	2.1274E-07	1.1072E+09	4.9418E-08	5.9260E+07	1.2551E-08	3.7620E+06	3.1954E-09	2.3944E+05	8.2238E-10	1.5406E+04
.750	2.1101E-07	9.9188E-01	4.9363E-08	9.9888E-01	1.2568E-08	1.0013E+00	3.2076E-09	1.0038E+00	8.2863E-10	1.0076E+00
.500	2.0914E-07	9.8308E-01	4.9349E-08	9.9859E-01	1.2599E-08	1.0038E+00	3.2288E-09	1.0104E+00	8.3880E-10	1.0200E+00
.250	2.0736E-07	9.7472E-01	4.9434E-08	1.0003E+00	1.2683E-08	1.0105E+00	3.2771E-09	1.0256E+00	8.5988E-10	1.0456E+00
.100	2.0699E-07	9.7299E-01	4.9716E-08	1.0060E+00	1.2872E-08	1.0255E+00	3.3679E-09	1.0540E+00	8.9459E-10	1.0878E+00
.080	2.0709E-07	9.7347E-01	4.9821E-08	1.0082E+00	1.2934E-08	1.0305E+00	3.3950E-09	1.0625E+00	9.0410E-10	1.0994E+00
.060	2.0730E-07	9.7441E-01	4.9986E-08	1.0115E+00	1.3024E-08	1.0376E+00	3.4327E-09	1.0743E+00	9.1689E-10	1.1149E+00
.040	2.0770E-07	9.7633E-01	5.0279E-08	1.0174E+00	1.3171E-08	1.0493E+00	3.4911E-09	1.0925E+00	9.3584E-10	1.1380E+00
.020	2.0884E-07	9.8166E-01	5.0964E-08	1.0313E+00	1.3481E-08	1.0741E+00	3.6039E-09	1.1279E+00	9.7123E-10	1.1810E+00
.010	2.1072E-07	9.9052E-01	5.1912E-08	1.0505E+00	1.3866E-08	1.1047E+00	3.7309E-09	1.1676E+00	1.0129E-09	1.2317E+00

depths. Linsky has also calculated a non-LTE version in which the abundance of the negative hydrogen ion is changed owing to the overpopulation of neutral hydrogen in its ground state; the principal change is that for non-LTE the outermost layers are more extended in height (Table V). However, the run of pressure versus optical depth is altered very little.

TABLE V

Departure Coefficients for Neutral Hydrogen

τ_{5000}(BCA)	x(km)	b_1	τ_{5000}(non-LTE)
0.000001	0.1500	5.8	0.0000004
0.00001	0.815	1.14	0.000009
0.0001	0.566	0.62	0.00011
0.001	0.408	0.53	0.0011
0.01	0.268	0.58	0.011
0.1	0.124	1.19	0.10
1.	0	1.00	1.0

The letter 'C' in the penultimate column of Table I indicates that the model is unstable against convection in the designated depth. Note that in Table IV the specific intensity is given per unit frequency, that is, in ergs cm^{-2} sec^{-1} $ster^{-1}$. In Table IV, the second column for each wavelength gives ϕ, the intensity ratio with respect to the center ($\mu = 1$), except for the initial number, which is the monochromatic specific intensity *per unit wavelength*.

The adopted temperature distribution of the BCA is graphically compared with previous models in Figures 1 and 2. The emergent intensities in the ultraviolet spectrum are shown in Figure 3.

4. Conclusions of the Working Group on the Temperature Model from Line Considerations

The spectral lines used so far by solar physicists may be roughly divided into three groups:

4.1. Lines for which a non-LTE theory is possible

These lines have a possible non-LTE theory in the sense that the relation between the electron temperature T_{el} and the excitation temperature T_{ex} can be derived from knowledge of a given model.

a. *The H and K lines of Ca II and Mg II*

From various investigations (see the Program for Tuesday 18 April, in Addendum I), a dilemma appears to exist: We have to assume either that the 'base' of the chromosphere is located at $\tau_{5000} \approx 10^{-4}$ with $T \leqslant 4200$–$4300\,°K$ (case for the thick chromo-

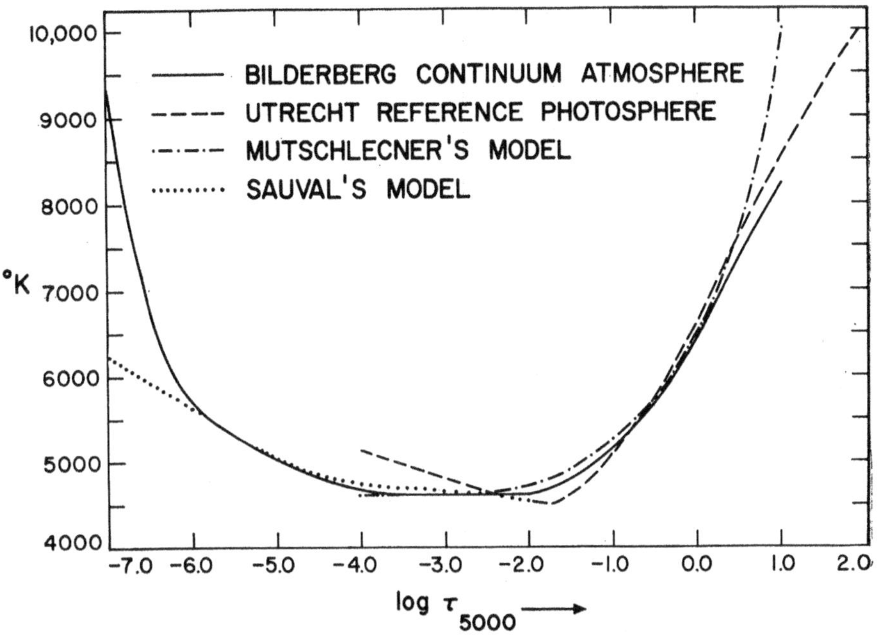

Fig. 1. The BCA temperature distribution compared to other models examined at the Bilderberg study week.

sphere), or that the base of the chromosphere occurs at $\tau_{5000} \ll 10^{-4}$ and $T_{min} > 4200$ °K. This latter situation is the 'optically thin' case, in which the optical depth of the chromosphere in the center of the K line of Ca II is of the order of 3; furthermore, $\Delta\lambda_D \approx 0.15$ Å.

In order to solve this dilemma, we must greatly improve our knowledge of the high chromosphere; we also need good photometry of the Ca II and Mg II lines, suitable for empirical analysis. We should mention here that the proposed BCA is in contradiction to both alternatives given above, but perhaps a further step forward may be achieved only when the inhomogeneities in the H and K lines are duly taken into account.

b. *The Balmer and Lyman Hydrogen Lines*

It is clear that the theory of the hydrogen lines is now in an adequate state. The wings of Balmer lines are suitably described by the URP model. It is not likely, however, that the minimum temperature in the low chromosphere can be deduced from an investigation of the Balmer lines, since the computations are rather insensitive to the temperature. On the other hand, the lines are so sensitive to the chromospheric model itself that undoubtedly the application of the theory of the hydrogen lines to an investigation of the line cores will greatly improve the model of the medium chromosphere.

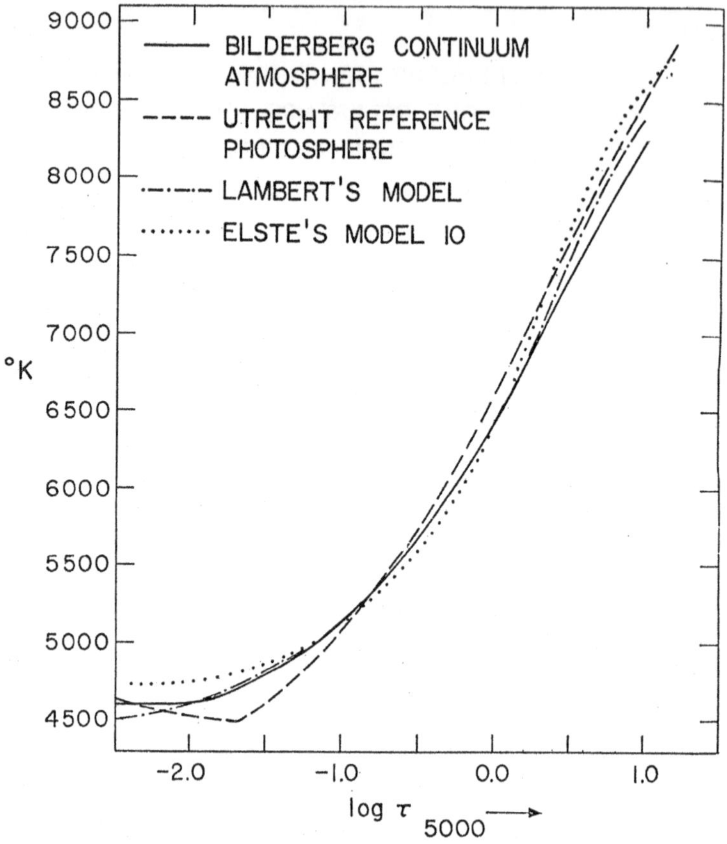

Fig. 2. The BCA photosphere temperature distribution compared to other models.

4.2. Metal and ion lines for which a relation between T_{el} and T_{ex} cannot be derived

(a) Holweger's distribution of T_{el} and his assumption that $T_{el} = T_{ex}$ appear to account for a very large number of observational data – on central intensities of Fraunhofer lines, with the exception of the resonance lines. However, noticeable differences, which appear mostly for $\log \tau_{5000} < -2$, still appear to occur between Holweger's source function and the source functions derived by De Jager and Neven (communication B13; see Addendum I). Part of these differences might be due to the fact that the measurements come from two different sets of observations. There is an obvious need for further accurate observations and for further theoretical investigations related to the source functions in the transition region between photosphere and chromosphere.

(b) The analysis of the detailed profiles of the NaD lines shows that for $\tau_{5000} \approx 10^{-5}$, $T_{el} \leqslant 4600° \pm 200°$. On the assumption of Holweger's model, this would mean that the microturbulent-velocity component $\zeta = 0.5$ km/sec at $\tau_{5000} \approx 10^{-5}$. This result

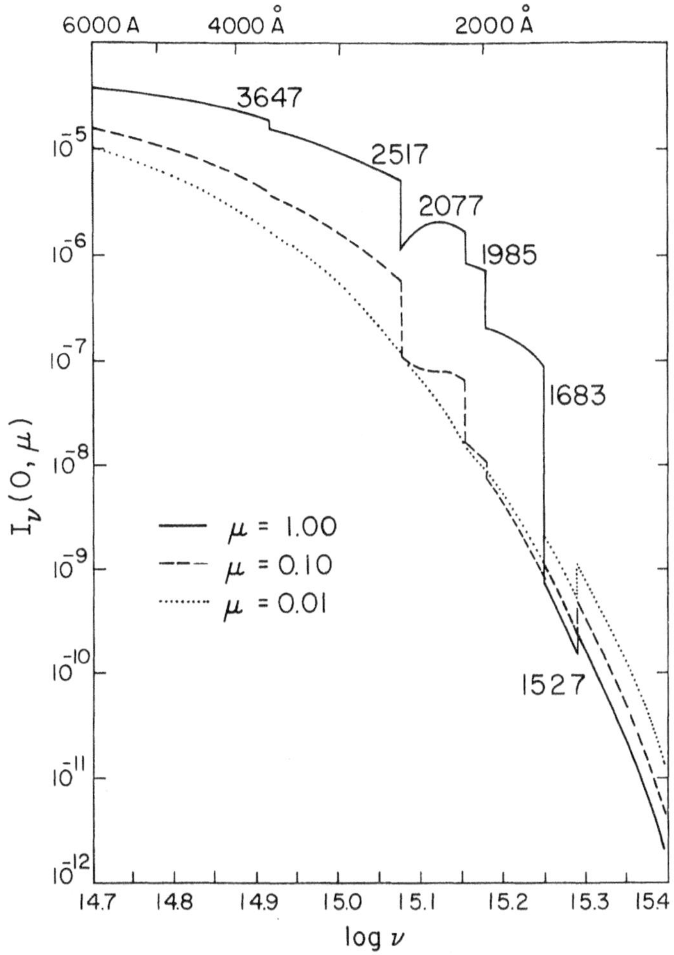

Fig. 3. Specific intensities in the ultraviolet region predicted from the BCA for the center ($\mu = 1$) and limb of the solar disk.

is based on a new approach, which requires further study. Similar empirical studies on very good material should be encouraged. Complete theoretical analyses of the non-LTE behavior of the NaD lines are being made in several institutes, and these results should be compared with those obtained from empirical investigations.

The working group felt that in order to decide, on a purely empirical basis, whether or not one has $T_{el} = T_{ex}$ between $\tau_{5000} \approx 10^{-3}$ and $\approx 10^{-5}$, further and more complete observations of the ultraviolet and infrared continuous spectra would be needed. These would allow a precise determination of $T_{el}(\tau_{5000})$ in this region. For this reason, it still seems premature to interpret the line data in terms of a $T(\tau)$ model in these layers of the low chromosphere.

4.3. Behavior of moderate intensity lines from ionized or neutral atoms in flash spectra

In order to understand Redman's observations about the occurrence of self-reversals in atomic and ionic lines of various excitation potentials in the flash spectrum, as reported by Houtgast (communication B14; see Addendum I), it seems necessary to assume that the different behavior of these lines is caused by the different behavior of the source functions at the same heights in the chromosphere. It may be necessary to assume that the temperature is increasing outward in the region about 500 km above the sun's limb.

5. Temperature Fluctuations, Velocity Fields in the Photosphere and Low Chromosphere

Although various valuable attempts have been made in the last fifteen years to determine the microscopic and macroscopic velocity fields and their variation with depth and direction, and to find the distribution function of the temperature fluctuations as a function of the optical depth, it was the consensus that these attempts did not yet yield reliable and generally acceptable results. The reason for these failures can easily be summarized.

5.1. Velocity fields

We can distinguish between the microturbulent velocity v_t with its vertical component ζ_t and the macroturbulent velocity field, which consists of a convective and a vibrational component v_c and v_v, respectively. These three quantities should be determined as a function of the optical depth τ_{5000} and of the angle θ with the vertical direction.

Little information is available in current literature about the relative contributions of v_c and v_v, so that we had better sum these two quantities into the macroscopic velocity $v_{\mathrm{ma}}(\tau_{5000}, \theta)$. Since the determination of the convective-velocity component is extremely sensitive to the assumed $\Delta T(\tau_0)$ relation, the difficulty in finding this latter function (see the second part of this section) makes it virtually impossible to find $v_c(\tau_{5000})$. Furthermore, the derived values of $v_t(\tau_{5000})$ are also directly dependent on the assumed values of $v_{\mathrm{ma}}(\tau_{5000})$, so that the difficulty in finding the latter function also makes it very hard to find the former. It seems probable that perhaps not much weight should be attached to already published values of these functions.

For the time being, we think it the safest assumption to take round values: $v_t = 1$ km/sec and $v_{\mathrm{ma}} = 2$ km/sec, *isotropic* and *independent of depth*. We should notice at the same time that there are clear indications that $v_{\mathrm{ma}} > 2$ km/sec at $\log \tau_{5000} \approx -6$, but the precise value is not yet known.

5.2. Temperature fluctuations

One of the important research programs for modern heliophysical research will be the determination of the temperature fluctuations ΔT as a function of the optical depth and also of the *distribution function* $\phi_{\Delta T}(\tau_{5000})$. This information can be obtained reliably only by means of very high-resolution photospheric pictures, at very different wavelengths, with a resolution exceeding the currently available ones by at least a factor of 3(0.″1). Present-day difficulties in finding $\Delta T(\tau_{5000})$ even for a simple two-stream model are partly due to insufficient photometric calibration of the existing material. The most reliable $\Delta T(\tau_{5000})$ function published so far is perhaps the one given by EDMONDS (1964).

Addendum I: Program of the Study Week on the Quiet Photosphere

A. Monday 17 April 1967; chairman: E. A. Müller

(1) Address of welcome.
(2) O. Namba: a study of the solar granulation.
(3) O. Namba: comment on a paper by Edmonds.
(4) D. Carbon, O. Gingerich, and R. Kurucz: line blanketing and solar windows.
(5) J. Houtgast: the near UV solar spectrum, and the Balmer jump.
(6) A. Sauval: limb darkening and photospheric model.
(7) D. L. Lambert: solar continuum and limb darkening between 5000 Å and 11 microns.
(8) R. M. Bonnet and J. E. Blamont: limb-darkening observations between 1800 and 2900 Å.
(9) H. Holweger: LTE model of the photosphere.
(10) J. P. Mutschlecner: the correlation of line-profile parameters and photospheric models.
(11) J. P. Mutschlecner: prediction of limb darkening and absolute intensity for several photospheric models.
(12) G. L. Withbroe: the center-limb variation of atomic and molecular solar lines.
(13) E. Müller, B. Baschek, and H. Holweger: center-to-limb observations of O I lines.
(14) E. Müller: analysis of selected lines of various elements in the solar atmosphere.
(15) C. de Jager: turbulence and convection from C I lines.

B. Tuesday 18 April 1967; chairman: R. W. Noyes

(1) J. T. Jefferies: a model based on the Na D lines.
(2) O. Gingerich and J. C. Rich: the far ultraviolet spectrum of the sun.
(3) A. Sauval: solar flux from 900 to 2000 Å.
(4) J. E. Blamont and C. Malique: observations of the variation of intensities at 1216 and 1300 Å near the solar limb.
(5) Y. Cuny: hydrogen lines.
(6) R. G. Athay and A. Skumanich: emission cores in the H and K lines.
(7) S. Dumont: étude de la zone de transition à partir des raies de resonance du Ca II.
(8) J. Linsky: the temperature distribution in the solar chromosphere as inferred from the calcium H and K lines.

(9) J. Zirker: empirical study of the H and K-line cores.
(10) H. A. Beebe and H. R. Johnson: predicted profiles of the Mg II resonance lines.
(11) P. J. Léna: observations of the center-to-limb variation of the solar brightness in the far infrared.
(12) R. W. Noyes, J. M. Beckers, and F. J. Low: center-to-limb variations in the far infrared and millimeter regions.
(13) C. de Jager and L. Neven: source functions in the cores of infrared Fraunhofer lines.
(14) J. Houtgast and C. de Jager: remarks on the model of the low chromosphere.

C. Wednesday morning 19 April 1967; chairman: J. T. Jefferies

First attempt to summarize: (1) Temperatures, (2) Turbulence, (3) Inhomogeneities.

D. Thursday 20 April 1967; chairman: J.-C. Pecker

(1) *Dynamical problems:*
O. Namba: granulation.
R. W. Noyes: supergranulation.
P. Delache: velocity fields.
(2) *A priori considerations on the construction of photospheric models:*
A. Skumanich: convection.
O. Gingerich: radiative equilibrium.
P. Souffrin: waves in the solar atmosphere.
(3) *Some problems of methodology:*
P. Delache: the Laplace transform.
J.-C. Pecker: roughness of the photosphere.
J. T. Jefferies: the determination of $\Delta\lambda_D$.
(4) *Meetings of working groups.*

Addendum II: Composition of Working Groups

(1) *Working group for the continuous radiation:* O. Gingerich (chairman), H. Holweger, D. L. Lambert, J. P. Mutschlecner, R. W. Noyes, A. Sauval.
(2) *Working group for spectral-line studies:* R. G. Athay, Y. Cuny, J. T. Jefferies, E. Müller, J.-C. Pecker (chairman).
(3) *Working group for temperature fluctuations and velocities:* R. M. Bonnet, P. Delache, C. de Jager, O. Namba, A. Skumanich (chairman), P. Souffrin, G. Withbroe.

Acknowledgments

We wish to thank Mr. Duane Carbon for carrying out the BCA computations and for preparing the accompanying graphs.

References

Bonnet, R. M. and Blamont, J. E.: 1968, *Solar Phys.* **3,** 64.
Carbon, D., Gingerich, O., and Kurucz, R. L.: 1968, *Solar Phys.* **3,** 55.
Detwiler, C. R., Garrett, D. L., Purcell, J. D., and Tousey, R.: 1961, *Ann. Geophys.* **17,** 263.
Edmonds, F.: 1964, *Astrophys. J.* **139,** 1358.
Gingerich, O.: 1963, *Astrophys. J.* **138,** 576.
Gingerich, O.: 1964, *Smithsonian Astrophys. Obs. Spec. Report* **167,** 17.
Gingerich, O. and Rich, J.: 1966, *Astron. J.* **71,** 161.
Gingerich, O. and Rich, J.: 1968, *Solar Phys.* **3,** 82.
Goldberg, L., Müller, E. A., and Aller, L. H.: 1960, *Astrophys. J. Suppl.* **5,** 1.
Heintze, J.: 1965, 'The Extreme Limb of the Sun', *Rech. Obs. Astron. Utrecht* **17** (2).
Heintze, J. R. W., Hubenet, H., and De Jager, C.: 1964, *Bull. Astron. Inst. Neth.* **17,** 442.
Holweger, H.: 1967, *Z. Astrophys.* **65,** 365.
Labs, D. and Neckel, H.: 1962, *Z. Astrophys.* **55,** 269.
Lambert, D. L.: 1967, Communication at the Bilderberg Conference.
Léna, P.: 1968, *Solar Phys.* **3,** 28.
Mutschlecner, J. P.: 1968, *Solar Phys.* **3** (in press). See also Müller, E. A. and Mutschlecner, J. P.: 1964, *Astrophys. J. Suppl.* **9,** 1.
Noyes, R. W., Beckers, J. M., and Low, F. J.: 1968, *Solar Phys.* **3,** 36.
Rich, J.: 1966, *Silicon and Carbon Monoxide Absorption in the Solar Ultraviolet Spectrum,* Thesis, Harvard University, Cambridge, Mass.
Sauval, A.: 1968, *Solar Phys.* **3,** 89.
Tousey, R.: 1963, *Space Sci. Rev.* **2,** 3.
Withbroe, G. L.: 1968, *Solar Phys.* **3,** 146.

SPECTRUM MEASUREMENTS OF THE SUN NEAR 1 CM WAVELENGTH

DAVID H. STAELIN, NORMAN E. GAUT, SARA E. LAW, and WOODRUFF T. SULLIVAN III*

M.I.T., Research Laboratory of Electronics, Cambridge, Mass., U.S.A.

(Received 13 June, 1967)

Abstract. The solar brightness temperature was measured at frequencies of 19.0, 21.0, 22.2, 23.5, and 25.5 GHz on 8 days in February and March, 1966. The observed average brightness temperatures were 10800, 10900, 11000, 10700, and 9800 °K, respectively. The daily values were close to these averages.

The solar brightness spectrum near 1 cm wavelength was measured on several days in the period January–March, 1966. The measurements were made with a 5-channel K-band radiometer mounted on the Massachusetts Institute of Technology, Lincoln Laboratory 28-ft millimeter antenna (Staelin and Barrett, 1966). The frequencies observed were 19.0, 21.0, 22.2, 23.5, and 25.5 GHz. The measurements were made by taking drift scans across the center of the sun every 10 min for a period of approximately 2 hours each day. The resulting measurements were corrected for atmospheric absorption by extrapolation to zero atmosphere of the observed dependence of solar intensity upon elevation angle.

The results were calibrated by comparison of the solar data with the average lunar-brightness temperature observed over a lunation. The comparison was made using the antenna patterns measured at each frequency (Staelin and Barrett, 1966). The true lunar brightness temperature was estimated from observations by many observers at many wavelengths (Moran, 1965). The errors involved in this comparison technique are very nearly the same at each frequency and thus the relative errors are small. The relative errors were dominated by the effective receiver noise and system-gain fluctuations present during the lunar and solar observations. If it is assumed that the true average lunar brightness is within 5% of the assumed 215 °K, then the absolute r.m.s. error at each frequency is approximately $\pm$ 700 °K.

The measured average solar brightness temperatures of the center of the sun are listed in Table I, together with the relative r.m.s. errors. Only the eight most accurate spectra were averaged. None of these spectra showed evidence of solar flares. The rest of the data are not included here.

This average spectrum is composed of data obtained on February 10, 18, 23, and 24, and March 3, 10, 11, and 14, 1966. The spectrum did not appear to vary much over this period.

* Presently at the University of Maryland.

TABLE I

Observed K-band Solar Spectrum

Frequency (GHz)	Central Solar Brightness Temperature (°K)	Relative rms Error (°K)
19.0	10800	±400
21.0	10800	±400
22.2	11000	±500
23.5	10700	±500
25.5	9800	±300

Acknowledgement

This work was supported principally by the National Aeronautics and Space Administration (Grant NsG-419).

References

MORAN, J. M.: 1965, *Radiometric Observations of the Moon near One-Centimeter Wavelength.* S. M. Thesis, Department of Electrical Engineering, Massachusetts Institute of Technology, Cambridge, Mass.

STAELIN, D. H., and BARRETT, A. H.: 1966, 'Spectral Measurements of Venus Near 1-Centimeter Wavelength', *Astrophys. J.* **144**, 352–363.

OBSERVATIONS OF THE CENTER-TO-LIMB VARIATION OF THE SOLAR BRIGHTNESS IN THE FAR INFRARED (10 TO 25 MICRONS)*

PIERRE J. LÉNA

*Université de Paris et Observatoire de Meudon, France***

(Received 25 April, 1967)

Abstract. Existing models of the solar atmosphere predict a limb brightening in the far infrared wavelengths. At shorter wavelengths this effect is confined to the extreme limb but at 25 μ it extends inward from the limb to $\cos\theta = 0.3$. Observations of $I(\mu)/I(1.0)$ were made with the McMath Solar Telescope through atmospheric windows at 10.4, 17.9, 20.4, and 24.2 microns, respectively. The measurements, after having been smoothed, are compared with the theoretical predictions taking into account the diffraction pattern of the telescope. The expected brightening does not appear.

1. Introduction

Different authors have emphasized the importance of measurements of the center-to-limb variation of the solar brightness at long wavelengths (PECKER, 1965; DE JAGER, 1963), in order to get more information on the transition layer between photosphere and the chromosphere, as well as on the absorption of the H^- ion. From different existing models of the solar atmosphere, NOYES *et al.* (1966) and LÉNA (1966) have computed expected center-to-limb variations at various wavelengths from the near-infrared to the submillimetric range. At the longer wavelengths, all the computations predict limb brightening at the extreme limb: this effect should be measurable at $\lambda = 20\mu$, for $\cos\theta \leqslant 0.3$.

Except for the millimeter to meter wavelengths, the previous observations of the relative intensities at different points on the solar disc in the continuum emission have been made at a wavelength $\lambda = 12.95\ \mu$ (SAIEDY, 1960), or below (PIERCE *et al.*, 1950), and extend from the center of the disc to, roughly, $\cos\theta = 0.4$ (1 min of arc from the limb). In a footnote, NOYES *et al.* (1966) report on an observation at 24.3 μ, $\cos\theta = 0.2$, in which neither brightening or darkening was observed. The present work extends previous observations; the center-to-limb variation has been measured at selected wavelengths between 10 and 25 μ from the center of the disc, $\cos\theta = 1.0$ to $\cos\theta = 0.14$, i.e. 11 sec of arc from the limb (see Note added in proof, p. 35). The results are presented herewith and compared with the theoretical predictions.

2. Observations

Measurements have been made with the McMath Solar Telescope at Kitt Peak National Observatory. The diameter of the main mirror of this telescope is 158 cm;

* Kitt Peak National Observatory Contribution No. 260. – Operated by the Association of Universities for Research in Astronomy, Inc., under contract with the National Science Foundation.
**Presently at Kitt Peak National Observatory, Tucson, Ariz., U.S.A.

the width at half intensity of the central lobe of the diffraction pattern is 3.4 sec of arc at $\lambda = 20\ \mu$. As this is not negligible it will be taken into account for the reduction of the results.

The measurements are relative: the intensity at various points along a solar radius is compared to the reference intensity of the center of the disc. Because of the vignetting in the telescope's focal plane, it has not been possible to measure *at the same time* the intensities at two widely different points in the sun's image. Rather, it has been necessary to intercompare the intensity at the center with a point near the limb by slowly moving the sun's image over the slit of the spectrometer and repeating this scan many times. However, by means of a mirror arrangement the detailed profile of the extreme limb has been obtained by rapid scanning of a small portion of the image, for which vignetting is negligible. The limb has been scanned using slits of different widths but of a constant length of 24 mm (55 sec of arc). The scans have been made at equatorial solar latitudes, either at the East or at the West limb. The scanned areas have been chosen without spots or visible solar activity.

Spectral isolation is obtained with a grating (Bausch & Lomb, 40 g/mm), plus one InSb or Ge interference filter, plus one or two Reststrahlen LiF reflection filters for wavelengths beyond 15 μ. In each case the amount of short-wavelength radiation reaching the detector has been measured and kept below 0.5% of the long-wavelength signal. The detector is a Golay cell which is made to act as a variable capacitor when placed in a measurement bridge driven at $f = .55$ Mc/s. The incident radiation is modulated at $f' = 10$ c/s. After suitable demodulation and filtering, the signal is recorded on a strip chart recorder or digitized and summed by a computer.

Except for possibly a few weak molecular lines, the spectrum of the sun is continuous in the range 10–25 μ. Since the phenomenon in which we are interested should not be strongly dependent on the wavelength, we are allowed to use a rather broad spectral interval; hence, in order to gain signal, an interval of $\Delta\lambda \simeq 1.4\ \mu$ was employed for most of the measurements.

The selection of the wavelengths in the spectrum cannot be arbitrarily decided because of the telluric absorption. The spectrum of the atmospheric absorption made by Farmer and Key (1965) was examined to decide which windows were to be used for this investigation. In spite of having selected the most open windows, the observed fluctuations of the atmospheric transmission are not negligible. For even on dry days, at wavelengths longer than 15 μ, a 30% variation in transmission in periods of about 10 min was noted. However, the frequency spectrum of this fluctuation varies roughly as $1/f$ in the vicinity of the frequency $f_0 = 1$c/s. If this is true then there is a gain in scanning the incoming data rapidly and summing repeated scans, as the following analysis shows.

Assume a radial scan of the sun and that this scan is sampled to get the measurement of the intensity in n points. Let T be the total duration of the scan. Each sampled point will correspond to the incident energy E integrated during a time T/n. The signal per point is then

$$S_1 = E\,T/n\,. \tag{1}$$

If the noise is assumed to have a $1/f$ *power spectrum*, the r.m.s. noise is, for sampling interval T/n,

$$N_1 = \frac{K}{\sqrt{f_0}}\left(\frac{T}{n}\right)^{\frac{1}{2}} = \frac{KT}{n^{\frac{1}{2}}} \tag{2}$$

where $(T/n)^{\frac{1}{2}}$ describes the random character of the noise and $f_0 = 1/T$ the frequency dependence. K is a constant.

Now assume that the radial scan is scanned N times faster and that N successive scans are added. The signal is

$$S_2 = \sum_N E\,T'/n = S_1 \tag{3}$$

where $T' = T/N$; the noise is

$$N_2^2 = \sum_1^N \frac{(KNT')^2}{f_0' n}, \tag{4}$$

where $f_0' = 1/T'$. Equation (4) then reduces to

$$N_2 = N_1 \cdot N^{-\frac{1}{2}}. \tag{5}$$

Thus the Signal/Noise ratio has been improved by a factor $(N)^{\frac{1}{2}}$.

The only limitation in this process has been the response time of the Golay cell (30 ms). The improvement by this technique has been good enough to reach a point where the predominant noise is no longer the atmospheric noise but the detector noise, which is mostly 'white' noise in this frequency range and which is not relevant in the above treatment.

In order to add repeated fast scans the SDS 910 computer was used. By suitable programming, the computer took care of scanning the image by the telescope, of the sampling and the numerical filtering of the data, and of the superposition and addition of the successive scans.

3. Reductions

Most of the observations have been reduced with a CDC 3200 computer in a four-step process. The Cooley and Tukey (1965) algorithm allows a versatile use of the Fourier transform for numerical deconvolution and filtering. The process here has been to

(1) Fourier transform the observed center-to-limb curve.

(2) Divide this transform by the transform of the scanning slit (which is the function $\operatorname{sinc} s = \sin \pi s/\pi s$); this is a deconvolution.

(3) Cut off the resultant transform above a frequency f_0, defined in such a way it retains in the original curve any structures wider than the slit width ($f_0 = 3F_0/4$, where F_0 is the first zero of $\operatorname{sinc} s$).

(4) Perform the inverse Fourier transform to get the reduced center-to-limb curve.

Figures 1 to 4 give the reduced observations at four different wavelengths: $\lambda = 10.4\ \mu$, $17.9\ \mu$, $20.4\ \mu$, $24.2\ \mu$, The noise level is the r.m.s. value of what remains

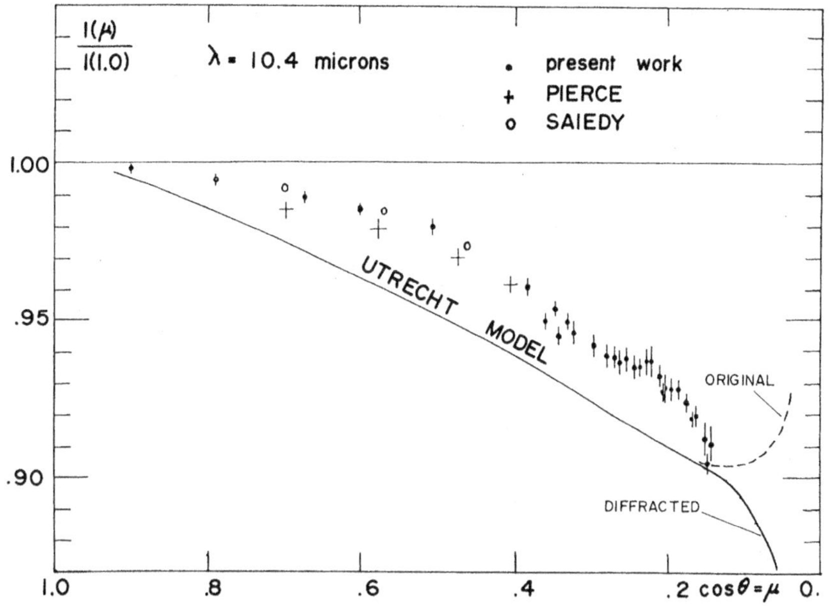

Fig. 1. Observed and computed limb darkening at 10.4 μ. The curve marked *original* is that calculated from the Utrecht model. The curve marked *diffracted* is the original curve modified by the diffraction pattern of the telescope.

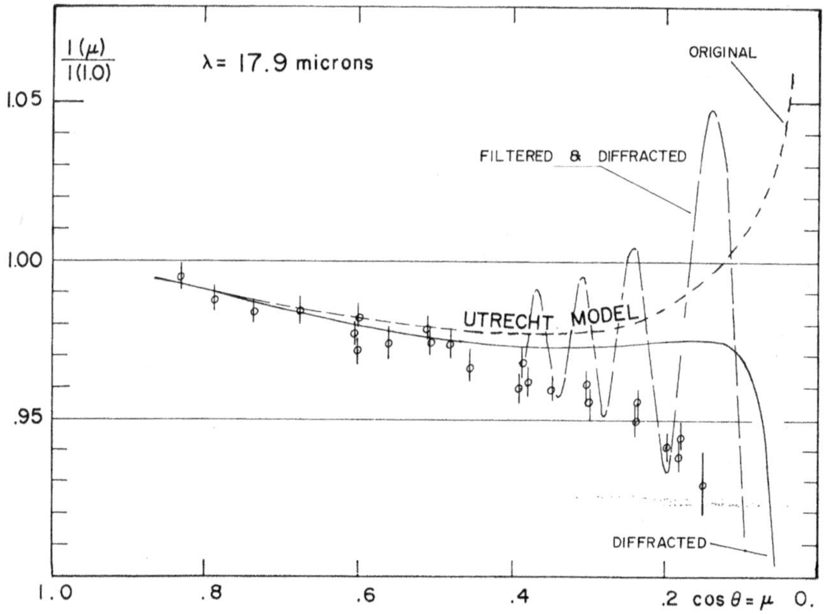

Fig. 2. Observed and computed limb darkening at 17.9 μ. The *original* curve is calculated from the Utrecht model. The *filtered and diffracted* curve is what one would observe of the original curve modified by the diffraction and strongly perturbed by the drastic suppression of its high frequencies.

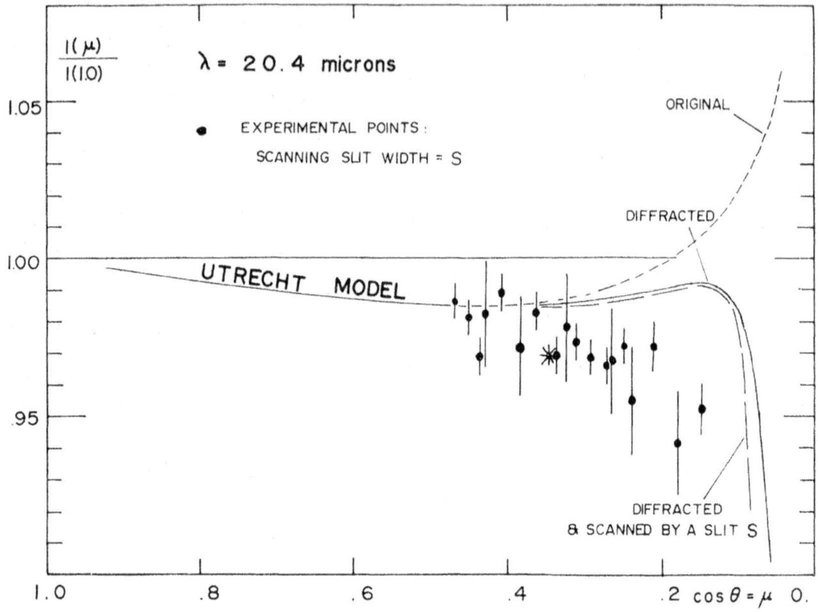

Fig. 3. Observed and computed limb darkening at 20.4 μ. The curve marked *diffracted and scanned by a slit S* is the theoretical curve that is to be compared with observation (see text). It is the original curve diffracted, filtered to avoid the Gibbs phenomenon, and convoluted with a slit of width *S*.

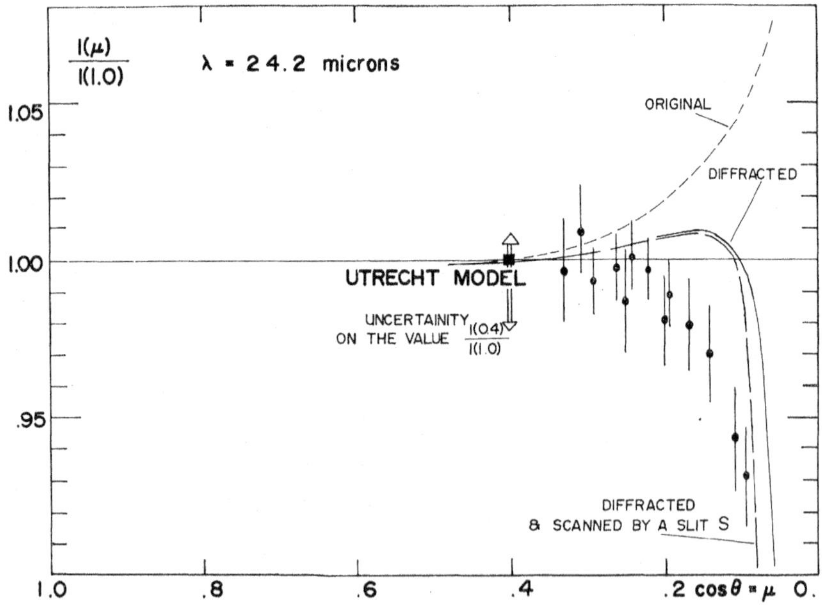

Fig. 4. Observed and computed limb darkening at 24.2 μ.

after filtering. The limb to which $\mu=0$ refers is the limb at the observed wavelength, arbitrarily fixed at the inflexion point of the observed curve. The scans at 20.4 and 24.2 μ cover only the limb area ($\cos\theta \leqslant 0.4$). To relate them with the reference central intensity, numerous systematic comparisons have been made between the center of the disc and one particular point of the limb, $\cos\theta=0.4$, chosen as a secondary reference.

The figures show four types of computed curves: Type (a): *Original* is the curve directly deduced from the model. Type (b): *Diffracted* is the original curve after diffraction by an objective of 158 cm and smeared by a slit, 2 sec of arc wide, parallel to the limb. (The smear simulates, on the model, some of those shifts which might have occurred during the superposition of successive scans.) Type (c): *Filtered and Diffracted* is Type (b) numerically filtered by the same process and with the same frequency cutoff as for the data. Figure 3 shows the Gibbs phenomenon from the strong frequency cutoff of the sharp limb. Type (d): *Diffracted and scanned by a slit S* means that Type (b) curve has been convoluted by a slit of the same width as the scanning slit, and the result filtered at a frequency f_1, high enough to just avoid the Gibbs phenomenon. The observational data on Figures 3 and 4 are to be compared with Type (d) curves because they have been filtered at the same frequency cutoff f_1, and were *not* deconvoluted.

To get an approximation of the diffraction pattern of the limb by the objective, the following analysis is made. The intensity distribution given by a point source in the focal plane of an objective (diameter D, focal length f) is the Airy function

$$f(m) = \frac{[2J_1(m)]^2}{m} \tag{6}$$

where $m=(\pi Dr)/f\lambda$ at the wavelength λ and at a point distant of r from the geometric image.

Note that the intensity distribution given by a long straight line source is

$$A(M) = 2\int_M^\infty \frac{mf(m)}{(m^2 - M^2)^{\frac{1}{2}}}\,dm\,, \tag{7}$$

which is the Abel transform of $f(m)$. We know that the one-dimensional Fourier transform of the function A is the two-dimensional Hankel transform of the function f (see BRACEWELL, 1956). Our case is particularly simple since the Hankel transform of $2J_1(m)/m$ is the function $\Pi(r/2)$ which is 1 for $r \leqslant 1$ and 0 elsewhere. Therefore, the Fourier transform of A is simply the convolution square of $\Pi(r/2)$, in two dimensions, according to the properties of the Hankel transforms. This square is the overlapping area of two equal circles and has the analytical expression:

$$A'(s) = \pi(s)^* \pi(s) = \frac{2}{\pi}\left[\mathrm{Arc}\cos(s) - s(1-s^2)^{\frac{1}{2}}\right]. \tag{8}$$

The relation between the uni-dimensional observed function $O(M)$ and the true original function $T(M)$ is thus

$$O(M) = T(M) * A(M)\,, \tag{9}$$

and performing the Fourier transform

$$O'(s) = T'(s) \cdot A'(s). \tag{10}$$

Since our observations go very close to the limb, it is important to know just how accurate are the above approximations.

As a check, we have first computed the diffracted aspect of a straight limb represented by a step-function. Then, we computed the exact two-dimensional diffraction pattern of a curved limb having the same step-function. In the instance of this particular and simple limb-profile, an exact computation is possible. The results compare within 0.4% for $\mu \geqslant 0.1$ at $\lambda = 25$ μ and below. The approximation is then legitimate in the range of our observations and within their accuracy. Thus, it is correct to start from the model T with Type (a) curves and to compute its Fourier transform. Next, one multiplies by the transform A', suitably scaled and then by a smearing function sinc s. Finally, one computes the inverse transform which gives curves of Type (b).

4. Discussion

Figures 1–4 show the theoretical limb-darkening and brightening curves based on the Utrecht Photospheric Model (Heintze *et al.*, 1964). The comparison between the observed and the computed data shows a discrepancy. Below about 15 μ, the computed curve is under the experimental points; this trend is reversed at 17.9 μ and longer wavelengths where the observed brightness is smaller than the predicted one. It is obvious that, to the accuracy of our measurements, the expected brightening does not appear. Either a modification of the photospheric model in its highest layers, or a modification of the coefficients of the H^- ion free-free absorption, or both, is required to fit observation with theory. In the transition zone between photosphere and chromosphere, where the temperature minimum is located, the thermal gradient might be lower than the value accepted in the Utrecht model, or *a fortiori*, in other models which give still stronger discrepancies at the long wavelengths.

As for future work, until observations can be made above the atmosphere, more extended and accurate sets of data are needed at 24 μ. The quasi-total absorption of the terrestrial atmosphere starts above 25 μ, except for a very weak window at 34 μ, noticeable on Farmer's spectra. This fact gives very little hope for observations from the ground at still longer wavelengths, where the phenomenon of limb brightening is accentuated. The present results need to be extended at the extreme limb by improvement of the scanning procedures and by reduction of the instrumental noise. Furthermore, comparisons should be made with measurements in the far ultraviolet, where the energy and diffraction problems do not exist to the same degree.

Acknowledgements

This investigation has been made possible through the encouragement of Professor J. C. Pecker and has been supported by the French Government. Appreciation is also expressed to Dr. N. U. Mayall and Dr. A. K. Pierce for their acceptance of the program and their subsequent assistance at Kitt Peak National Observatory. The author is, moreover, indebted to Dr. J. W. Brault and Mr. Richard S. Aikens, who gave time and help on computer problems and the Fourier transform program.

Note added in proof. Recent improvements of experimental technique allow to scan up to 2 sec of arc from the limb, with a good stability of the image. The results will be published later.

References

Bracewell, R. N.: 1956, *Australian J. Phys.* **9,** 198.
Cooley, J. W. and Tukey, J. W.: 1965, *Math. of Computation* **19,** 297.
De Jager, C.: 1963, *Mém. Soc. Roy. Sci. Liège* (Fifth Series) **9,** 151.
Farmer, C. B. and Key, P. J.: 1965, *Appl. Opt.* **4,** 1051.
Heintze, J. R. W., Hubenet, H., and De Jager, C.: 1964, *Bull. Astron. Inst. Neth.* **17,** 442.
Léna, P. J.: 1966, *Ann. Astrophys.* **29,** 361.
Noyes, R. W., Gingerich, O., and Goldberg, L.: 1966, *Astrophys. J.* **145,** 344.
Pecker, J. C.: 1965, in *The Solar Spectrum* (ed. by C. de Jager). D. Reidel Publishing Co., Dordrecht, Holland, p. 29.
Pierce, A. K., McMath, R. R., Goldberg, L., and Mohler, O. C.: 1950, *Astrophys. J.* **112,** 289.
Saiedy, F.: 1960, *Mon. Not. Roy. Astron. Soc.* **121,** 483.

OBSERVATIONAL STUDIES OF THE SOLAR INTENSITY PROFILE IN THE FAR INFRARED AND MILLIMETER REGIONS

R. W. NOYES

Smithsonian Astrophysical Observatory, Cambridge, Mass., U.S.A.

J. M. BECKERS

Sacramento Peak Observatory, Sunspot, N.M., U.S.A.

and

F. J. LOW

University of Arizona, Tucson, Ariz., U.S.A.

(Received 30 June, 1967)

Abstract. Observations of the intensity distribution near the solar limb at 2.43 and 22.5 μ, show the absence of limb brightening to within 1 or 2 arc sec of the limb. Observations at 1.2 mm indicate limb brightening at this wavelength. These results are compared with the Utrecht Reference Photosphere and with existing data on the solar flux in the millimeter range, and suggest that the temperature minimum is broad and extends above $\tau_{5000} = 2 \times 10^{-3}$. A sharp rise of temperature is required above $\tau_{5000} = 10^{-5}$.

In this paper we shall discuss a series of investigations that we have carried out during the past year to study the solar intensity distribution in the far infrared and millimeter regions. Part of the work has been reported previously (NOYES, GINGERICH, and GOLDBERG, 1966; NOYES, 1966; NOYES *et al.*, 1966), but some of the results are new.

In February and March of 1966 we made preliminary measurements of the solar limb intensity at 24.3 μ in an attempt to see the brightening predicted by several current models of the solar atmosphere (e.g., NOYES, GINGERICH and GOLDBERG, 1966). The results were negative to within the 10-arcsec resolution limit of our measurements. This observation is described in Section 1.

Next we measured the infrared limb profile with higher spatial resolution during the November 12, 1966, total solar eclipse. This experiment has indicated that limb darkening at 22.5 μ persists to the limit of our resolution, about 1 or 2 arcsec from the limb. We discuss the experiment and preliminary results in Section 2.

At the same time, we have analyzed several drift curves of the sun made at a wavelength of 1.2 mm, and found definite evidence for limb brightening at this wavelength. We describe these data and their analysis in Section 3.

Finally, in Section 4 we discuss the implications of these various results for the temperature structure of the low chromosphere.

1. Measurements of the Solar Limit Profile at 24.3 μ

In February and March of 1966 we made several measurements of the solar limb intensity distribution, using the 5-foot $f/14$ telescope of the Lunar and Planetary

Laboratory of the University of Arizona. A carefully selected sheet of black polyethylene 0.004″ thick was placed in front of the telescope to prevent excessive heating of the optics. The transmission of the polyethylene, while very low in the visible, was about 40% at 24 μ. Infrared scans of Venus with and without the polyethylene in place showed no noticeable degradation of the image. The infrared radiation passed through a narrow-band interference filter with its peak transmission at 24.3 μ and a halfwidth of 0.9 μ, and then through a 5-arcsec circular aperture at the detector. The detector was a germanium bolometer cooled by liquid helium to 4 °K. The detectivity of such a system is actually far greater than the noise introduced by other sources, principally sky transmission fluctuations, but this equipment was already in use for stellar work and was readily available.

Fluctuations in the emission and transparency of the daytime sky were largely, but not completely, removed by chopping between two regions of the sun about

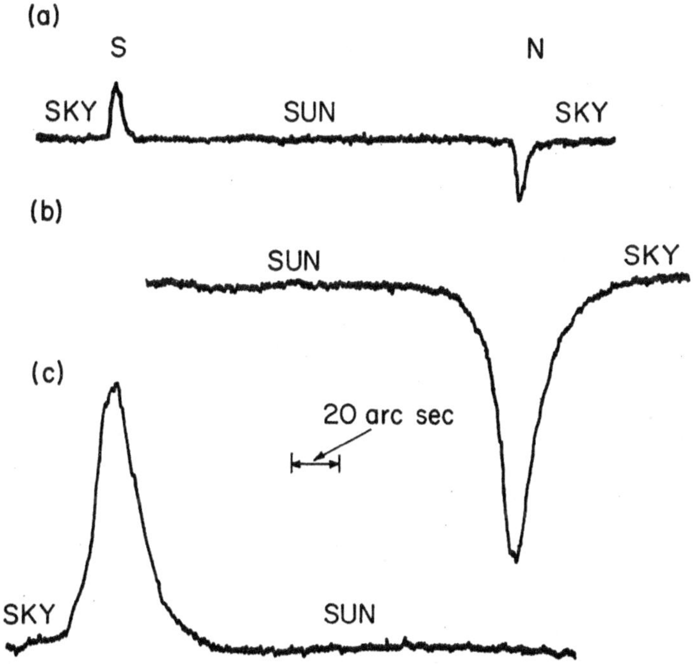

Fig. 1. Differential chopper scans of the sun at 24.3 μ, March 22, 1966. Chopper separation: 18 arcsec. (a): entire sun, (b) North limb, (c) South limb. (b) and (c) are at higher gain and higher dispersion. The scale for (b) and (c) is given in the figure.

18 arcsec apart, at the rate of 10 cps. Scans were made across the sun in the direction (N–S) of the line between the two regions being chopped, so that the signal approximated the spatial derivative of the intensity. Figure 1 shows a typical scan of the entire sun, with separate scans of the limb region at higher gain and time resolution. The flatness of the 25-μ sun is at once evident.

Reasonably good data were obtained for 18 sets of limb scans. After averaging scans of the North and South limbs to remove the effects of an asymmetry in the instrumental profile, we were able to draw two conclusions:

(a) At a position 40 arcsec inside the limb, where the effect of the instrumental profile on the measurements is small, the slope of the intensity is $-0.06\% \pm 0.07\%$ arcsec. In terms of $\mu = \cos\theta$, at $\mu = 0.28$, $\mathrm{d}I\,\mathrm{d}\mu = -0.16 \pm 0.20$.

(b) At a distance 10 arcsec from the limb ($\mu = 0.14$), which lies within the instrumental profile of the apparatus, the instrumentally smoothed profile is asymmetric, even after averaging the North and South limbs. This presumably is caused by an asymmetry in the original unsmoothed derivative of the intensity about the limb position. The direction of the asymmetry at 10 arcsec from the limb is such as to imply limb darkening at that position. Thus we have somewhat indirect evidence that the darkening persists to within at least 10 arcsec of the limb.

2. Eclipse Measurement at 22.5 μ

In order to attain better spatial resolution near the limb, we attempted a similar observation during the November 12, 1966, total solar eclipse. This eclipse had the appeal of passing over the Andes, where high-altitude sites are reasonably accessible. We set up a 12-inch $f/8$ cassegrain telescope at an elevation of 14000 feet on the centerline of the eclipse north of Arequipa, Peru. A prefilter of high-quality black polyethylene, similar to that described above, was stretched over the entrance aperture of the telescope. At the cassegrain focus we placed a rectangular slit, 1 arcmin by 3.5 arcmin, with its long axis parallel to the apparent motion of the moon. A rotating two-level mirror displaced the image by 4.2 arcmin in the direction of the long axis of the slit, at a frequency of 12 cps. This allowed us to chop between a region including a 1-arcmin slice of the solar crescent and a region on the dark lunar disk. The chopping eliminated sky-emission fluctuations, but not transparency fluctuations. The solar radiation passed into a liquid helium dewar through a window of Irtran 6, then successively through the focal plane slit, through a 22.5-μ interference filter (Figure 2), and through a KRS-5 Fabry lens, which finally imaged the objective onto the detector. The detector was a germanium bolometer especially constructed for the experiment at the Lunar and Planetary Laboratory of the University of Arizona, and had a noise-equivalent power of $1.6 \times 10^{-12} w$ at 4°K.

The entire slit-filter-lens-detector assembly was cooled to 4°K by a liquid helium bath.* The signal from the detector was fed into a low-noise parametric amplifier, a phase-lock differential amplifier, an integrating digital voltmeter, and a digital printer.

On the morning of the eclipse thin clouds covered the sky, gradually thinning out as totally approached. At totality, the sun appeared to be uniformly covered by very thin cirrus, although examination of a large-scale photograph of the sun obtained by

* The liquid helium, as well as liquid nitrogen for precooling, was airshipped from the United States, then transported by train and truck to the site, which was 40 miles horizontally and 6000 feet vertically from the nearest paved road. It was only through the efforts of many people, especially the Smithsonian Observing Station in Arequipa, Peru, that the logistic difficulties were overcome.

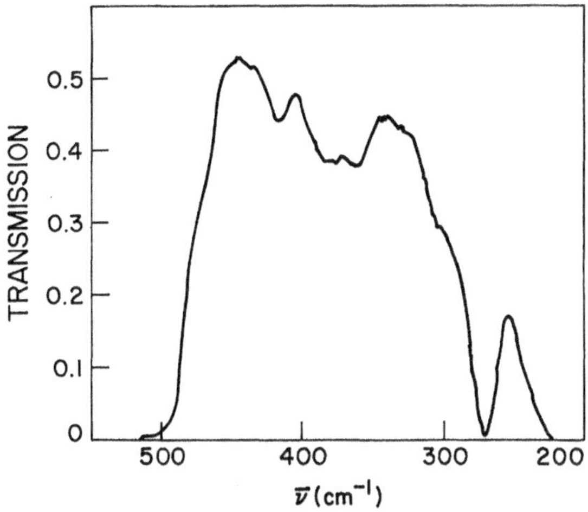

Fig. 2. Transmission curve of the filter used for the eclipse observation. The filter is completely blocked below 555/cm. (The short wavelength energy passed by the system was measured to be less than 0.1% of that passed in the 20-μ region.)

a Kitt Peak National Observatory expedition at the same site does not show the cirrus. Shortly after totality the sky became very cloudy.

Before second and third contacts, we attempted to set the slit on the lunar limb at those position angles that were calculated by one of us (BECKERS, 1966) to cause a minimum of resolution loss due to irregularities of the lunar limb and distance off axis from the sun–moon centerline. Unfortunately, flexure in our guide telescope caused an error in positioning the slit, and we missed these optimum positions. Even more serious, at third contact the limb was placed quite close to the end of the slit, where the sensitivity drops rapidly with position. Calibration curves of the variation of slit sensitivity with position were obtained the previous day; these indicate that at third contact a sizeable correction to the data is necessary.

The data near second and third contacts are shown in Figures 3 and 4. Figures 3a and 4a show the raw data, i.e., the integrated intensity $I(x)$ between the lunar and solar limbs, as a function of the distance x in arcsec between the lunar and solar limbs. There is considerable noise in the data, consisting of two types: (a) high-frequency ($\geqslant 1$ cps) noise, which is proportional to the signal $I(x)$ and is almost certainly due to rapid transparency fluctuations, and (b) low-frequency ($\leqslant 0.1$ cps) variations, which are presumably due to slower drifts in sky transparency. Fortunately, during the few seconds preceding second contact and following third contact, there is no evidence for drifting in sky transparency.

In Figures 3b and 4b we have plotted the 'average remaining intensity' $\bar{J}(x) = I(x)/x$. The run of these points itself should be an index of whether the limb region shows brightening or darkening. The fluctuations in these data make clear the difficulty of deriving a very precise estimate of the limb darkening from them. Actually, the data for $\bar{J}$ suffer from other uncertainties, which are not evident in Figures 3b and 4b: (a) there is an uncertainty of about ± 0.1 arcsec in the location of the limb, which causes a fractional uncertainty of about $0.1/x$ in $\bar{J}$ at a distance x arcsec from the limb, (b) the third contact data have been corrected for the variation of sensitivity with position in the detector slit, described above. This makes these data considerably less reliable than the second contact data.

Nevertheless, there does appear a tendency for the mean remaining intensity $\bar{J}$ to drop slightly toward the limb. This tendency may be followed to within about 3 arcsec of the limb before second contact, where the apparent influence of irregularities of the lunar profile becomes noticeable, and to within about 1 or 2 arcsec of the limb in the less reliable third contact data. We must emphasize, however, that noise in the data, uncertainties in the calibration, and possible transparency drifts make such a conclusion a probability rather than a certainty.

Several curve-fitting procedures have been applied to the data in Figures 3 and 4 in an attempt to get the function $J(x) = \mathrm{d}I(x)/\mathrm{d}x$ directly. These include least-squares

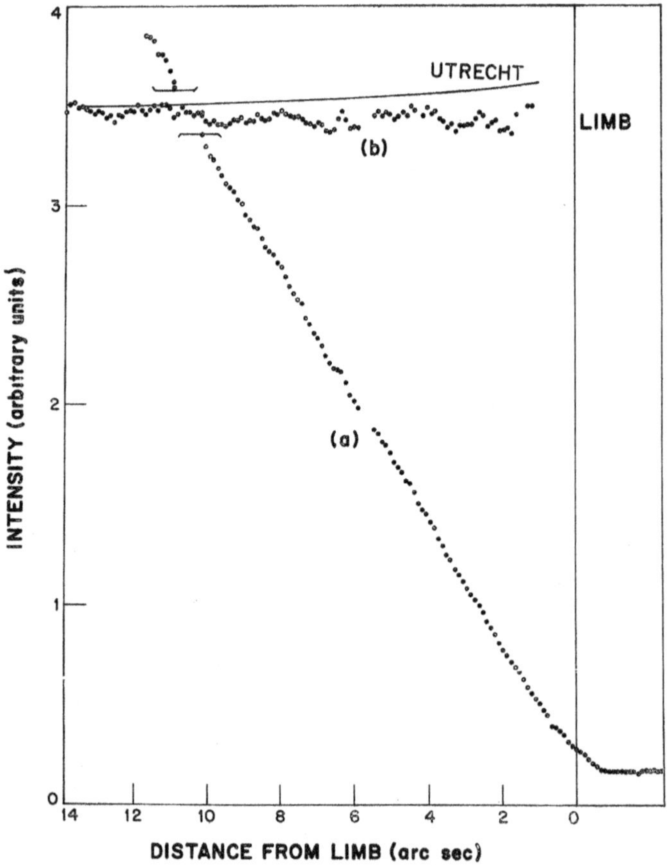

Fig. 3. (a) Integrated intensity $I(x)$ before second contact. (b) Average remaining intensity $\bar{J} = I(x)/x$, where $x =$ distance remaining to limb. Solid line is the prediction of the Utrecht Reference Photosphere (HEINTZE, HUBENET, and DE JAGER, 1964).

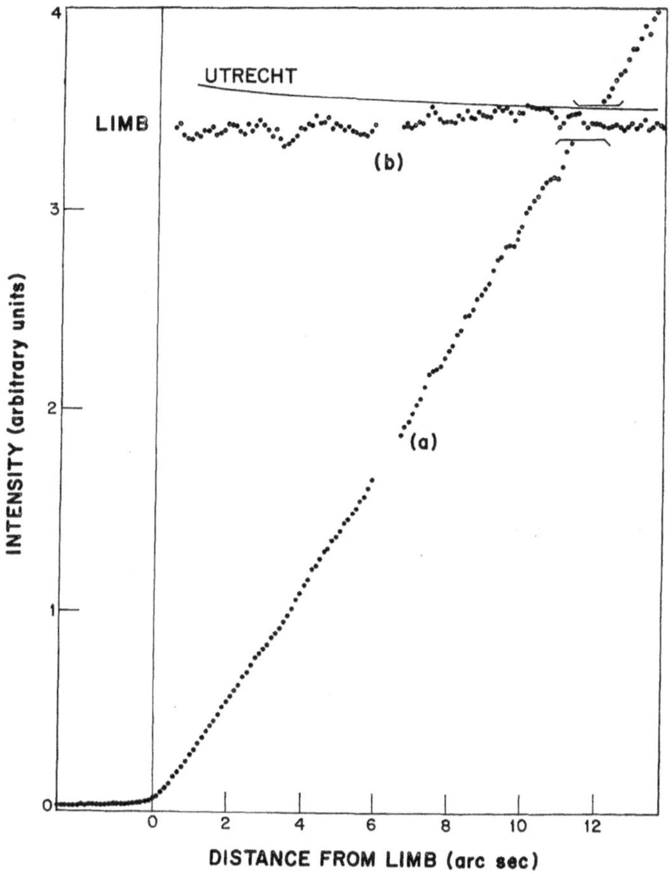

Fig. 4. (a) Integrated intensity $I(x)$ after third contact. (b) Average remaining intensity $\bar{J} = [1/\bar{S}(x)]$ $I(x)/x$, where $\bar{S}(x)$ is the average slit sensitivity in the region between 0 and x. Solid line is the prediction of the Utrecht Reference Photosphere (HEINTZE, HUBENET, and DE JAGER, 1964).

fits of polynomials directly to the entire data string near each contact, and polynomial fits at each point to those points in its immediate neighborhood. We also tried other numerical techniques that suppress the high-frequency noise relative to the lower frequency components of the signal.

For the second contact data, the various methods concur in yielding either a flat distribution or slight darkening near the limb. (Further than 10 arcsec inside the limit some apparent brightening occurs, but we believe that this is due to a slow drift in sky transparency; for the last 10 arcsec, the time of about 25 sec remaining until contact is sufficiently short to make the sky drift less important.) The amount of darkening near the limb is rather sensitive to the fitting technique used. None of the techniques used produced a brightening at the limb, although the data are not good enough to rule out a very small brightening. The amount of brightening predicted by the Utrecht Reference Model (HEINTZE, HUBENET, and DE JAGER, 1964) seems outside the limits of error, however. The measured darkening for $x < 10$ arcsec is $(1/J)(dJ/dx) = -0.005 \pm 0.005$/arcsec, where the uncertainty simply includes the spread among different curve-fitting techniques.

The third contact data give identical results, but are unfortunately less reliable, due to the calibration problem mentioned above. However, they can be followed to within about 1 arcsec of the limb, in distinction to the second contact data, and yield no evidence of a brightening even this close to the limb.

3. Millimeter Region

The millimeter data consist of three drift curves of the sun obtained in February 1965 at the National Radio Astronomy Observatory, using a 5-foot spun plastic antenna with a conical feed into a germanium bolometer. The effective wavelength was 1.2 mm. In addition, both drift curves and North–South scans of the full moon were obtained. The intensity scans of the moon in the North–South and East–West directions appeared very similar.

Figure 5 shows drift curves of the sun and the moon and their derivatives, with the East and West limbs averaged. Although the solar drift curves do not explicitly show limb brightening, we may infer its presence from the asymmetry of the drop-off of the smoothed intensity at the limb. If the intensity distribution were flat, we would expect its derivative to be symmetric about the limb position. The lower left curve of Figure 5 shows the observed derivative for the sun. We see that: (a) the tail to the

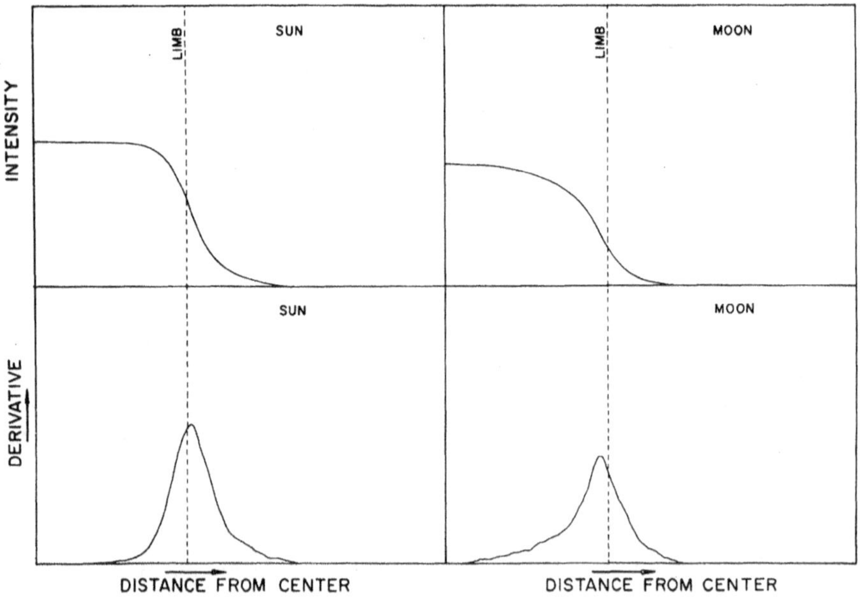

Fig. 5. Upper left: Solar drift curve at 1.2 mm, averaged over East and West limbs. Upper right: Lunar drift curve at 1.2 mm, averaged over East and West limbs. Lower cuıves are the spatial derivatives of the upper ones.

derivative is smaller on the side toward the disk center, and (b) the peak of the derivative is moved outward somewhat from the limb position. Both effects are easily explained by the presence of limb brightening which, while not large enough to survive as explicit brightening after instrumental smoothing, does slow down the drop-off of the instrumentally smoothed profile at and just inside the limb.

By way of comparison, the lower right-hand curve of Figure 5 shows the analogous curve for the moon, which is known to have considerable limb darkening at 1.2 mm. Here the effects are just the opposite: the increased slope on the side toward the center of the moon due to the lunar limb darkening is clearly evident, as well as a shift of the position of maximum slope inside the limb.

A rather crude attempt has been made to correct for the effects of instrumental smoothing, using BRACEWELL's (1955) chord-construction technique. This method suffers from our lack of detailed knowledge of the antenna pattern. We use the equation

$$I_{\text{corr}}(x) = 2I_{\text{obs}}(x) - \tfrac{1}{2}\left[I_{\text{obs}}\left(x - \frac{a}{2}\right) + I_{\text{obs}}\left(x + \frac{a}{2}\right)\right], \tag{1}$$

where I_{obs} and I_{corr} are the observed and corrected intensities, respectively (BRACEWELL, 1955). This equation assumes that the antenna profile looks like a single-slit diffraction pattern of halfwidth a. Although we do not know the 'best' value for a, it certainly is at least as large as the halfwidth of the lower left curve in Figure 5, which gives a crude measure of the instrumental profile. (Assuming that the true intensity near the solar limb is nearly a step function. Note that limb brightening can only serve to decrease the apparent halfwidth over the true antenna profile halfwidth.) This minimum halfwidth is $a = 4.9$ arcmin. Figure 6 illustrates the effect of applying Equation (1) to the data for several values of a, and shows explicit brightening for values of a greater than 4.0 arcmin. Thus we have further evidence that the inferred brightening is real.

Calculations were made of the area under the upper left curve of Figure 5, compared with the area under a step-function distribution extending to the white light limb, where we assume circular symmetry in both distributions. The ratio of these areas should equal the excess flux emitted by the 1.2-mm sun over that of a uniform disk of emissivity equal to that at the solar disk center. We find this ratio to be 1.14 ± 0.03, where the last figure gives the internal consistency of the three drift curves. Systematic errors due to the lack of circular symmetry of the sun or the antenna patterns are possible, but hard to estimate. In addition, the ratio would be an overestimate if the antenna smoothing lowered the intensity at the center of the disk, but the spread of the inferred antenna profile seems too narrow to allow this.

4. Discussion

The persistence of darkening at 22.5 μ to within 2 arcsec of the limb requires that the temperature minimum in the low chromosphere lie above $\tau_{5000} = 2 \times 10^{-3}$. The pres-

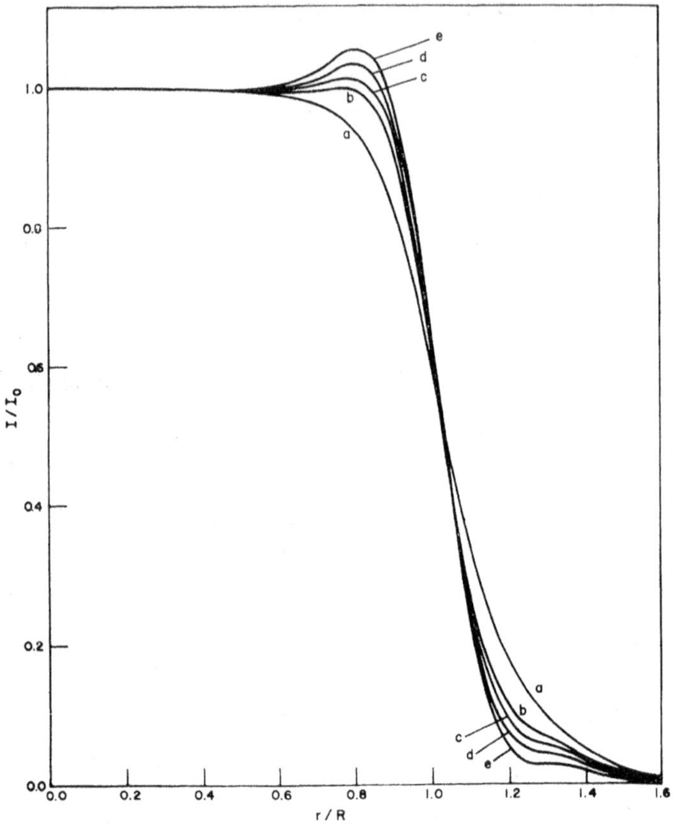

Fig. 6. 1.2-mm limb profile as restored from the original data using BRACEWELL's (1955) technique: (a) original data; (b) restoration with $a = 4.0$ arcmin; (c) restoration with $a = 4.6$ arcmin; (d) restoration with $a = 5.1$ arcmin; (e) restoration with $a = 5.7$ arcmin.

ence of limb brightening at 1.2 mm requires that the temperature minimum lie below $\tau_{1\,\mathrm{mm}} = 1$, or $\tau_{5000} \sim 1.5 \times 10^{-5}$. The present data do not place any further restrictions on the model. However, it is interesting to consider this data in the light of our knowledge of the solar intensity distribution with wavelength in the millimeter region. Figure 7 (see also Table I) shows the various published measurements of brightness temperature T_B in the far infrared and millimeter range, plotted versus wavelength in millimeters. The solid curve in Figure 7 is the prediction of the Bilderberg continuum atmosphere* (GINGERICH and DE JAGER, 1968), and is only one of many plausible fits to these rather widely scattered data, but any fit will have to show a rather sharp rise in the 1 to 10-mm range. In order to fit the brightness temperatures observed in the visible and infrared, a rather broad low-temperature region is necessary.

* Extrapolated upward according to $T(\tau_1) = 4820 + 215/\sqrt{\tau_1}$ above $\tau_1 = 10^{-3}$, where τ_1 is the optical depth at 1 mm. Since the millimeter data alone determined the model in this height range, the agreement between the model and the data is good.

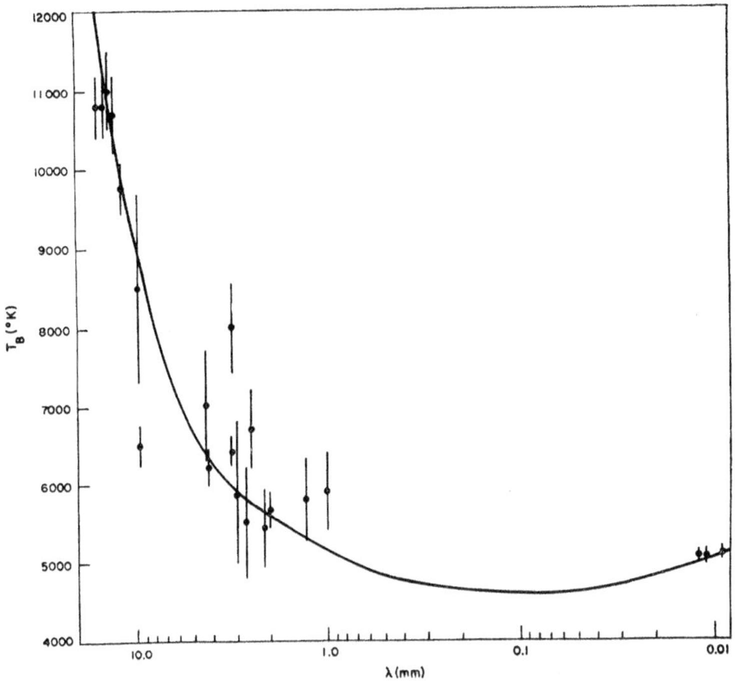

Fig. 7. Data from Table I on solar brightness temperature T_B in the sub-millimeter and millimeter regions, plotted against wavelength λ. The solid line is the predicted brightness temperature from the Bilderberg model (GINGERICH and DE JAGER, 1968).

TABLE I

Brightness temperature measurements in the submillimeter and millimeter ranges

λ_{mm}	T_B	ΔT_B	Reference
0.0086	5160	40	SAIEDY, 1960
0.0111	5040	30	SAIEDY, 1960
0.0120	5050	80	SAIEDY, 1960
1.0	5900	500	LOW and DAVIDSON, 1965
1.3	5800	500	BASTIN *et al.*, 1964
2.0	5670	230	WORT, 1962
2.15	5430	500	TOLBERT and STRAITON, 1961
2.5	6700	500	BASTIN *et al.*, 1964
2.73	5500	700	TOLBERT and STRAITON, 1961
3.0	5870	950	TOLBERT and STRAITON, 1961
3.2	6400	200	SIMON, 1965
3.2	8000	560	TOLBERT, 1966
4.3	6200	310	TOLBERT, 1966
4.3	7000	700	COATES, 1958
8.6	6500	260	TOLBERT, 1966
8.6	8500	1200	COATES, 1958
11.8	9800	300	STAELIN, 1968
12.8	10700	500	STAELIN, 1968
13.5	11000	500	STAELIN, 1968
14.3	10800	400	STAELIN, 1968
15.8	10800	400	STAELIN, 1968

If we accept a homogeneous model of the chromosphere, then, it is clear that a sharp temperature rise must occur above about $\tau_{1\,\mathrm{mm}}=1$, or $\tau_{5000}\sim 10^{-5}$. In view of the known inhomogeneous structure (spicules, etc.) above this level, however, such a conclusion seems unwarranted. For instance, the increased brightness temperature past 1 mm could be caused by the increased optical thickness of the hot chromospheric structures, while the mean atmosphere, i.e., the interspicular regions, remain rather cool.

Acknowledgements

We are grateful to Mr. Arnold W. Davidson for construction of the 22.5-μ bolometer, and for the major effort in securing the 1.2-mm solar drift curves. We thank Mr. James M. Percy for outstanding assistance in the eclipse expedition. The DuPont Company of Wilmington, Delaware generously supplied the special polyethylene prefilters for the infrared work.

References

Bastin, J. A., Gear, A. E., Jones, G. O., Smith, H. J. T., and Wright, P. J.: 1964, *Proc. Roy. Soc* **A278,** 543.
Beckers, J. M.: 1966, *Sac. Peak Research Note No. 25,* A.F.C.R.L. Office of Aerospace Research.
Bracewell, R. N.: 1955, *Australian J. Phys.* **8,** 200.
Coates, R. J.: 1958, *Astrophys. J.* **128,** 83.
Gingerich, O. and De Jager, C.: 1968, *Solar Phys.* **3,** 5.
Heintze, J. R. W., Hubenet, H., and De Jager, C.: 1964, *Bull. Astron. Inst. Neth.* **17,** 442.
Low, F. J. and Davidson, A. W.: 1965, *Astrophys. J.* **142,** 1278.
Noyes, R. W.: 1966, *Infrared Astronomy Conference*. Goddard Institute for Space Sciences New York.
Noyes, R. W., Gingerich, O., and Goldberg, L.: 1966, *Astrophys. J.* **145,** 345.
Noyes, R. W., Beckers, J. M., Low F., J., and Davidson, A. W.: 1966, *Astron. J.* **71,** 866.
Saiedy, F.: 1960, *Monthlty Notices Roy. Astron. Soc.* **121,** 483.
Simon, M.: 1965, *Astrophys. J.* **141,** 1513.
Staflin, D. H., Gaut, Norman E., Law, Sara E., and Sullivan, Woodruff T.: 1968, *Solar Phys.* **3,** 26.
Tolbert, C. W.: 1966, private communication.
Tolbert, C. W. and Straiton, A. W.: 1961, *Astrophys. J.* **134,** 91.
Wort, D. J. H.: 1962, *Nature* **195,** 1288.

THE HIGH-DISPERSION CONTINUOUS ULTRAVIOLET SOLAR SPECTRUM AND THE BALMER-JUMP

J. HOUTGAST

Utrecht University Observatory 'Sonnenborgh'

(Received 23 May, 1967)

Abstract. Among the intensities, determined at about 200 wavelengths between λ 3000 and λ 4100 Å in the spectrum of the centre of the sun's disk (Houtgast, 1965), the 32 highest ones (windows) were plotted and compared with absolute intensities given by other authors.

The intensities in between the Fraunhofer lines from λ 3600 to λ 4000 Å, as determined here for the first time with high dispersion, reveal a detailed picture of several absorption features, one of which can be attributed to a Balmer jump of 0.03, a value in accordance with that found for stars and in agreement with the strengths of the high Balmer lines.

The much higher value of the Balmer jump for the sun, as quoted in literature, in reality refers to the total intensity jump between λ 4000 and λ 3600 Å, which is mainly due to the crowding of Fraunhofer lines.

1. Comparison of Results

Absolute intensities in the solar spectrum were determined by several authors in different ways. As fundamentally different one can consider the measurements made with low and high resolving power. My measurements were made with the photoelectric equipment of Dr. E. W. Mitchell, attached to the Snow telescope and spectrograph at Mount Wilson; they have a resolution better than 0.1 Å. The measured points were chosen between the Fraunhofer lines, which does not mean that all these points are windows, where the intensity of the continuous spectrum is reached, since this may be prevented by crowding of lines in many spectral regions. Considering the total of 200 measured points, 32 of them, given in the last column of Table I, are distinctly high ones in their surrounding spectral regions. A plot of these points will give us, for the time being, the best impression of some characteristics of the continuous spectrum between λ 3000 and λ 4000 Å.

In Table I the results of several authors are given. Great differences are shown, but the best observations agree in more than one respect, e.g. a sudden rise in intensity from λ 3700 to λ 4000 Å and the absolute intensities between λ 4000 and λ 4100 Å. Five of the series of observations are plotted in Figure 1.

All but one series of absolute intensities were taken directly from the publications quoted. The one that was corrected by us is that of Labs and Neckel (1962). These authors describe exactly how and what they have measured, namely integrated intensities in wavelength bands of 20.0 Å. Afterward they made a correction for the influence of the Fraunhofer lines to obtain the intensity of the continuous spectrum by adding to the measured intensities the equivalent widths (from the *Utrecht Photometric Catalogue*, 1960) of the lines within each interval to the measured intensities. This,

TABLE I

Intensities I in 10^{14} erg $sec^{-1} cm^{-2} sterad^{-1}$, $\Delta\lambda = 1$ cm for the centre of the sun's disk at selected wavelengths between λ 2920 Å and λ 4100 Å

λ(Å)	M* 1939	P 1940	Ch. 1950	L-N 1962	M-S 1963	Ma 1957	$\overline{\text{Ma}}$ 1963	Ma 1965	H 1965
2920		0.31_5							
77.3									2.13
3000	1.10	0.91_5							
52.7									2.48
84.6									2.56
3100	1.30								
08.3									2.78
11.5/6								1.40	2.67
24.4								1.46	
49.1								1.42	2.72
67.5								1.28	
89.1/2								1.40	2.84
3200	1.72	2.12	2.52						
04.6/7								1.26	2.83
33.4								1.27	
59.3									2.88
62.6								1.28	
94.4								1.31	
3300.0/3/4	2.28	2.36						1.28	3.03
25.8								1.35	
55.6									3.08
81.8					2.47				
87.0					2.66			1.32	
3400.0/3/4	2.76	2.61	2.90				2.88	1.31	3.08
19.4					2.38				
29.2					2.64				
34.6								1.29	
70.9									3.09
99.7					2.61				
3500	3.08	2.53					3.04		
01.3						1.10			
04.1					2.82				3.13
39.3					2.86			1.20	
49.6									3.21
57.5						1.60			
95.5									3.28
3600	3.37	2.97	3.24				3.23		
12.3					2.67				
52.8/9					2.74	2.64			
58.6									3.41
59.0					2.85				
78.4/6					2.60	3.36			
91.8					2.80			2.39	
98.8						3.51			
99.3									3.40
—			3.39						

Table I (continued)

λ(Å)	M* 1939	P 1940	Ch. 1950	L-N 1962	M-S 1963	Ma 1957	$\overline{\text{Ma}}$ 1963	Ma 1965	H 1965
3700	3.64	2.95					3.31		
+			4.52						
55.6/8						3.73		2.62	3.44
76.8									3.66
80.0					3.42				3.86
82.9									3.93
84.9					3.42				
3800	3.91	3.05	4.56				3.84		
11.5						3.87		2.50	3.91
52.0						3.80			3.94
63.2									3.94
83.9					3.49				
3900	4.19	3.91					4.28		
01.3								3.20	
09.2									3.92
14.8									3.92
88.1									4.22
99.8/9						4.33		3.91	4.44
4000	4.48	4.44	4.60				4.61		
12.9								4.16	
13.2					4.70				
10–30				4.65					
20.7									4.47
29.0					4.75				
30–50				4.50					
50.6						4.70			
51.6									4.45
50–70				4.45					
69.8						4.51		4.04	
68.8–88.8				4.52					
81.5/6					4.81				4.43
4100	4.68	4.52					4.83		

* The characters at the head of the table refer to: M: Mulders, smoothed values; P: Pettit, smoothed values; Ch: Chalonge, inter- and extrapolated (at 3700 + and 3800 Å) values; L-N: Labs and Neckel: in bands of 20 Å, corrected for Fraunhofer lines; M-S: Murasheva and Sitnik; Ma 1957: Makarova, corrected for Fraunhofer lines; $\overline{\text{Ma}}$: Makarova, averages of different authors; Ma: Makarova; H: Houtgast.

however, is not correct, they should have taken the total amount of absorption from the *Utrecht Atlas* (1940) in the wavelength regions considered. That the two procedures do not yield the same result lies in the fact that for blended lines the sum of the equivalent widths is greater than their total area of absorption. For this reason a correction for this was applied in the four regions of overlap with their measurements lying between λ 4030 and λ 4088.8 Å. In applying this correction, the height of the Atlas continuum was determined by making use of the intensities of the points at $\lambda\lambda$ 4020.7, 4051.6 and 4081.5, taken from Table I, column H.

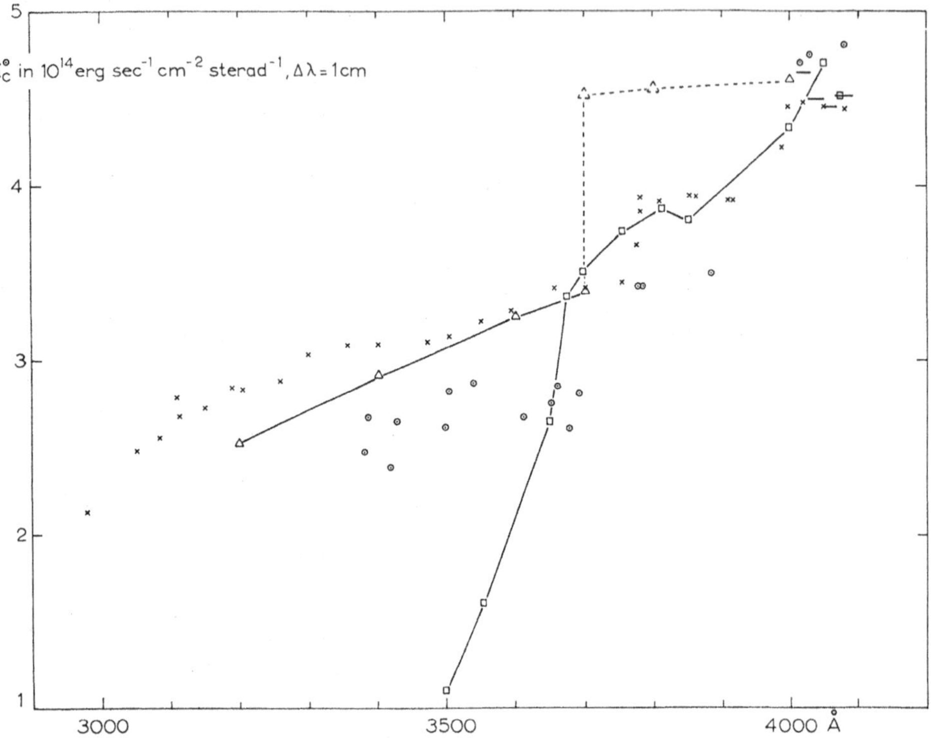

Fig. 1. Observed intensities for the centre of the sun's disk in absolute units against wavelength, from different authors: △ CHALONGE *et al.* (1950), extrapolated points △; — LABS and NECKEL (1962), corrected (see text); ⊙ MURASHEVA and SITNIK (1963); □ MAKAROVA (1957); × HOUTGAST, this paper, Table I.

2. The Previous Balmer Jump

The Balmer jump at λ 3640 Å in the solar spectrum was hitherto determined by using the spectral regions $\lambda > 4000$ Å and $\lambda < 3700$ Å. The extrapolations from continuum points from both regions towards λ 3700 Å, show different intensities, indicated respectively as I_{3700^+} and I_{3700^-}. The Balmer jump is defined as $D = \log I_{3700^+} - \log I_{3700^-}$ for which the following values were obtained, quoted from LABS (1957):

D. Chalonge	(1946, 1950)	0.125;
R. Michard	(1950)	0.14;
D. Labs		0.13.

The spectral region between λ 3700 Å and λ 4000 Å was not taken into account because there the observed intensity points between the Fraunhofer lines do not define a run that can be described by an unique colour- or partition temperature, as do the points for $\lambda > 4000$ Å and $\lambda < 3700$ Å.

The graph in Figure 2 shows the good agreement between LABS and NECKEL's (1962) results and those described here. The Balmer jump, determined by the extra-

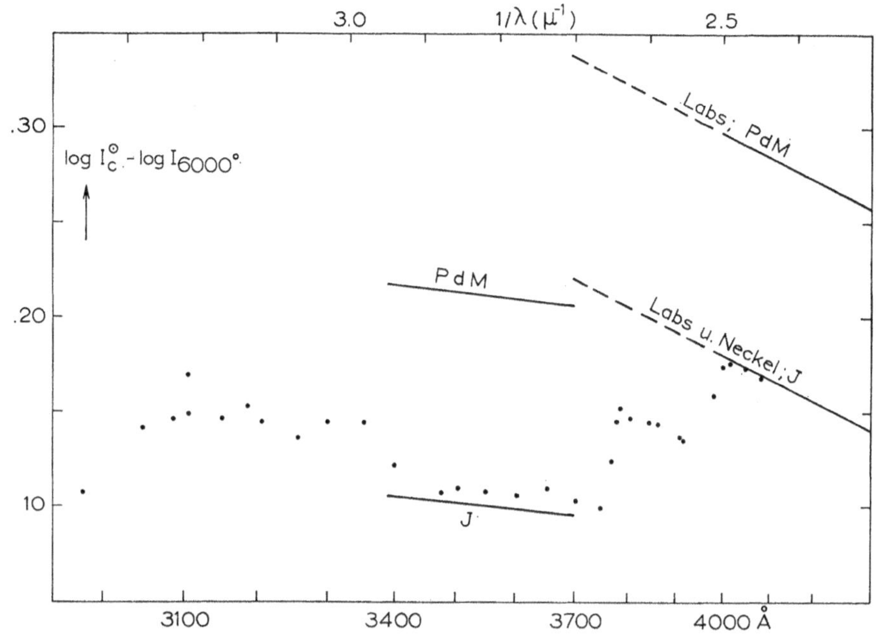

Fig. 2. Comparison of observational results in the region of the Balmer-jump. PdM: LABS (1957), from observations at the Pic-du-Midi; J: LABS and NECKEL (1962), from observations at Jungfraujoch; • HOUTGAST, this paper. The jump at $1/\lambda = 2.7 (\lambda = 3700$ Å) between the extrapolation from the region $1/\lambda < 2.5$ and that from $1/\lambda > 2.7$ amounts to $\Delta \log I = 0.13$.

polation as indicated in the figure, that is to say without taking into account the points between λ 3700 and λ 4000 Å, has the value of 0.13.

3. The Real Balmer Jump

Now we will consider in more detail all the intensity points which are available. They are compared to Wien's law:

$$\log \frac{I_\lambda \cdot \lambda^5}{2C_1} = -\frac{0.621}{\lambda T},$$

which for the wavelengths and temperatures considered here, does not deviate for more than a fraction of a percent from Planck's law. A plot of $\log (I_\lambda \cdot \lambda^5)/2C_1$ against $1/\lambda$ yields a straight line, the height and the slope of which are determined by the temperature. The lines for $T = 6000°$, 6300°, and 6500° are drawn in Figure 3. We also plotted the measured intensities in the same way. The height of each point is a measure for the radiation temperature for that wavelength, the slope of the series of points, if they lay on a straight line, determines the colour temperature in that wavelength region. Interesting conclusions in this respect may be drawn from Figure 3, but this is not our purpose here.

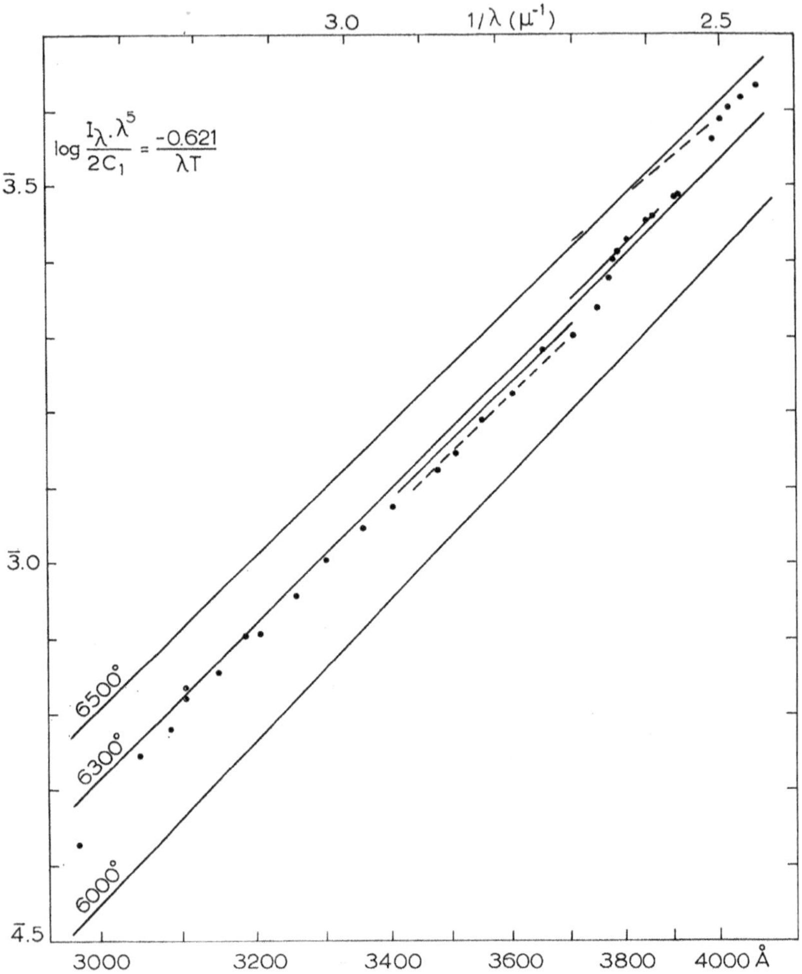

Fig. 3. Straight line relation for Wien's law. The points represent the values from the last column in Table I. The shorter lines determine at λ 3700 Å the values 0.13 (old ----) and 0.03 (new —) for the Balmer-jump.

Below λ 3900 Å till λ 3800 Å the continuum is depressed by many Fraunhofer lines, not due to hydrogen, but the gradient (colour temperature) is constant. From here, below λ 3800 Å, the depression by the wings of the overlapping higher Balmer lines sets in. At λ 3640 Å we expect the Balmer jump, but obviously around this wavelength there is also a depression of the true continuum by overlapping lines. We see this from the point at λ 3658 Å, which from all points in these surroundings seems the least influenced point. It is this point that indicates the intensity level which may be influenced only by the Balmer absorption. If we now make the extrapolation from the points around λ 3800 Å (no Balmer influence) and from the Balmer continuum below λ 3700 Å, we find at λ 3700 Å a jump in $\log I$ of 0.03, as indicated in Figure 3.

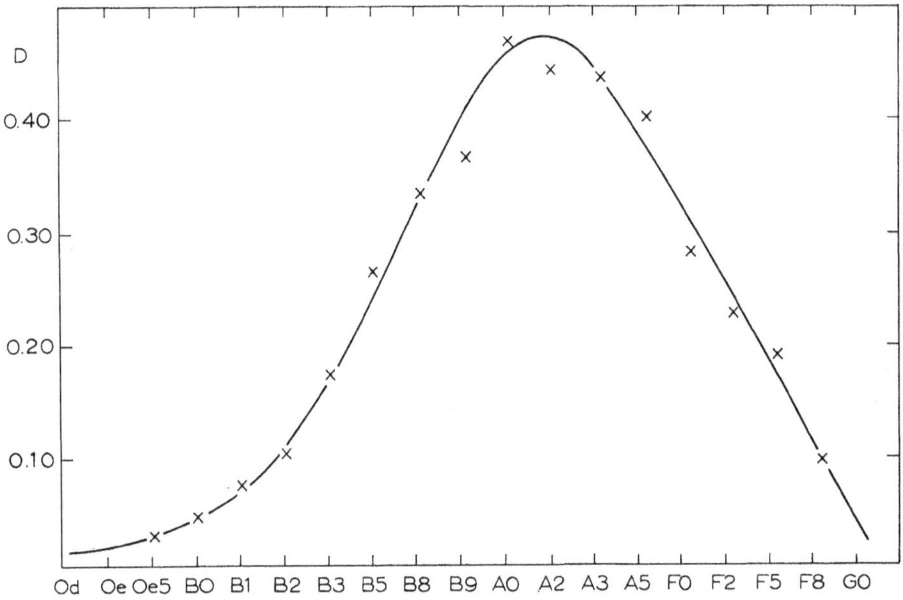

Fig. 4. The Balmer-jump for dwarf stars as a function of spectral type (BARBIER and CHALONGE, 1939). For the spectrum of the centre of the sun's disk, being G0, the value of D is in agreement with the determination of this paper.

This value, deviating considerably from the earlier value of 0.13, fits remarkably well into the values obtained for stars as a function of spectral type. As an illustration of this data of BARBIER and CHALONGE (1939) are reproduced here in Figure 4.

There is however more: the height of the Balmer-jump has to agree with the strengths of the high Balmer lines. With the data given by UNSÖLD (1955) we can check whether this condition is satisfied. From the Balmer lines he derives the number of hydrogen atoms in the second quantum state and finds $\log N_{0.2}H = 15.7$. The absorption coefficient per hydrogen atom at the Balmer limit is $R = 1.38 \times 10^{-17}$, resulting in a total absorption coefficient of 0.07.

If we write with Unsöld for the spectral depression R caused by an absorption coefficient κ:

$$\frac{1}{R} = \frac{1}{\kappa} + \frac{1}{R_c},$$

in which R_c is the maximum depression in saturated hydrogen lines in the wavelength region considered, for which we may take 0.80, then $R = 0.065$, which in terms of the Balmer-jump D can be written as $1–10^{-D}$. It follows that $D = 0.03$, in full agreement with the observational value as derived in this communication.

References

BARBIER, D. and CHALONGE, D.: 1939, *Ann. Astrophys.* **2**, 254.

Canavaggia, R., Chalonge, D., Egger-Moreau, M., Oziol-Peltey, H.: 1946, 1950, *Ann. Astrophys.* **9**, 143; **13**, 355.
Houtgast, J.: 1965, *Proc. Kon. Nederl. Akad. Wetensch. Amsterdam*, series B **68**, No 5.
Labs, D.: 1957, *Z. Astrophys.* **44**, 37.
Labs, D. and Neckel, H.: 1962, *Z. Astrophys.* **55**, 269.
Makarova, E.A.: 1957, *Astr. Zhu.* **34**, 539; *Soviet Astr.* **1**, 531.
Makarova, E.A.: 1963, *Observatory* **83**, 183. (Also in 1964, *Astr. Zhu.* **41**, 288; *Soviet Astr.* **8**, 222.)
Makarova, E.A.: 1965, *Astr. Zhu.* **42**, 681; *Soviet Astr.* **9**, 525.
Michard, R.: 1950, *Bull. Astr. Inst. Neth.* **11**, 227.
Mulders, G.F.W.: 1939, *Publ. Astr. Soc. Pacific* **51**, 220.
Murasheva, M.S. and Sitnik, G.F.: 1963, *Astr. Zhu.* **40**, 819. (Also in 1964, *Soviet Astr.* **7**, 623.)
Pettit, E.: 1940, *Astrophys. J.* **91**, 159.
Unsöld, A.: 1955, *Physik der Sternatmosphären.* Springer-Verlag, Berlin, p. 481.
Utrecht Photometric Atlas of the Solar Spectrum: 1940, Amsterdam.
Utrecht Preliminary Photometric Catalogue of Fraunhofer Lines, 1960; Rech. Astr. Obs. Utrecht, XV. (Also in Moore, Ch.E., Minnaert, M.G.J., Houtgast, J.: 1966, *The Solar Spectrum 2935 Å to 8770 Å,* Second Revision of Rowland's Preliminary Table of Solar Spectrum Wavelengths, National Bureau of Standards Monograph 61, Washington, D.C.)

EFFECTS OF LINE BLANKETING ON THE SOLAR WINDOWS

DUANE CARBON, OWEN GINGERICH, and ROBERT KURUCZ

Smithsonian Astrophysical Observatory and Harvard College Observatory, Cambridge, Mass., U.S.A.

(Received 18 July, 1967)

Abstract. The increasingly high flux predicted to the violet of 4500 Å by many model solar atmospheres stands in contradiction to the observations. Since one possible cause of the disagreement is that the solar 'windows' by which the observed continuum is established might be obscured by line wings, we have made detailed calculations of these narrow spectral regions. With the exception of a few windows affected by the wings of Balmer lines, those redward of the Balmer discontinuity appear free of line blanketing. Even the assumption that the ultraviolet continuum is depressed 5% by unseen lines not included in our calculations leaves substantial disagreements between the models and observations. The discrepancies could perhaps be explained by a veil of weak lines across the ultraviolet spectrum.

Our calculations indicate that the windows become narrower at shorter wavelengths. Many of the ambiguities to the violet of 3600 Å would be resolved if spectrophotometric tracings with a band pass of 10 mÅ were available.

1. Introduction

One of the most remarkable discrepancies between many model solar atmospheres and the observations is the increasingly high flux predicted to the violet of λ 4500 Å. (In the Bilderberg Continuum Atmosphere (GINGERICH and DE JAGER, 1968), this disagreement is less pronounced, but nevertheless present beyond the Balmer discontinuity.) This spectral region is graphed in Figure 1. The observed values quoted by

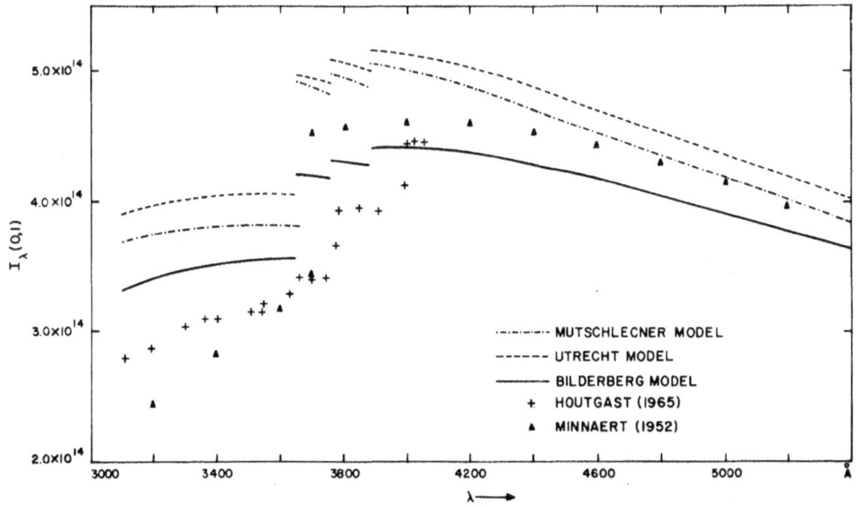

Fig. 1. Predicted and observed intensities at the center of the solar disk ($\lambda\lambda$ 3100–5400 Å).

MINNAERT (1953) are based on a re-evaluation of earlier observations, especially those of CANAVAGGIA *et al.* (1950). The values of LABS and NECKEL (1962) redwards of 4010 Å are in reasonable agreement with those of Minnaert.

The 'observed' Balmer discontinuity given by Minnaert depends on the values of 3800 Å and 3700^+ Å, which have been extrapolated on the assumption that the color temperature remains the same between 3700^+ Å and 5000 Å. If an additional opacity source becomes important around 3900 Å and increases to the violet, then the foregoing assumption is faulty and the observed Balmer discontinuity is only an upper limit. The observations of HOUTGAST (1965) suggest that this might be the case; in this paper we shall investigate the alternative possibility that the observed windows have been depressed by line blanketing.

The models based on empirical $T(\tau)$ relations such as the Mutschlecner model (MÜLLER and MUTSCHLECNER, 1964), the Utrecht Reference Photosphere (URP) (HEINTZE, HUBENET, and DE JAGER, 1964), and the Bilderberg Continuum Atmosphere (BCA) predict significantly higher intensities in the near ultraviolet than the observed values. Theoretical models driven to radiative equilibrium show similar disagreement with the observations in this wavelength region (GINGERICH, 1966; SWIHART, 1966).

Attempts to reduce the disagreement by changing the model atmospheres have not produced encouraging results. Although an increase in the efficiency of convective energy transport brings the predicted intensities into slightly better agreement, the improvement is inadequate. Furthermore, the URP and BCA models already take into account the convective gradient. While another continuous opacity source in the ultraviolet could reduce the emergent intensities, it would also diminish the size of the Balmer discontinuity in an unacceptable manner. To avoid the ambiguity of the observed Balmer discontinuity, we define a related parameter $I_\lambda(4000)/I_\lambda(3600)$. Table I shows that the predicted discontinuities are already too small.

TABLE I

Observed and predicted log $I_\lambda(4000)/I_\lambda(3600)$

Observed:	0.160	Minnaert
	0.136	Houtgast
Predicted:	0.117	Mutschlecner model
	0.100	Utrecht Reference
	0.093	Bilderberg Atmosphere

Finally, non-LTE effects in the major sources of continuous opacity – neutral hydrogen and H^- – do not help. For stars of solar type, departures from LTE in neutral hydrogen lead to an over-population of the $n=3$ level and an under-population of the $n=2$ level (KALKOFEN, 1968). This will lead to an increased flux through the Balmer continuum and to a smaller Balmer discontinuity. Both results will increase the disagreement between theory and observation. Although non-LTE over-population of the bound state of H^- would improve the situation, recent estimates of the collisional cross-section of H^- for associative detachment indicate that H^- will be in LTE at the continuum-forming depths (SCHMELTEKOPF, FEHSENFELD, and FERGUSON, 1967; DALGARNO and BROWNE, 1967).

Because of the discrepancy between the models and the observations, we have re-examined the observational data. The observed values for the emergent intensity depend, of course, upon isolation of the continuum. The results quoted by CANAVAGGIA *et al.* (1950) and by LABS and NECKEL (1962) were obtained by correcting measurements taken at medium and low resolution for the effects of absorption lines. HOUTGAST's (1965) results were obtained at much higher resolution and are believed to represent continuum intensities directly (HOUTGAST, 1967). Because the absorption lines become more and more crowded violetward of 5000 Å, the possibility arises that at these shorter wavelengths there are no intervals entirely free of the effects of line absorption. If the so-called solar 'windows' (CANAVAGGIA and CHALONGE, 1946; CANAVAGGIA *et al.*, 1950) used in the determination of the continuum intensities are significantly blanketed by the opacities in the wings of neighboring absorption lines, then the observed continuum will be depressed below the 'true' (unblanketed) continuum. Such blanketing could help to explain the discrepancies between the model-atmosphere results and the observations.

2. The Blanketing Calculations

In order to estimate the extent of the line blanketing, we have attempted to calculate the detailed spectrum in the vicinity of the solar windows. We have concentrated our efforts on the windows listed by CANAVAGGIA *et al.* (1946, 1950), since these have been used most extensively by observers for establishing the level of the continuum. The atmospheric model used in our calculations was the BCA with the adopted microturbulent velocity of 2 km/sec (independent of depth). GOLDBERG, MÜLLER, and ALLER's (1960) abundances were adopted for the metals. A He/H ratio of 0.10 by number was used. Unless otherwise noted, all results were obtained on the basis of this model. The continuous opacity sources included were neutral hydrogen, H^-, H_2^+, electron scattering, Rayleigh scattering, and bound-free transitions of silicon and magnesium. The parameters used for the calculation of the line opacities were taken primarily from the extensive table compiled by STROM and KURUCZ (1966). This table provided gf values and excitation potentials for approximately two-thirds of the lines indicated by the Revised Rowland Table (MOORE, MINNAERT, and HOUTGAST, 1966) in the vicinity of the windows. With the exception of unidentified weak lines, the remaining third of the lines listed in the Revised Rowland Table fell into two groups. The first and larger group consisted of metallic lines for which no gf values, either theoretical or experimental, are available. The fewer remaining identified lines were due to CN, CH, and NH.

In the case of the metallic lines without gf values, preliminary gf values were assigned. These were subsequently adjusted in a fashion to be described later. Since our computer programs have not yet been adapted for molecular calculations, the lines of CN, CH, and NH were approximated by artificial lines of C I having 7.7 eV excitation potential. Neutral carbon was chosen, since it would represent an upper limit on the abundances of the radicals, and because it has roughly the same mass and therefore, doppler width as CH and NH. The high-excitation potential was chosen to force the line formation down to the large optical depths indicated by the results of COHEN and STROM (1968) and of WITHBROE (1967). Since lines of low-excitation potential can be observed easily in the laboratory, most unidentified lines must be either higher-excitation lines of abundant metals or lines of rare elements. For this reason the unidentified lines in the vicinity of the windows were represented by artificial Fe I lines of 3.3 eV excitation potential. We feel that these very crude approximations did not seriously influence the results of the investigation. The artificial lines were not great in number, generally weak in strength, and, in most cases, not close enough to any windows to influence the level of the continuum.

Since the solar windows were chosen far from strong lines, any blanketing should be due to weaker lines in the immediate vicinity of the windows. We have, therefore, attempted to reproduce the detailed spectrum for 1.25 Å on either side of the window wavelengths given by Canavaggia *et al.* To do this, a list was made containing every line inside a 4-Å interval centered on the window wavelength. At every 0.01 Å in this 4-Å interval, the opacity from each line in the interval was calculated using a Voigt profile and the atomic parameters described above. For the windows redward of the Balmer discontinuity, the opacities from the wings of the nearer hydrogen lines were also calculated. If the opacity due to any line at a particular point exceeded 0.0001 times the local continuous opacity, it was included in the calculation of the total opacity at that point; otherwise, it was neglected. This procedure was repeated at each of the 400 wavelength points for 30 depths in the model atmosphere (from $\tau_{5000} = 10^{-6}$ to $\tau_{5000} = 17.8$). These opacities were then used in an LTE calculation of the emergent intensity $I_\lambda(0,1)$ at every 10 mÅ for the central 2.50 Å of the 4-Å interval.

We then compared the emergent spectrum obtained in this manner with available atlases of the solar spectrum. These were the two unpublished high-dispersion atlases, the McMath-Hulbert Atlas of the Solar Spectrum, and the Atlas of the Solar Spectrum from 3000 to 7500 Å (Delbouille, Neven, and Roland, to be published) as well as the lower dispersion *Photometric Atlas of the Near Ultraviolet Solar Spectrum* (BRUCKNER, 1960) and the *Utrecht Photometric Atlas* (MINNAERT, MULDERS, and HOUTGAST, 1940). Since the instrumental profiles were known for the McMath-Hulbert Atlas (GATHIER, 1962), the Brückner Atlas, and the Utrecht Atlas, direct comparions could be made with these atlases simply by folding the calculated spectra through the instrumental profiles. Comparing the calculated spectra with the observed provided a check on the preliminarily assigned gf values. Whenever the calculated line strengths differed significantly from the observed, the gf values were altered to bring the strengths into better agreement. For the lines from the tabulation by Strom and Kurucz, any changes in gf values found necessary were usually within the tentative error brackets of 0.3 in log gf assigned by them. The spectrum was then recalculated with the new gf values, and another comparison made. Quite reasonable agreement between the observed and calculated spectra could usually be achieved after three to six trial calculations. Figures 2 and 3 show two examples of solar windows. In these graphs the intensity at the center of the disk has been normalized to the true continuum.

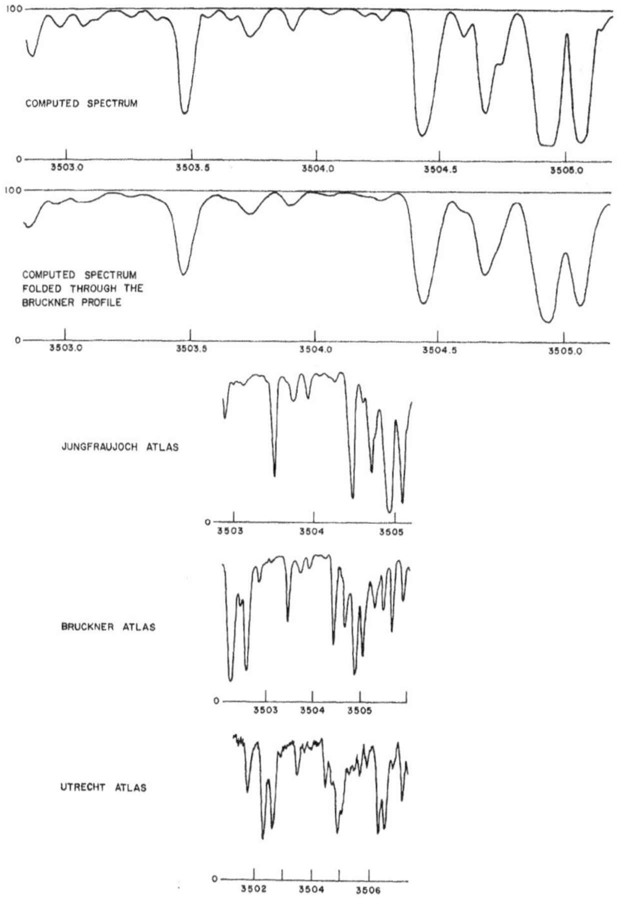

Fig. 2. Computed and observed spectra near the solar window at 3504 Å. The Utrecht Atlas trace is for the integrated disk; the remaining traces represent intensities at the center of the disk.

We wish to note explicitly several interesting results obtained in the process of performing these calculations. In our initial calculations we included no microturbulence in the line broadening. We found that the predicted line profiles were noticeably narrower and deeper than was observed. In subsequent calculations we used the 2 km/sec microturbulent velocity distribution adopted at the Bilderberg Conference. As may be seen from Figures 2 and 3, the line profiles calculated with this velocity distribution are in reasonable agreement with the observed profiles. Additional experimentation showed that a microturbulent velocity distribution that is increasing with depth provides even better agreement when calculated line wings are compared

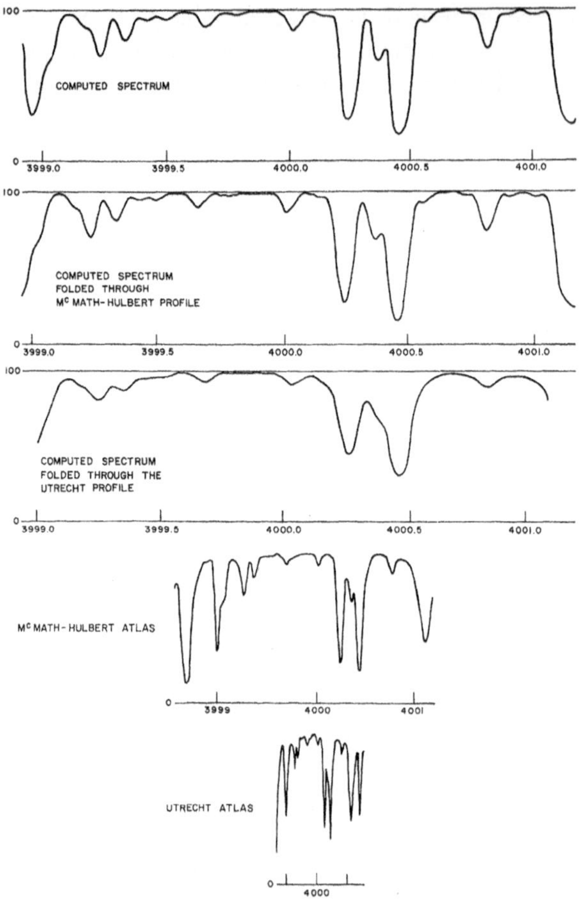

Fig. 3. Computed and observed spectra near the solar window at 4000 Å. All traces represent intensities at the center of the disk.

with the observed. However, since there seems to be no firm basis for choosing any particular depth-dependent microturbulent velocity distribution, we have chosen the 2 km/sec distribution for our calculations.

Second, we note that the profiles of lines of moderate equivalent width calculated using the URP $T(\tau)$ relation show prominent central reversals even at the center of the disk. These reversals are not apparent in any of the atlas traces of the lines. The BCA model, with its much broader temperature minimum, predicts no reversals for the lines included in our calculations.

A third point of interest concerns the strengths of the wings of the Balmer lines at the window wavelengths. We used the most recent broadening theory due to Griem (1967) in our calculations. The degree of depression of the continuum caused by hydrogen-line blanketing in the BCA is shown in Table II, column two, for those windows that were significantly affected.

TABLE II

Depression of continuum by hydrogen lines

Window (Å)	BCA (%)	Mutschlecner (%)	URP (%)
3659	7	13	9
3665	10	15	13
3692	12	20	15
4000	< 1	< 1	< 1
4117	< 1	2	< 1
4365	< 1	< 1	< 1
4880	< 1	2	< 1

It must be stressed that the degree of depression of the continuum is dependent upon the choice of model atmosphere. As shown in columns three and four, both the Mutschlecner and the URP $T(\tau)$ relations predict greater depressions than the BCA. The difference between the BCA and the Mutschlecner models arises because the temperatures where much of the continuum is formed ($\tau_{5000} \approx 1–2$) are higher in the latter model. Hence it might be more appropriate to speak of the increase in the continuum rather than the depression of the hydrogen wings.

3. Conclusions

The windows that we examined were measured by CANAVAGGIA *et al.* (1946, 1950), and they also correspond to the highest points reported by HOUTGAST (1965). We have found that the windows redward of the Balmer discontinuity are free of line blanketing, with the exception of those listed in Table II.

In the region to the red of 3900 Å, the McMath-Hulbert Atlas, with its very high resolution, allowed us to compare the calculated and observed spectra with very little difficulty. On the violet side of the discontinuity the situation is slightly ambiguous. The lower dispersion of the available atlases for the ultraviolet made the comparison between predicted and observed spectra troublesome. Only lines that appeared in the atlas traces could be taken into account in the calculations. The low resolution of the available atlases leaves open the possibility that the apparent windows are blanketed by crucially placed, unseen lines of a few milliangstrom equivalent width. Such lines could perhaps introduce as much as 5% ambiguity into the level of the apparent continuum. Within this uncertainty, however, we find that the windows in the ultraviolet are essentially unblanketed by recognized lines.

Even assuming that the continuum is depressed 5% by blanketing in the ultraviolet still leaves a large discrepancy between the models and observations. We feel that we cannot justifiably account for these disagreements on the basis of blanketing from the lines included in our calculations. These results contradict the investigation mentioned by Pecker (in the discussion following GINGERICH, 1966). Pecker and Gökdogan found a correlation between abundance and wavelength from a curve-of-growth analysis of solar iron lines; this discrepancy could be removed by assuming a higher true continuum (and therefore larger equivalent widths) in the ultraviolet region. However, their indirect procedure was probably biased by systematic normalization errors in the gf values for ultraviolet iron lines.

Although introducing more microturbulence into the models at large depths would put more blanketing opacity into the line wings, enough microturbulence to blanket the windows significantly would grossly distort line profiles that now agree fairly well with the observations.

Both our calculations and the high-dispersion atlases indicate that the windows become narrower at shorter wavelengths. Because the intervals of continuum at shorter wavelengths are so narrow, it is possible that the highest points registered in the Utrecht Atlas are not representative of true continuum, but are depressed by instrumental broadening. If this is the case, then the corrections for blanketing applied by Canavaggia *et al.* (1950) are not great enough. Since the resolving power of 10^5 reported by Houtgast (1965) is comparable to that of the Utrecht Atlas, his results may also represent a continuum depressed by instrumental broadening. It appears possible to estimate the importance, if any, of this effect by comparing the blanketing corrections calculated on the basis of the Utrecht Atlas with those obtained from the high-resolution atlases.

The discrepancies could be explained by postulating a veil of weak lines across the ultraviolet spectrum. Since these lines must be very closely spaced to produce the required effect, the veil would act as an additional continuous opacity source similar to bound-free absorption. If the distribution in wavelength of such weak lines resembled the distribution of stronger lines (that, is concentrated toward the violet), it might be possible to circumvent the difficulty mentioned earlier, namely, that an additional opacity source would tend to diminish the $I_\lambda(4000)/I_\lambda(3600)$ ratio.

Acknowledgments

The authors wish to thank Deane Peterson for providing the computer subroutines needed to calculate the hydrogen-line opacities. They also wish to thank Dr. G.L. Withbroe for bringing our attention to the very useful Jungfraujoch Atlas.

References

Brückner, G.: 1960, *Photometric Atlas of the Near Ultraviolet Solar Spectrum*. Vandenhoeck & Ruprecht, Göttingen.

Canavaggia, R. and Chalonge, D.: 1946, *Ann. Astrophys.* **9**, 143.

Canavaggia, R., Chalonge, D., Egger-Moreau, M., and Oziol-Peltey, H.: 1950, *Ann. Astrophys.* **13**, 355.

Cohen, J.G. and Strom, S.E.: 1968, *Astrophys. J.* (in press).

Dalgarno, A. and Browne, J.C.: 1967, *Astrophys. J.* **149**, 231.

Gathier, P.J.: 1962, *Bull. Astron. Inst. Neth.* **16**, 128.

Gingerich, O. J.: 1966, *J. Quantit. Spectrosc. Radiat. Transfer* **6**, 609.

Gingerich, O.J. and De Jager, C.: 1968, *Solar Phys.* **3**, 5.

Goldberg, L., Müller, E.A., and Aller, L.H.: 1960, *Astrophys. J. Suppl.* **5**, 1.

Griem, H.R.: 1967, *Astrophys. J.* **147,** 1092.
Heintze, J.R.W., Hubenet, H., and De Jager, C.: 1964, *Bull. Astron. Inst. Neth.* **17,** 442.
Houtgast, J.: 1965, *K. Ned. Akad. Wet. Proc. Sect. Sci. Ser. B* **68,** 306.
Houtgast, J.: 1967, private communication.
Kalkofen, W.: 1968, *Astrophys. J.* (in press).
Labs, D. and Neckel, H.: 1962, *Z. Astrophys.* **55,** 269.
Minnaert, M.: 1953, 'The Photosphere', in *The Sun* (ed. by G.P. Kuiper) (Vol. I in the series 'The Solar System'). Univ. of Chicago Press, Chicago, p. 88.
Minnaert, M., Mulders, G.F.W., and Houtgast, J.: 1940, *Photometric Atlas of the Solar Spectrum.* Utrecht Observatory, Amsterdam.
Moore, C.E., Minnaert, M., and Houtgast, J.: 1966, *Nat. Bur. Stand., Monogr. No. 61.*
Müller, E.A. and Mutschlecner, J.P.: 1964, *Astrophys. J.*, Suppl. **9,** 1.
Schmeltekopf, A.L., Fehsenfeld, F.C., and Ferguson, E.E.: 1967, *Astrophys. J.* **148,** L155.
Strom, S.E. and Kurucz, R.L.: 1966, *J. Quantit. Spectrosc. Radiat. Transfer* **6,** 609.
Swihart, T.L.: 1966, *Astrophys. J.* **143,** 358.
Withbroe, G.L.: 1967, *Astrophys. J.* **147,** 1117.

LIMB-DARKENING OBSERVATIONS BETWEEN 1800 AND 2900 Å

R. M. BONNET and J. E. BLAMONT

Service d'Aéronomie du C.N.R.S., Réduit de Verrières, Verrières-le-Buisson, France

(Received 13 June, 1967)

Abstract. We succeeded in several attempts at measuring the solar limb-darkening in the continuous spectrum between 1800 Å and 2900 Å. Good spectroheliograms were obtained during a rocket flight in November 1964 and during a balloon flight in October 1966 at 2190 Å, 2665 Å, 2885 Å, and 1980 Å, 2235 Å respectively. A rocket flown at the beginning of 1967 provided excellent spectra allowing the measurement of center-to-limb variation in the continuum and lines from 1800 Å to 2800 Å.

The first series of pictures yielded absolute values of the central intensity leading to a minimum temperature close to 4700°K. The continuous opacity of the photosphere layers has been deduced from the two first experiments. It seems that a non-metallic source of continuous absorption must be taken into account between 2500 Å and 3000 Å. Moreover, as the source functions for the shortest wavelengths show a very flat variation near the minimum temperature, it seems that the transition zone between the photosphere and the chromosphere must be very gradual.

These results are discussed in the light of the latest results deduced from the new spectra obtained.

1. Introduction

In order to measure the ultraviolet solar limb-darkening we have launched a series of four balloon and rocket experiments. All of them make use of photographic film. Three consisted in taking spectroheliograms within narrow bands of the continuum. The other one consisted in recording a stigmatic spectrum of the disk. Table I is a general view of the results obtained in each experiment. The different instrumental specifications are also included. Since two of the experiments have just been launched at the beginning of 1967, the results presented in this paper refer mainly to the two earliest.

2. Observations (Experiments of November 1964 and October 1966)

Several spectroheliograms were obtained in November 1964 with a multiple band-pass filter (Bonnet and Courtes, 1962) aimed at the sun by a biaxial pointing system built at the University of Colorado (Boulder, USA). The wavelengths were chosen in the 'windows' of the solar spectrum as free as possible from absorption lines at 2190 Å, 2665 Å, and 2885 Å. The adjustable band-width was taken equal to 43 Å. The resolution achieved on the pictures is always better than 15 seconds of arc (Figure 1). Another experiment of exactly the same type was launched with the biaxial pointing system developed at the Service d'Aéronomie on a 38000 m^3 balloon in October 1966. The wavelengths were chosen for this second experiment at 1960 Å and 2280 Å in order to frame the silicon discontinuity at 1990 Å. The band-pass was 80 Å.

Due to the vicinity of the atmospheric window at 2100 Å, both bands were shifted

TABLE I

General View of the UV Limb-darkening Results

Date of launching	Space Vehicle used	Sun follower	Pointing accuracy during flight	Type of experiment	Spectral resolution	Wavelengths studied	Spatial resolution	Exposure times and photographic emulsion	Main Results
November 8, 1964	Veronique Rocket	University of Colorado	±1′	spectro-heliograms	43 Å	2190 Å	10″	0.03 sec and 0.06 sec	Continuum limb-darkening curves
						2665 Å	10″	0.03 sec and 0.06 sec	Continuum absolute intensities
						2885 Å	15″	0.03 sec III-0, O, UV (16 mm)	
October 3, 1966	38000 m³ Balloon	Service d'Aéronomie	±1′ 30″	spectro-heliograms	80 Å	1980 Å	15″ to 20″	0.08 sec and 0.11 sec 103a-0, UV 35 mm	Continuum limb-darkening curves and facular structure
						2235 Å	15″ to 20″	0.08 sec and 0.11 sec III-0, UV 35 mm	
January 13, 1967	Veronique Rocket	S.P.C. 300 Ball Brothers Research Corporation	±7′	stigmatic spectra	0.4 Å	Range 1800 Å to 2800 Å	15″ to 20″	0.1 sec and 0.5 sec 103a-0, UV 16 mm	Continuum and lines limb-darkening curves Absolute intensities
March 22, 1967	38000 m³ Balloon	Service d'Aéronomie	±2′	spectro-heliograms	80 Å	2040 Å	1′ to 2′	0.08 sec and 0.11 sec 103a-0, UV 35 mm	Facular structure
						2215 Å	30″	0.08 sec and 0.11 sec 103a-0, UV 35 mm	Limb-darkening curves

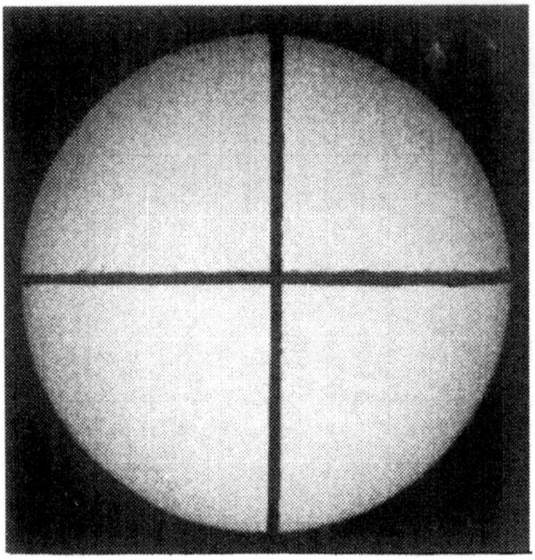

Wavelength = 2885 Å Film: Kodak III-0, UV
Band = 43 Å Image diameter = 5 mm
Exposure time = 25 m.s.

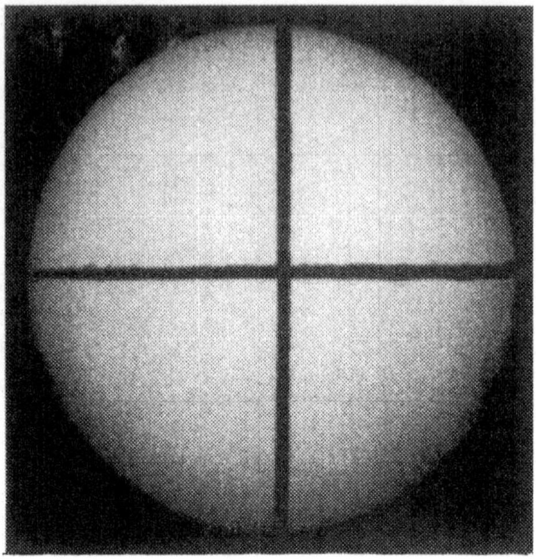

Wavelength = 2665 Å Film: Kodak III-0, UV
Band = 43 Å Image diameter = 5 mm
Exposure time = 60 m.s.

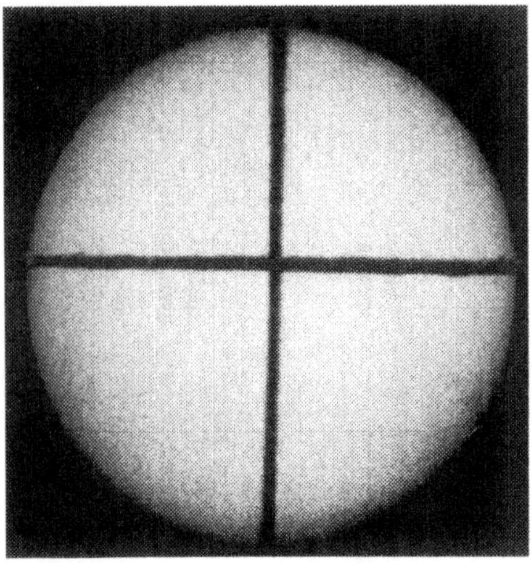

Wavelength = 2190 Å Film: Kodak III-0, UV
Band = 43 Å Image diameter = 5 mm
Exposure time = 60 m.s.

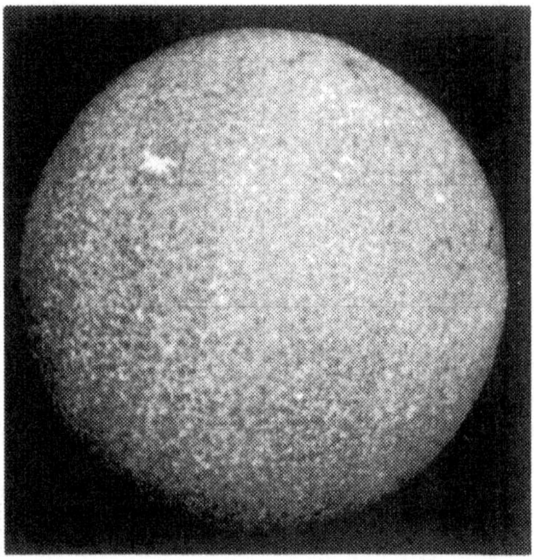

Spectroheliogram in the Calcium K line
Picture of November 8, 1964
(Observatoire de Meudon)

Fig.1. Rocket spectroheliograms of November 8, 1964.

toward this wavelength and the pictures actually correspond to 1975 ± 5 Å, and 2235 ± 5 Å. On these pictures too, the resolution is of the order of 15″ (Figure 2).

Absolute photometric calibration allowed us to measure the central intensity averaged over the band-pass of the filter in the rocket experiment. The results are reported in column 2 of Table II. In the balloon experiment, absolute measurements

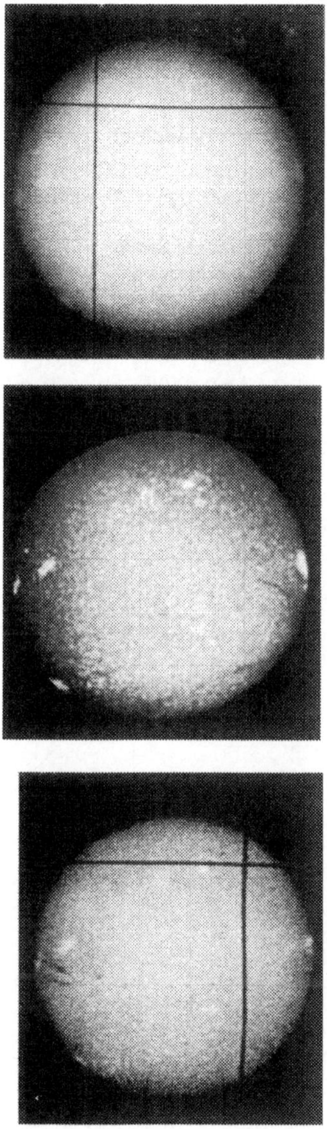

Wavelength = 2235 Å
Band = 80 Å
Exposure time = 110 m.s.
Film = Kodak III-0, UV
Image diameter = 14 mm

Spectroheliogram in the Calcium K line
Picture of October 3, 1966
(Observatoire de Meudon)

Wavelength = 1980 Å
Band = 76 Å
Exposure time = 80 m.s.
Film = Kodak 103a-0, UV
Image diameter = 8 mm

Fig. 2. Balloon spectroheliograms of October 3, 1966.

are impossible because we do not know the actual ozone thickness above the balloon. The results obtained at 2190 Å, 2665 Å, 2885 Å have already been published (BLAMONT and BONNET, 1966). A strong difference was noticed between the 2190 Å and 2235 Å data, obtained during the rocket and the balloon flights respectively (BLAMONT and BONNET, 1967). After a careful analysis, it was found that the photometric calibration used to process the results of the 2190 Å pictures was not reliable. New calibrations were undertaken, which led to the final result shown in Figure 3. It can be seen that the curves are very similar. Thus, for further computations we have taken only one curve, since the differences between them lie within the experimental errors, estimated at 5%.

TABLE II

Different Determinations of the Central Intensity at 2190 Å, 2665 Å, 2885 Å, averaged over 40 Å given in ergs cm^{-3} sec^{-1} steradian^{-1}

Wavelength	Our rocket measurements	Computed from Tousey's measurements of the flux	Computed for		
			$T=4500$	$T=4600$	$T=4700$
2190 Å	9.9×10^{12}	1.4×10^{13}	5.0×10^{12}	6.95×10^{12}	9.4×10^{12}
2665 Å	7.2×10^{13}	6.0×10^{13}	3.75×10^{13}	4.86×10^{13}	6.25×10^{13}
2885 Å	1.8×10^{14}	1.2×10^{14}	7.3×10^{13}	9.3×10^{13}	1.2×10^{14}

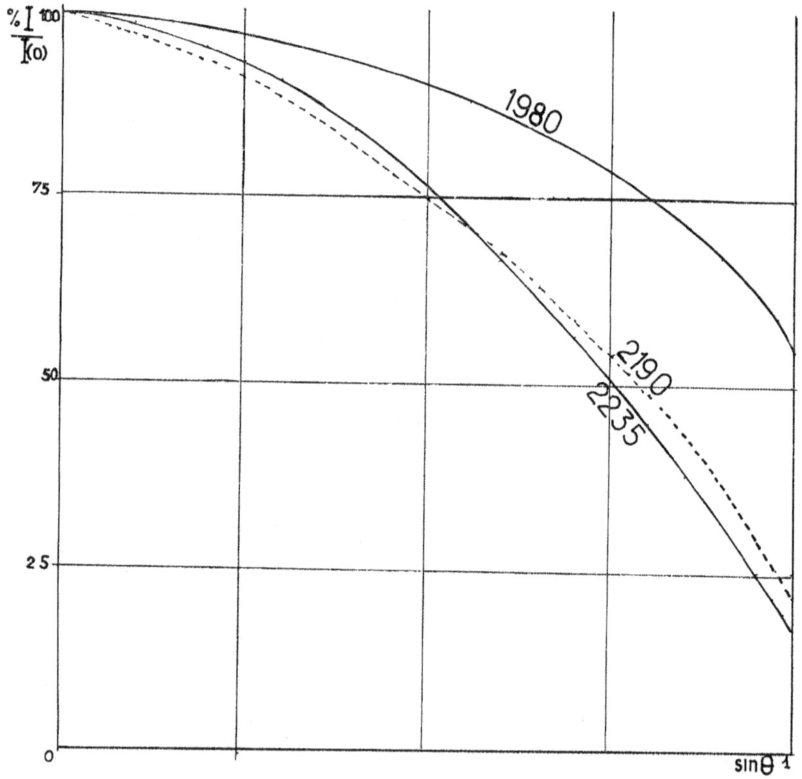

Fig. 3. Comparison of the solar limb-darkening at 2190 Å and 2235 Å plotted versus sinθ.

The four curves used in our computation are plotted in Figure 4 versus cosθ. We have included the curve obtained from photoelectric ground-based observations by PEYTURAUX (1955). It will be noticed that there is a huge discontinuity between 2190 Å and 1980 Å in the solar limb-darkening. This is also observed in Figure 2, where the sun looks very different from one wavelength to the other. The similarity between the 1980 Å picture and the Ca K spectroheliogram is especially striking.

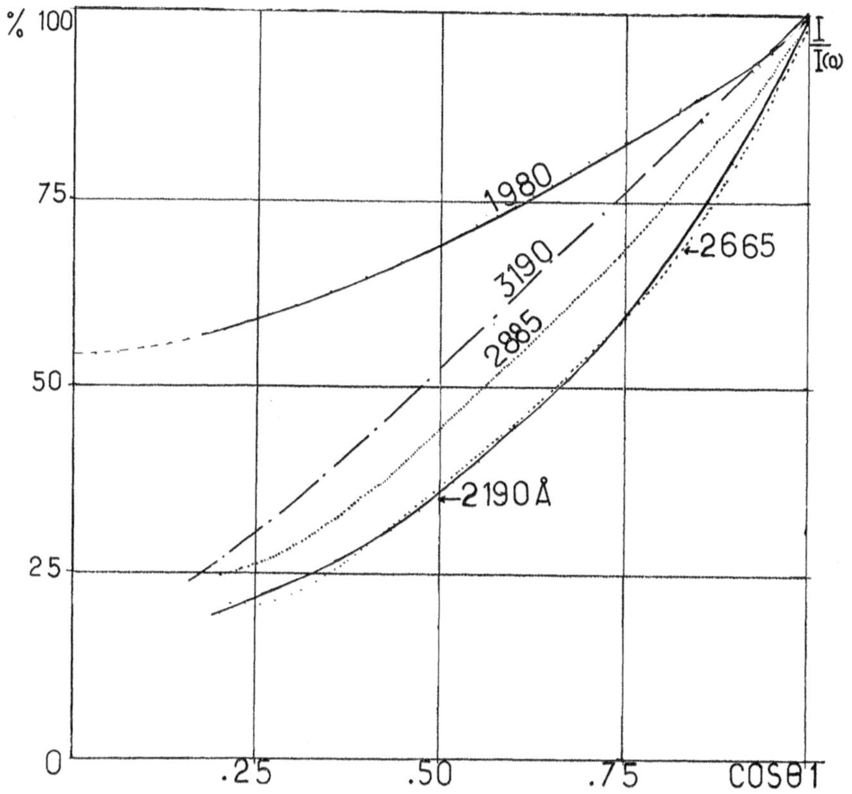

Fig. 4. Solar limb-darkening at 1980 Å, 2190 Å, 2665 Å and 2885 Å, plotted versus $\cos\theta$. The curve corresponding to 3190 Å is taken from PEYTURAUX (1955).

3. The Source-Function

We have used an analytical method to invert the Laplace Integral:

$$\frac{I_\lambda(\mu)}{I_\lambda(\mu=1)} = \int_0^\infty \frac{S_\lambda(\tau_\lambda)\, e^{-\frac{\tau_\lambda}{\mu}}}{I_\lambda(\mu=1)} \frac{d\tau_\lambda}{\mu} \tag{1}$$

where μ is $\cos\theta$, τ_λ the optical depth, and $S_\lambda(\tau_\lambda)$ the source-function, these last two quantities referring to the single wavelength λ.

The information one can get on $S_\lambda(\tau_\lambda)$ is entirely concentrated in the $(I_\lambda(\mu))/(I_\lambda(\mu=1))$ curve. It is clear that it decreases exponentially with increasing optical depths. Therefore, instead of integrating Equation (1) over τ_λ one can assume to integrate it over intervals containing the same amount of information, i.e. to take $x_\lambda = e^{-\tau_\lambda}$.

Hence, Equation (1) becomes:

$$\frac{I_\lambda(\mu)}{I_\lambda(\mu = 1)} = \int_0^1 S'_\lambda(x_\lambda)\, x_\lambda^{\frac{1}{\mu}-1} \frac{dx_\lambda}{\mu}. \tag{2}$$

The inversion of Equation (2) is then made by means of a least-squares method in expanding $S'_\lambda(x_\lambda)$ in a power series in x_λ.

Applied to several test problems, this method gave good results (BONNET and LAGO, 1966). For experimental problems, the agreement between the observed and computed values of $(I_\lambda(\mu))/(I_\lambda(1))$ is very good and of the order of a few percents. Nevertheless, since we must introduce negative powers of x_λ in the expansion of $S'_\lambda(x_\lambda)$ the values of $S_\lambda(\tau_\lambda)$ for large values of τ_λ are certainly overestimated. The variation of $S_\lambda(\tau_\lambda)$ with τ_λ is reported in Figure 5 for the wavelengths we are dealing with.

Using the expansion of $S_\lambda(\tau_\lambda)$ one can compute the monochromatic flux as a

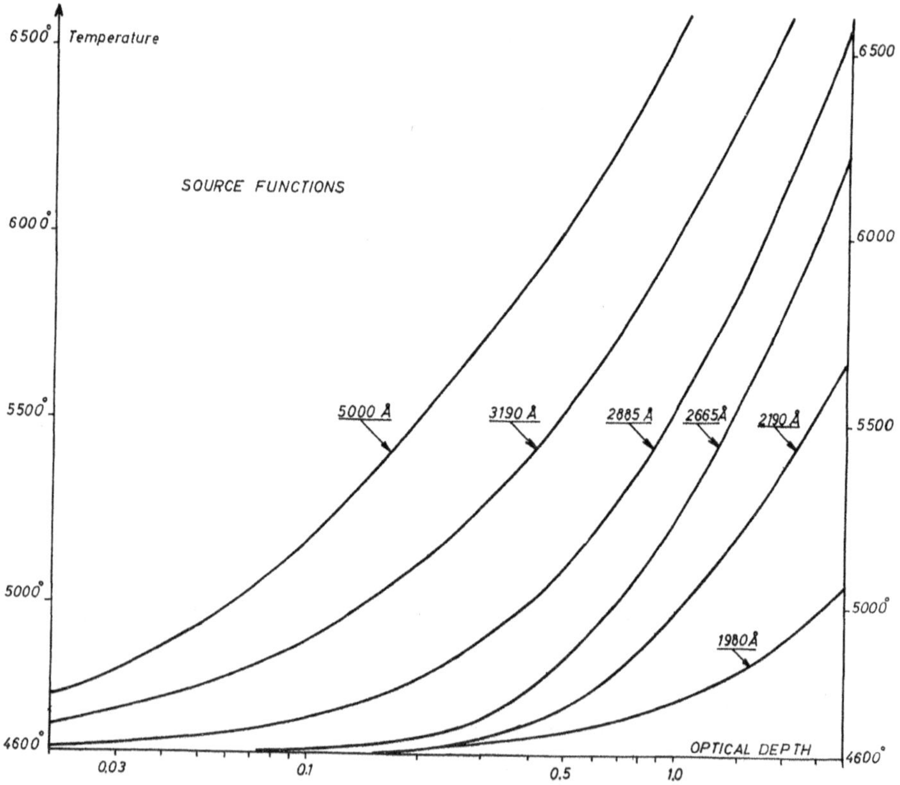

Fig. 5. The empirical source functions deduced from limb-darkening measurements assuming a unique surface temperature of 4600°K.

function of τ_λ from:

$$\frac{F_\lambda(\tau_\lambda)}{I_\lambda(\mu=1)} = 2\int_{\tau_\lambda}^{\infty} \frac{S_\lambda(t_\lambda)}{I_\lambda(\mu=1)} E_2(t_\lambda - \tau_\lambda)\,\mathrm{d}t_\lambda - 2\int_{0}^{\tau_\lambda} \frac{S_\lambda(t_\lambda)}{I_\lambda(\mu=1)} E_2(\tau_\lambda - t_\lambda)\,\mathrm{d}t_\lambda .$$

From this, one can get the flux at $\tau_\lambda = 0$. If we set $F_\lambda(\tau_\lambda)$ to be equal to the U.V. flux values given by TOUSEY (1963), we are able to determine the central intensity $I(\mu=1)$. The result of this computation is reported in column 3 of Table II.

4. The Minimum Temperature (assuming LTE)

Assuming local thermodynamic equilibrium, we can set $S_\lambda(\tau_\lambda)$ to be equal to the blackbody radiation function at the wavelength λ, and thus get a relation between τ_λ and T. In particular we obtain the temperature corresponding to $\tau_\lambda = 0$, the so-called boundary temperature T_0 or minimum temperature as far as a minimum can be revealed in the variation of $S_\lambda(\tau_\lambda)$ with τ_λ.

In columns 4, 5, and 6 of Table II, we report the values of the central intensity which give a surface temperature T_0 successively equal to 4500°, 4600°, 4700°. Our measurements, together with the determinations from the U.V. fluxes, lead to a surface temperature higher than 4700°.

As far as LTE is assumed, this result is in good agreement with the determinations of Matsushima, who found from visible limb-darkening observations a surface temperature close to 4650° for wavelengths longer than the Balmer discontinuity of hydrogen and a temperature greater than 4700° below this discontinuity (MATSUSHIMA and TERASHITA, 1967).

Nevertheless, experimental errors are important and lead to a rather broad uncertainty for T_0. Let us put

$$X = \frac{I(\mu=0)}{I(\mu=1)} \quad \text{and} \quad Y = I(\mu=1).$$

We have:

$$X \cdot Y = \frac{C_1}{\lambda^5} \frac{1}{e^{\frac{C_2}{\lambda T_0}} - 1} \cong \frac{C_1}{\lambda^5} e^{-\frac{C_2}{\lambda T_0}}.$$

Through logarithmic derivation we obtain:

$$\Delta T_0 \cong \frac{T_0^2 \lambda}{C_2}\left(\frac{\Delta X}{X} + \frac{\Delta Y}{Y}\right).$$

If we assume that $T_0 = 4700°$, and for $\lambda = 3000$ Å, we can write:

$$\Delta T_0 \cong 450°\left(\frac{\Delta X}{X} + \frac{\Delta Y}{Y}\right).$$

We include in X and Y the experimental errors and the incertainty due to the averaging over the band of the filter. Experimental errors on X can be estimated to 5% in the best conditions. But the error due to the inversion of the Laplace Integral cannot be estimated with any certainty.

Errors on Y are certainly more important since absolute photographic photometry is not very accurate; we think that they must lie between 15% and 20% at best.

The fact that the measurements are not made in the true continuum but are averaged over a finite spectral range including a lot of absorption lines, have opposite effects in modifying X and Y. X is certainly overestimated since the center-to-limb contrast is lower in the lines than in the continuum, as we shall see later, and therefore will lead to a value of T_0 too high. Conversely, Y is underestimated since the absorption by the lines decreases the continuum intensity. Which of these two effects is the most important is difficult to assert! A computation of the limb-darkening and of the blanketing of the lines for all those included in the bands of the filter is the only way to answer the question. This is in progress now, but not entirely finished.

If the experimental errors were only taken into account, T_0 could be determined with an accuracy of $\pm$ 100°.

5. Empirical Opacity

The analytical derivation of $\tau_\lambda(T)$ leads to $d\tau_\lambda/dT\,(T)$ which, with respect to the same quantity at a given reference wavelength λ_0, gives the ratio x_λ/x_0 of the corresponding absorption coefficients. As usual, we have taken $\lambda_0 = 5000$ Å. The $\tau_0\,(T)$ relation for this wavelength has been deduced from Pierce and Waddell's limb-darkening observations (PIERCE and WADDELL, 1961). We note that the model atmosphere given by these authors is in better agreement with the observations at 5000 Å than the Utrecht Reference Model Atmosphere or U.R.M.A. (HEINTZE, HUBENET, and DE JAGER, 1964). In the same way the empirical model of MÜLLER and MUTSCHLECNER (1964) agrees very well with the observations at 5000 Å. Consequently, assuming that there must be a unique surface temperature we have chosen 4600°, which is the boundary of the Müller and Mutschlecner model. It leads to central intensities at 1980 Å and 3190 Å respectively equal to:

$$\left.\begin{matrix} 1.015 \times 10^{12} \\ 2.020 \times 10^{14} \end{matrix}\right\} \text{ergs/cm}^3\text{/steradian/sec.}$$

As shown by BARBIER (1946) the fact that T_0 is not accurately defined is not of great importance for the determination of the empirical opacity.

In Figures 6, 7, and 8 we have reported the results obtained for 4800°, 5200°, 7000°, and 8000°. They are compared to a theoretical computation made at Meudon following Gingerich's evaluation of the continuous absorption by various hydrogenic sources and by metals such as magnesium, silicon, carbon, and aluminium (GINGERICH, 1965).

It can be noticed that:

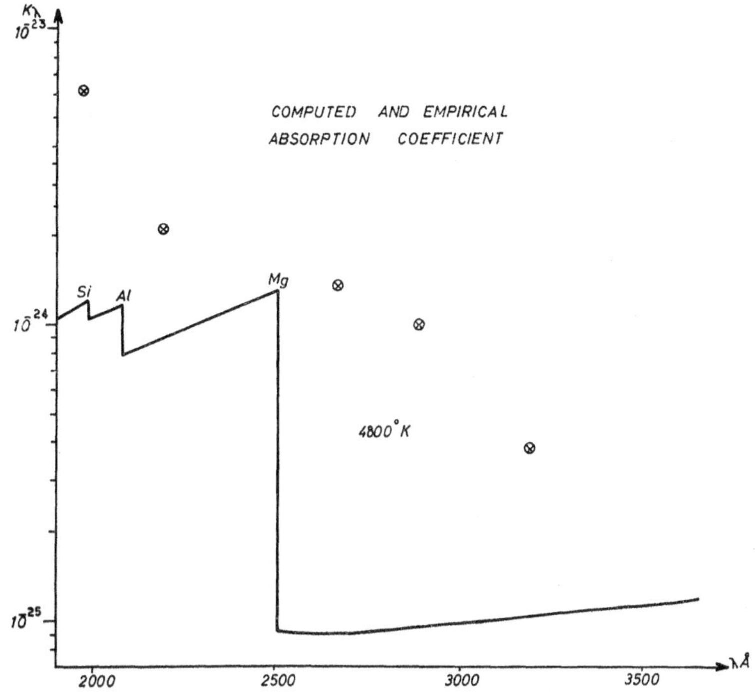

Fig. 6. Comparison between the empirical opacity deduced from limb-darkening observations and the computation made at Meudon following GINGERICH (1965) for an electron temperature of 4800 °K. The full line refers to the computation, the circles with crosses represent our determinations.

(1) In the transition zone, i.e. $T \simeq 4800°$, the empirical opacity is always higher than the computed one, especially between 3190 Å and 2500 Å where its variation with λ is entirely different.

(2) The discontinuity due to magnesium does not appear so strong in the empirical determination as it does in the theoretical one.

(3) There is a large increase in the opacity between our two measurements at 2190 Å and 1980 Å as could be deduced from the aspect of the limb-darkening curves.

(4) At higher temperatures, the agreement between measurements and computations including metals is considerably better in the range 2500 Å, 3190 Å, but the measured opacity at 1980 Å is in excess by a factor of 3 to 5 with respect to the computed one.

DISCUSSION

That the disagreement is greater for low temperatures is not surprising. Given the profile of our source-functions in the corresponding layers, a slight variation in the temperature induces a large variation in the optical depth. More curious is the variation of the empirical opacity compared to that of the computed one. Several factors can be taken into account to explain the discrepancy.

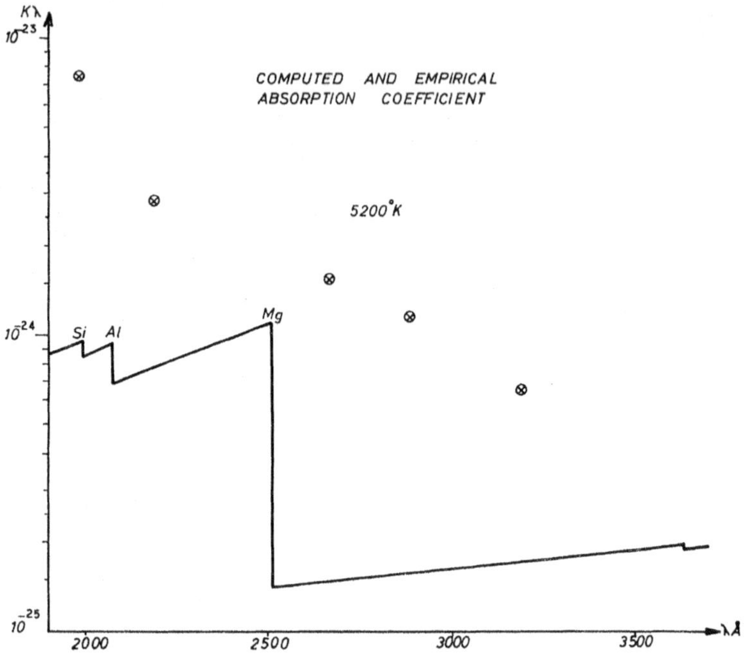

Fig. 7. Comparison between the empirical opacity deduced from limb-darkening observations and the computation made at Meudon following GINGERICH (1965) for an electron temperature of 5200 °K. The full line refers to the computation, the circles with crosses represent our determinations.

1. *Effect of the Lines*

As seen earlier, lines intercepted by the filter have the effect of decreasing the center-to-limb contrast. The result is an apparent increased continuous opacity, and the more numerous the lines, the greater this effect. We think that this effect is very important and our values of the continuous opacity are certainly an upper limit of what can be found in the solar atmosphere.

2. *Effect of Inhomogeneities*

That the sun, even with a resolution of some 15″, cannot be considered as homogeneous is shown in Figure 2: the 1980 Å picture looks very much like the Ca K spectroheliogram and, at least for this wavelength, smoothing the limb-darkening curve certainly leads to a poor approximation of the source-function. But as we shall see in Section 6 and as it can be observed on the other pictures of Figures 1 and 2, the inhomogeneities of dimensions greater than 15″ are very diffuse for wavelengths higher than 2190 Å. Hence, the disagreement appears in a region where this effect must not play the prime role.

Although these two effects must modify the final result once they have been taken into account, we suggest that a non-metallic source of continuous opacity contributes to the solar absorption coefficient in the range 2500–3000 Å. This result confirms that of MATSUSHIMA and TERASHITA (1967), deduced from observations in the visible. The variation of this opacity with temperature and wavelength suggests that it could be the quasi-molecule H_2 (ERKOVITCH, 1960). As pointed out by Dalgarno (PAGEL, 1959), the rate coefficient for the association detachment reaction $H + H^- \rightarrow H_2 + e^-$ is of the order of 10^{-10} cm^3 sec^{-1}.

It is possible that the H_2 molecule thus produced might be a quasi-molecule contributing to the UV opacity (BELY, 1967).

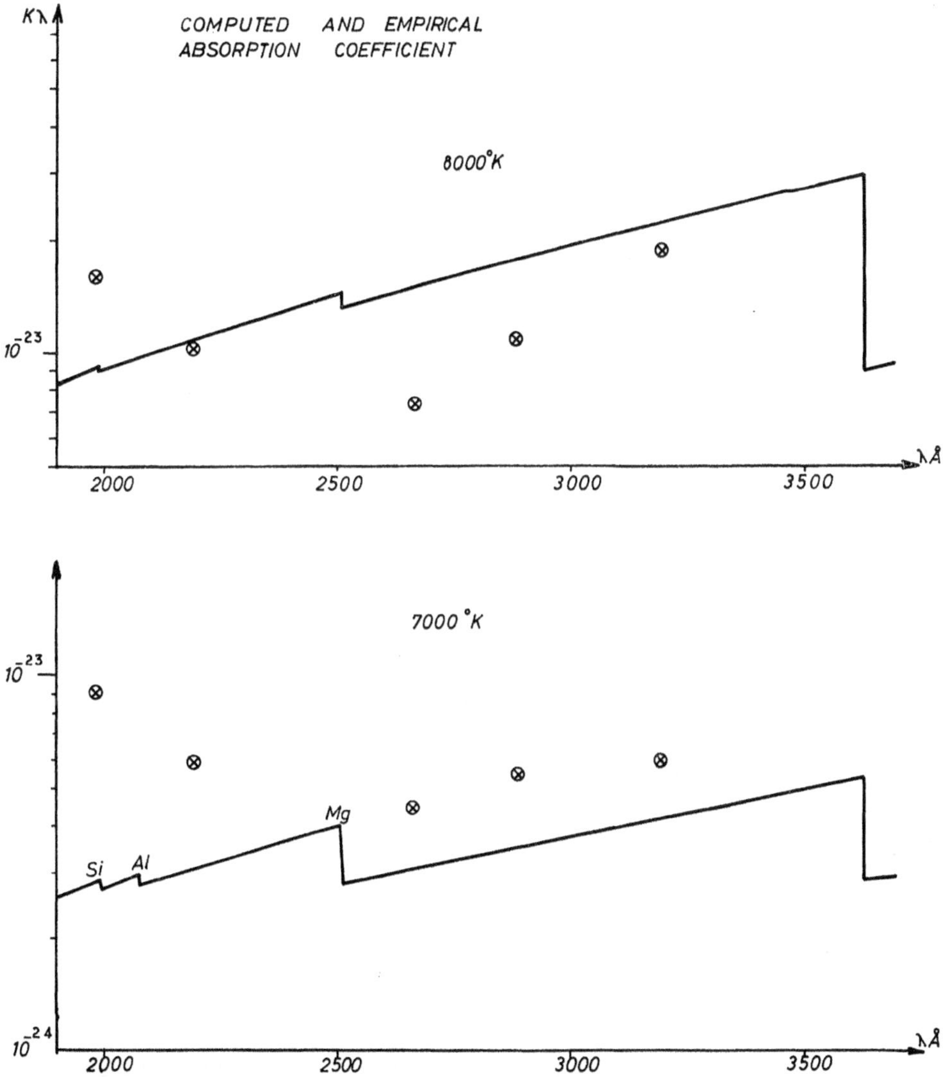

Fig. 8. Comparison between the empirical opacity deduced from limb-darkening observations and the computations made at Meudon following GINGERICH (1965) for an electron temperature of 7000 °K and 8000 °K. The full lines refer to the computation, the circles with crosses represent our determinations.

3. *Departures from Local Thermodynamic Equilibrium*

Let us assume that local thermodynamic equilibrium cannot hold in the photospheric layers studied, and that the source functions cannot be identified exactly to the Planck function $B_\lambda(T)$ but to (PAGEL, 1963):

$$\frac{B_\lambda(T)}{\bar{b}} = B_\lambda(T_{ex}) \quad \text{with} \quad \bar{b} < 1 .$$

Hence we have $T < T_{ex}$. We have introduced in our computation b factors $\simeq \frac{1}{2}$ in order to lower the surface temperature close to the minimum of 4500° of the Utrecht Model 1964. The source functions thus obtained have been compared to the τ_{5000} (T) relation given by the average column of this model. The result is a general slight increase of the empirical opacity. We have not evaluated what the effect would be of introducing departures from LTE in the computation of metallic absorption but if Ni^* is the number density of any metallic atoms involved, computed in LTE, $\alpha_{\lambda i}$ the corresponding absorption cross-section, and bi the coefficient of departure from LTE. The actual number density of the absorbing atom i is $Ni = biNi^*$, and we have:

$$\frac{x_\lambda}{x_{5000}} \cong \frac{\sum_i biNi^*\alpha_{\lambda i}}{N_{\mathrm{H}}^* - \alpha \mathrm{H}_{5000}^-} = \bar{b} \cdot \frac{\sum_i Ni^*\alpha_{\lambda i}}{N_{\mathrm{H}}^* - \alpha \mathrm{H}_{5000}^-} = \bar{b} \, \frac{x_\lambda^*}{x_{5000}} .$$

The asterisk refers to computation in LTE. Since $\bar{b}$ is lower than unity, we see that the computed opacity in the case of LTE departure is lowered.

Consequently the effect of introducing LTE departures is to increase the discrepancy existing between computed and measured opacity.

6. Latest Observations (Experiments of January 1967 and March 1967)

In order to study in more detail the observed discontinuity in the balloon experiment of October 3, 1966 between 1980 Å and 2235 Å, we launched on January 13, 1967, a Veronique rocket equipped with a stigmatic double dispersion spectrograph and a biaxial SPC 300 sun follower built by Ball Brothers Research Corporation. The spectral range covered was 1800 Å, 2800 Å, with a spectral resolution varying from 0.37 Å near 2300 Å to 0.7 Å at the extremities.

Excellent spectra were obtained corresponding to exposures of 0.1 sec, and 0.5 sec. Given the very good pointing accuracy, details of angular dimensions lower than 20″ could be resolved in the direction of the slit, which cut across the disk at a distance less than 1′ from its center. Figure 9 represents two such exposures corresponding to approximately the same solar diameter. It can be seen that the facular plages appear strongly in the regions where the absorption is important and in the chromospheric emission lines.

Numerous data were obtained on center-to-limb variations both in the continuum and absorption lines. Figures 10 and 11 represent two low-resolution densitometer

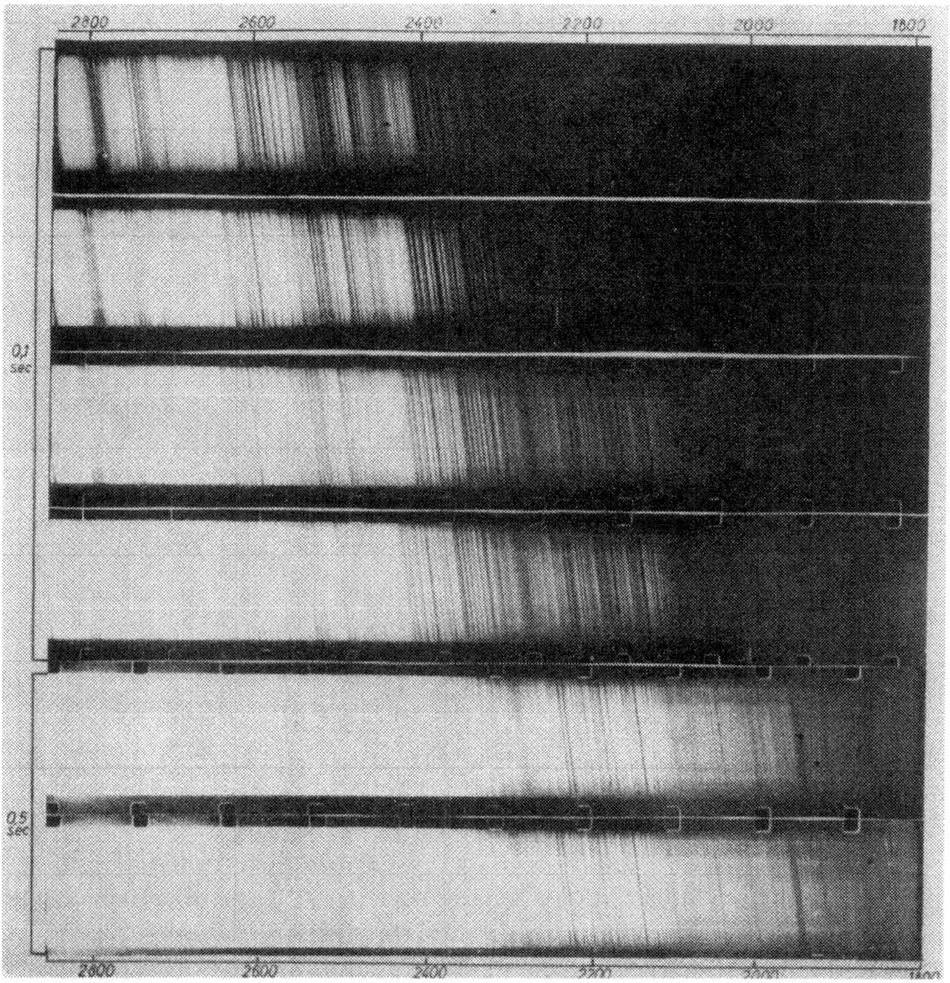

Fig. 9. Solar disk stigmatic spectrum 1800 Å–2800 Å. Veronique flight of 13th January 1967. Mean spectral resolution 0.5 Å, angular resolution 20 seconds of arc.

tracing referring to the center and limb of the disk. Althougth the corrections due to instrumental transmission and film sensitivity are not yet finished, the most striking features of the solar spectrum can be revealed in terms of central intensity and limb-darkening.

The most conspicuous one occurs at 2076 Å which corresponds to the ionization limit of Al I. The central intensity varies by a factor of 5 to 7 from one part to the other of this wavelength. Apart from the discountinuity of Ca I at 2028 Å and that of Si I at 1990 Å which is hardly seen on our spectra, it is the only feature which can explain the very different structures observed on our 1980 Å and 2235 Å spectroheliograms.

In order to be sure of this, we have taken spectroheliograms with wavelengths peaked at 2040 Å and 2215 Å, during a balloon flight in March 1967. Despite the fact

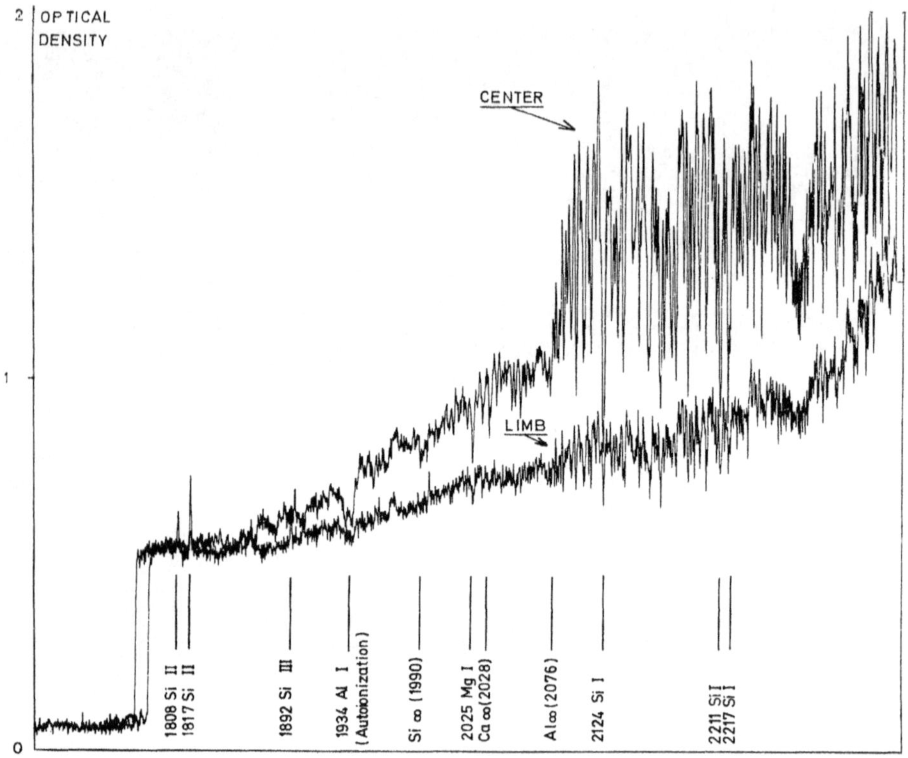

Fig. 10. Low-resolution densitometer tracing for the center and limb of the solar disk. The ordinates represent the optical density of a 0.5 sec exposure.

that the angular resolution was very poor (of the order of 1 min of arc), the sudden change in the solar structure from one picture to the other is well shown.

Anyhow, theoretical computations of metallic absorption must inevitably include a new source to explain the variation of the central intensity at 2076 Å. As already shown by Kodaira (1965), Aluminium could be that source.

The discontinuity due to the photo-ionization of Mg I which theoretically occurs at 2510 Å is less easily observed, owing to the presence of a strong blend of Fe lines. Nevertheless, the variation of the central intensity by a factor of 2 to 3 at 2650 Å and 2500 Å can be more likely attributed to magnesium than to any other absorber.

The discontinuity observed at 2412 Å corresponds to a change in the film sensitivity.

Another important feature of the spectra is the Al I auto-ionization doublet at 1934 Å. As shown by Withbroe (1966), the center-to-limb variation of this line brings information on the solar model. Withbroe computed this variation using two different models, the Utrecht Reference Model and the Model of Müller and Mutschlecner. The first predicts a strong limb brightening while the latter leads to the exactly opposite conclusion. The difference is due to the presence of a chromosphere in the first Model, while in the second one the electron temperature decreases monotically with de-

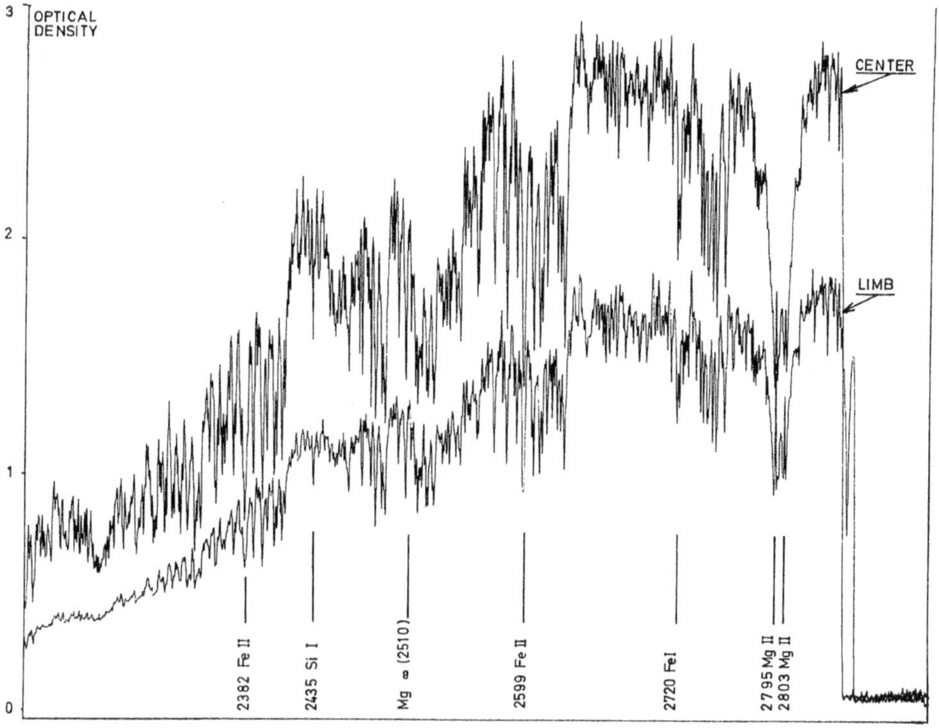

Fig. 11. Low-resolution densitometer tracings for the center and limb of the solar disk. The ordinates represent the optical density of a 0.1 sec exposure.

creasing optical depths. We have made several tracings of the center-to-limb variation of the line: no limb brightening can be revealed, but the observation corresponds fairly well to the variation predicted by the computation using Müller and Mutschlecner's model.

7. The Photospheric Model

The preceding observations and computations are often in disagreement with some characteristics of the Utrecht Reference Model Atmosphere 1964. On the contrary, the model of Pierce and Waddell and that of Müller and Mutschlecner seem to fit quite well visible and UV observations.

A. CRITICISM OF THE UTRECHT REFERENCE MODEL 1964

1. *Visible Observations*

The τ_{5000} (T) relation deduced from visible limb-darkening (PIERCE and WADDELL, 1961) is not in agreement with the same relation given in the average column of the U.R.M.A., especially for optical depths lower than 0.1. As a matter of fact, the variation of the temperature with τ_λ is considerably more gradual for the empirical

τ_{5000} (T) relation. As pointed out by the authors themselves, the computed radiation intensity is 10% too small near 5000 Å.

The determination of the surface temperature through extrapolations of the τ_λ (T) relations toward $\tau_\lambda = 0$ yields to a minimum temperature of 4600°–4650° (MATSUSHIMA and TERASHITA, 1967).

2. *Ultraviolet Observations*

Since the solar ultraviolet radiation comes from the upper layers of the photosphere near the transition zone, they give more information on these layers than the visible observations (Figure 12).

The gradual variation of the temperature near the transition is confirmed:

(1) by the smooth limb-darkening profile below 2076 Å;

(2) by the center-to-limb variation of the auto-ionization doublet of Al I;

(3) by the very slow variation of the brightness temperature between 2000 Å and 1500 Å which never falls under 4700 °K (TOUSEY, 1963).

The minimum temperature as determined from our measurements would be close to 4700 °K. Nevertheless, as seen earlier this determination is not very precise and the extrapolation of T near $\tau_\lambda = 0$ is certainly affected by inhomogeneities. Therefore, a minimum temperature of 4500 °K is not incompatible with our observations. In the

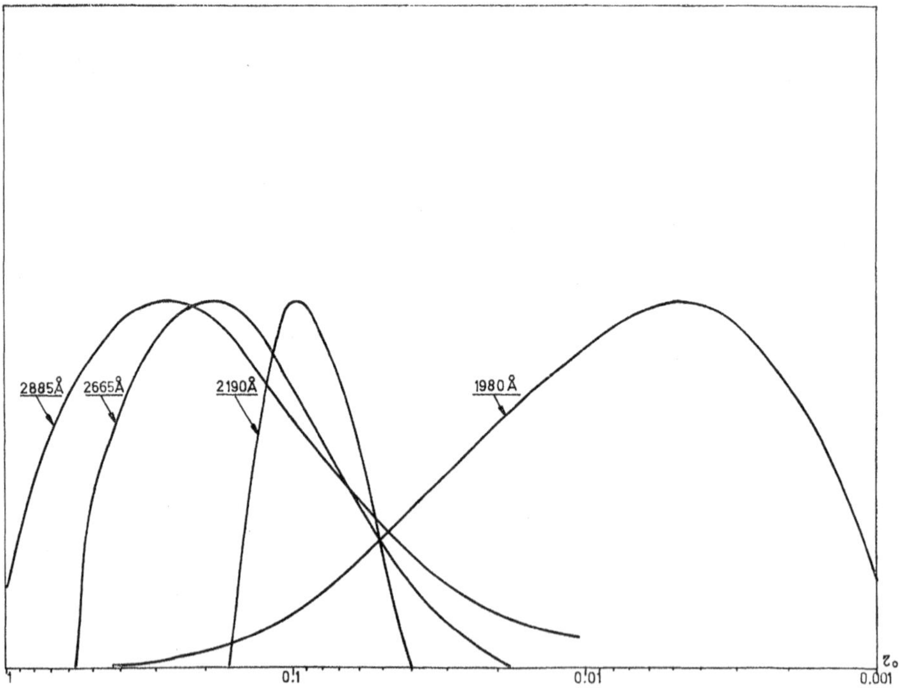

Fig. 12. Contribution functions plotted *vs.* the optical depth at 5000 Å.

assumption of L.T.E., it seems to be in contradiction with the fact that the transition is gradual or even flat with a brightness temperature always higher than 4700 °K.

3. *Eclipse Observations*

The gradient $dT/d\tau_{5000}$ of the U.R.M.A. has been deduced from eclipse observations of the intensity decrease at the extreme limb. RODDIER and DELACHE (1967) pointed out that eclipse determinations of the $dT/d\tau_{5000}$ gradient are always overestimated. This could explain the remarkable stressed transition of the U.R.M.A.

B. REQUIREMENTS FOR A PHOTOSPHERIC MODEL

The major criticism that we address to the U.R.M.A. concerns the transition between the photosphere and the chromosphere. The average empirical models of Pierce and and Waddell on the one hand and Müller and Mutschlecner on the other hand, which do not present any chromospheric temperature rising and therefore can be considered as models with very low temperature gradient, are in good agreement with visible and UV observations. The boundary temperature of these models is of some 4600 °K, which is a good compromise between the 4500 °K of the U.R.M.A. and our overestimated determinations of more than 4700 °K.

References

BARBIER, D.: 1946, *Ann. Astrophys.* **9**, 173–198.
BELY, O.: 1967, Private Communication.
BLAMONT, J. E. and BONNET, R. M.: 1966, *C.R. Ac. Sciences* **262B**, 152.
BLAMONT, J. E. and BONNET, R. M.: 1967, *C.R. Ac. Sciences* **264B**, 1158.
BONNET, R. M. and COURTES, G.: 1962, *Ann. Astrophys.* **25**, 367–373.
BONNET, R. M. and LAGO, B.: 1966, *C.R. Ac. Sciences* **262B**, 1636.
ERKOVITCH, S. P.: 1960, *Optics and Spectroscopy* **8**, 162.
GINGERICH, O.: 1965, Contribution to the 120th meeting of the American Astronomical Society, Berkeley, Calif., December, 1965.
HEINTZE, J. R. W., HUBENET, H., and DE JAGER, C.: 1964, *Bull. Astron. Inst. Neth.* **17**, 442.
KODAIRA, K. : 1965, *Z. Astrophys.* **60**, 240.
MATSUSHIMA, S. and TERASHITA, Y.: 1967, *Ann. Astrophys.* **30**, 183.
MÜLLER, E. A. and MUTSCHLECNER, J. P.: 1964, *Astrophys. J. Suppl.* **9**, 1.
PAGEL, B. E. J.: 1959, *Monthly Notices Roy. Astron. Soc.* **119**, 609.
PAGEL, B. E. J.: 1963, *Planetary Space Sci.* **2**, 333.
PEYTURAUX, R.: 1955, *Ann. Astrophys.* **15**, 302.
PIERCE, A. K. and WADDELL, J. H.: 1961, *Mem. Roy. Astron. Soc.* **68**, 89.
RODDIER, F. and DELACHE, P.: 1967, Private Communication.
TOUSEY, R.: 1963, *Space Sci. Rev.* **2**, 3.
WITHBROE, G. L.: 1966, *Astrophys. J.* **146**, 295.

THE FAR ULTRAVIOLET SPECTRUM OF THE SUN

OWEN GINGERICH

Smithsonian Astrophysical Observatory, Cambridge, Mass., U.S.A.

and

JOHN C. RICH

Harvard College Observatory, Cambridge, Mass., U.S.A.

(Received 26 June, 1967)

Abstract. Predictions of solar flux and limb darkening calculated from the Utrecht Reference Model, the Mutschlecner model, and the Bilderberg Continuum Atmosphere are compared with the rocket ultraviolet observations. The calculations employ recent experimental and theoretical determinations of metallic photo-ionization cross-sections. The spectral region between the continuous absorption edge from the first excited level of silicon at 1680 Å and the ground-state absorption edge at 1525 Å proves particularly significant for an investigation of the solar temperature minimum. The analysis indicates that the solar temperature minimum is relatively broad and flat, having a nearly constant temperature over somewhat more than a scale height, and that the chromospheric rise occurs just above $\tau_{5000} = 10^{-4}$ or so. Within the assumptions of the present models, a satisfactory interpretation of the ultraviolet spectrum can be reached with a temperature minimum of 4600° ± 100 °K.

1. Introduction

The solar ultraviolet spectrum shows a pronounced change from limb darkening (above about 1700 Å) to limb brightening (below about 1525 Å). That the limb-brightened spectrum below 1525 Å reflects the chromospheric temperature rise was pointed out by Tousey in the initial discussions of the 1962 rocket spectrum obtained by the U.S. Naval Research Laboratory (Tousey, 1963, 1964). Since that time numerous workers have emphasized how the change from limb darkening to limb brightening reflects conditions near the limb of the sun where the temperature must pass through the minimum value.

The continuous absorption in this ultraviolet region has been discussed by a number of workers (De Jager, 1963; Zwaan, 1962). Photo-ionization continua of neutral metals appear to make the most important contributions to the opacity. Bode (1965), improving the earlier work of Vitense (1951), has provided tables and graphs of the total continuous coefficient resulting from the sums of the various contributors in the stellar-like mixture.

Unfortunately, many analyses have neglected the explicit consideration of the low-lying excited atomic levels. The explicit inclusion of many levels of each atomic system is limited by practical matters of computing time and effort, but our results show that more than one level must be included. Silicon and magnesium are cases of immediate interest, and whereas the 2510-Å edge of the excited $^3P^0$ level of magnesium appears explicitly in most analyses, the 1680-Å and 1980-Å edges of the excited levels of silicon have generally been grouped with the ground level.

By explicitly including the ^1D excited level of silicon with its edge at 1680 Å, we can show that the continuous spectrum of the sun from 1690 Å to 1525 Å must arise from the region of the solar temperature minimum. Since it is difficult in the photometry to disentangle from the continuum the many emission lines in this spectral region, it is not easy to establish a precise numerical value for the minimum temperature. A minimum temperature of 4500 °K is entirely consistent with the observations, but the low-temperature region must be considerably more extended than that given in the Utrecht Reference Model (HEINTZE, HUBENET, and DE JAGER, 1964).

2. Silicon and Metallic Photo-Ionization Cross-Sections

Recent theoretical and experimental work in photo-ionization cross-sections has provided improved bound-free absorption cross-sections for those elements that, because of abundances and ionization, appear to provide most of the total absorption coefficient in this region of the spectrum – that is, for silicon, magnesium, and aluminium. Table I summarizes these cross-sections. The column labeled 'hydrogenic' contains those values calculated from Kramer's formula with an effective nuclear charge (cf. UNSÖLD, 1955); although manifestly unreliable for low-lying levels, they nevertheless provided some rough indication about possibly important absorbers, and hence furnished considerable incentive for the laboratory work carried out last year by RICH (1966, 1967). The column labeled 'quantum defect' contains those values calculated by various workers who have used BURGESS and SEATON's (1960) quantum-defect method.

TABLE I

Photo-ionization cross-sections (Series limit values in megabarns (10^{-18} cm^2))

		limit (Å)	hydrogenic	quantum defect	experimental
Si I	^3P	1525	4.8	34. (a) 26. (b)	37. (a)
	^1D	1680	5.4	48. (a) 23. (b)	35. (a)
	^1S	1980	6.3	46. (a) 39. (b)	
Mg I	^2S	1620	4.8	2.6 (c)	1.2 (a)
	$^3P^0$	2510	7.4	19. (e)	45. (f)
Al I	^2P	2071	6.7	18. (g) 28. (h) 19. (e)	22. (i) 21. (j)

References: (a) RICH (1966); (b) BODE (1965); (c) BURGESS and SEATON (1960); (d) DITCHBURN and MARR (1953); (e) PEACH (1962); (f) BÖTTICHER (1958); (g) BURGESS, FIELD, and MICHIE (1960); (h) VAINSTEIN and NORMAN (1960); (i) KELM: see BODE (1965); (j) PARKINSON and REEVES (1967).

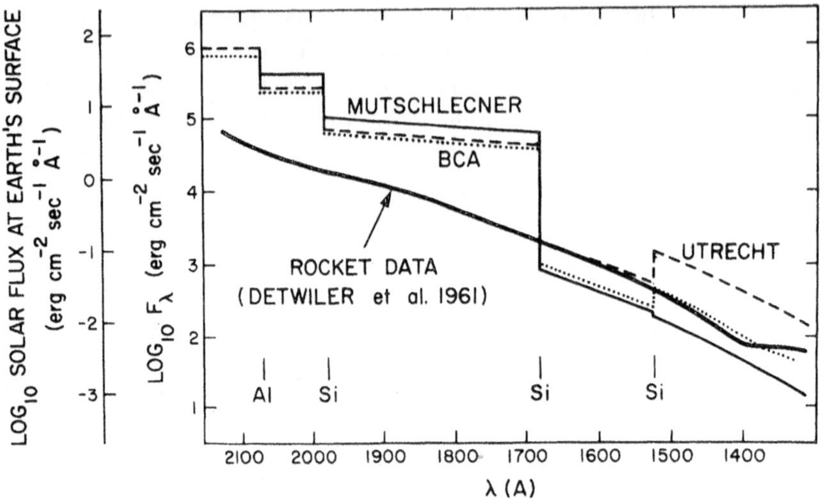

Fig. 1. Calculated ultraviolet solar flux. Opacity sources include metal continua, H^-, H, H_2^+, and Rayleigh scattering. Solid line is that calculated from the Mutschlecner model, which does not have a chromospheric temperature rise; dashed line is that from the Utrecht model. Flux is defined such that $\int_0^\infty F_\lambda \, d\lambda = \sigma T_e^4/\pi$.

3. Comparison of Two LTE Models with Observations

If we adopt photospheric abundances (GOLDBERG, MÜLLER, and ALLER, 1960) and if we assume, perhaps naively, that local thermodynamic equilibrium (LTE) holds throughout these regions in the high photosphere and low chromosphere, we can calculate the ultraviolet flux and limb darkening for a specified temperature distribution. We have carried out such computations with a computer program described elsewhere by GINGERICH (1961, 1963). We assume that the outer atmosphere is in hydrostatic equilibrium.

Figure 1 compares the results of three such LTE-flux calculations with data from a rocket experiment by the U.S. Naval Research Laboratory. The three models we use for purposes of illustration are the Utrecht Reference Model, designated 'Utrecht', the Mutschlecner model (MÜLLER and MUTSCHLECNER, 1964), and the Bilderberg Continuum Atmosphere (BCA). We include all the known significant sources of opacity: the bound-free metallic and silicon cross-sections (experimental values whenever available), neutral hydrogen, H^-, and H_2^+ (GINGERICH, 1964), the quasi-H_2 molecule (SOLOMON, 1964), and Rayleigh scattering. The discontinuities in the calculated continuum result from the idealized metallic and silicon photo-ionization edges.

The most striking feature of this figure is the jump at 1680 Å due to the ^1D edge of excited silicon. Although its discontinuity does not appear in the comparatively rough data of DETWILER *et al.* (1961), which have been averaged over 50-Å intervals, it is clearly visible in the 1962 rocket spectrum (Figure 2). The region near 1680 Å

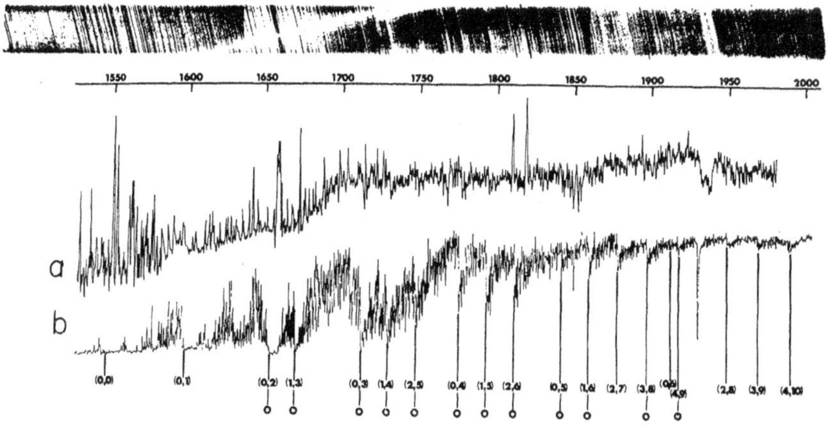

Fig. 2. Above: Part of the 1962 solar rocket ultraviolet spectrum obtained by TOUSEY *et al.* (1964). (a) The tracing of the spectrum (PORTER, TILFORD, and WIDING, 1967) showing the decrease in continuum intensity near 1680 Å. (b) The absorption curve of carbon monoxide obtained in the Harvard Shock-Tube Laboratory (GOLDBERG, PARKINSON, and REEVES, 1965). Illustration courtesy of the U.S. Naval Research Laboratory.

marks a transition from an absorption-plus-emission spectrum on the longer wavelength side to an almost pure emission spectrum below 1680 Å. Hence, in the region above 1680 Å, where absorption lines distort the continuum level, the true continuum should be significantly higher than the experimental data shown, whereas in the region below 1680 Å the converse applies. Consequently, if the lines could have been eliminated from the photometry, a small discontinuity would appear in the observed continuum. (Similarly, the silicon ground-state emission edge at 1525 Å does not show in the coarse photometry, but is clearly seen upon inspection of the 1962 rocket spectrum.)

Nevertheless, the predicted discontinuity of two orders of magnitude at 1680 Å is too large. The blanketing by the stronger silicon and iron lines on the redward side of this edge, as well as very many CO lines, can probably amount to only about half an order of magnitude. One important reason that the discontinuity is so large is that a substantial amount of flux arises from deeper layers because of the rapid increase in the Planck function in this wavelength region. Hence, an absorption that plays only a minor role elsewhere might be important in this limited spectral region from 1680 Å to 1800 Å. For example, absorption from quasi-molecular hydrogen has a small but noticeable effect on the predicted continuum.

In all models calculated the Sc I ^{1}S jump at 1980 Å and the Al I jump at 2071 Å appear to be of about the same magnitude, yet the observations indicate a significant discontinuity only for the aluminum at about λ 2085. This fact may have an explanation through a number of non-model effects (such as inadequate observations, cross-section and abundance errors, or a misidentification of the absorber at 2071 Å

TABLE II

Theoretical ultraviolet limb-darkening and limb-brightening predictions (out to $\mu = 0.1$)

λ(Å)	observed	Utrecht	Mutschlecner*	BCA
1700	pronounced darkening	pronounced darkening	pronounced darkening	pronounced darkening
1600	apparently neutral	pronounced brightening	neutral	slight brightening
1500	pronounced brightening	pronounced brightening	neutral	pronounced brightening

* The Mutschlecner model is extrapolated out to $\tau_{5000} \approx 10^{-6}$.

(Parkinson, 1967)), but it could be indicative of the manner in which the temperature first begins to flatten out as one moves higher into the very upper photosphere.

Table II summarizes qualitatively the limb darkening and limb brightening predicted by three models at three representative wavelengths. Obviously, the new Bilderberg Continuum Atmosphere is more satisfactory than the previous models. The Utrecht Reference Model predicts limb brightening at 1600 Å, contrary to observations, and the extrapolated Mutschlecner model does not predict limb brightening or an emmision edge at 1500 Å, again contrary to the observations.

4. Requirements for a Model

Of all the ultraviolet spectrum, the region between 1525 Å and 1680 Å is the most important for the interpretation of the solar temperature minimum. The absorption coefficients show that if any section of the ultraviolet continuum arises primarily from the temperature minimum, this is it. Furthermore, the fact that this region shows neither pronounced limb darkening nor pronounced limb brightening strengthens this interpretation.

Because of the great strength of the silicon absorption, optical depth unity corresponds to around $\tau_{5000} = 0.001$, and therefore the temperature minimum must extend farther out than represented, for example, in the Utrecht model. In other words, the temperature minimum must be rather broad, perhaps constant from $\tau_{5000} = 10^{-3}$ to 10^{-4}; otherwise, the models will predict limb brightening in the region.

The radiation temperature for this region, as measured by the NRL group, is 4700 °K. If the measured continuum is raised by emission lines, the radiation temperature for the true continuum would be somewhat less. The minimum temperature could actually be somewhat lower than the radiation temperature of this region, provided that significant additional radiation were contributed both by the chromosphere and by the photosphere.

On the other hand, if this critical region from 1680 Å to 1525 Å does have some limb brightening, then the spectrum could arise from the low chromosphere and redwards of 1680 Å from the high photosphere. In that case, the temperature minimum could not be directly observed in the ultraviolet. The observational evidence is sufficiently ambiguous, so that this possibility cannot be ruled out. Clearly, it would be immensely desirable to have additional calibrated spectrum and limb observations in this region, especially with somewhat higher resolution.

Recently, PORTER, TILFORD, and WIDING (1967) have argued similarly for contrary reasons that nowhere in the ultraviolet solar spectrum does the continuum arise from the region of the temperature minimum. They believe that the presence of CO band heads in absorption in the 1680-Å to 1525-Å region requires that this spectrum be formed in deeper and hotter photospheric layers. Since the CO lines can be formed by scattering, this argument has not convinced us, particularly in view of the heavy silicon absorption in this region. Unless the first excited level of silicon is enormously depleted by some sort of non-LTE mechanism, it is difficult to construct any reasonable model in which the continuum in this region can arise from a layer as deep as the upper photosphere.

Because the analysis of the calcium H and K lines by various investigators working independently indicates a minimum temperature in the vicinity of 4100 °K, we must now inquire whether such a low minimum is compatible with any interpretation of this spectral region. It would apparently require that the NRL calibration of 4700 °K as the radiation temperature near 1600 Å is in error for some unknown reason. Note that at 1600 Å a change from 4700 °K to 4200 °K in the radiation temperature corresponds to an order of magnitude decrease in the emergent flux. But a further difficulty arises in explaining the intensity of the carbon monoxide lines. The amount of carbon monoxide depends particularly critically on temperature around 4500 °K, and even if the minimum is 4500 °K the weakness of the CO lines above 1680 Å (RICH, 1966) or in the infrared (NEWKIRK, 1966) can hardly be accounted for. With a minimum temperature of 4100 °K, the CO spectrum would be entirely inexplicable.

Within the assumptions of the present models, a satisfactory interpretation of the ultraviolet spectrum can be reached with a temperature minimum of 4600 °K ± 100 °K.

Acknowledgment

We wish to thank Dr. Richard Tousey for providing special prints of the ultraviolet solar spectrum and Mr. Duane Carbon for carrying out the BCA computations.

References

BODE, G.: 1965, Ph.D. Dissertation, University of Kiel.
BÖTTICHER, W.: 1958, *Z. Physik* **150**, 336.
BURGESS, A. and SEATON, M. J.: 1960, *Monthly Notices Roy. Astron. Soc.* **120**, 121.
BURGESS, A., FIELD, G. B., and MICHIE, R. W.: 1960, *Astrophys. J.* **131**, 529.
DE JAGER, C.: 1963, *Bull. Astron. Inst. Netherlands* **17**, 209.
DETWILER, C. R., GARRETT, D. L., PURCELL, J. D., and TOUSEY, R.: 1961, *Ann. Géophys.* **17**, 263.
DITCHBURN, R. W. and MARR, G. V.: 1953, *Proc. Phys. Soc.* **A66**, 655.
GINGERICH, O.: 1961, Ph.D. Dissertation, Harvard University.

Gingerich, O.: 1963, *Astrophys. J.* **138**, 576.
Gingerich, O.: 1964, in *Proceedings of First Harvard-Smithsonian Conference on Stellar Atmospheres* (Smithsonian Astrophys. Obs. Spec. Rep., 167), p. 17.
Goldberg, L., Müller, E. A., and Aller, L. H.: 1960, *Astrophys. J. Suppl.* **5**, 1.
Goldberg, L., Parkinson, W. H., and Reeves, E. M.: 1965, *Astrophys. J.* **141**, 1293.
Heintze, J. R. W., Hubenet, H., and De Jager, C.: 1964, *Bull. Astron. Inst. Netherlands* **17**, 442.
Müller, E. A. and Mutschlecner, J. P.: 1964, *Astrophys. J. Suppl.* **9**, 1.
Newkirk, G.: 1966, private communication.
Parkinson, W.: 1967, private communivcation.
Parkinson, W. and Reeves, E.M.: 1967, private communication.
Peach, G.: 1962, *Monthly Notices Roy. Astron. Soc.* **124**, 371.
Porter, J. R., Tilford, S. G., and Widing, K. G.: 1967, *Astrophys. J.* **147**, 172.
Rich, J. C.: 1966, Ph.D. Dissertation, Harvard University.
Rich, J. C.: 1967, *Astrophys. J.* **148**, 275.
Solomon, P. M.: 1964, *Astrophys. J.* **139**, 999.
Tousey, R.: 1963, *Space Sci. Rev.* **2**, 3.
Tousey, R.: 1964, *Quart. J. Roy. Astron. Soc.* **5**, 123.
Tousey, R., Purcell, J. D., Austin, W. E., Garrett, D. L., and Widing, K. G.: 1964, in *Space Research*, IV, North-Holland Publ. Co., Amsterdam, p. 703.
Unsöld, A.: 1955, *Physik der Sternatmosphären*, 2nd ed., Springer-Verlag, Berlin.
Vainstein, L. A. and Norman, G. E.: 1960, *Opt. Spektr.* **8**, 79.
Vitense, E.: 1951, *Z. Astrophys.* **28**, 81.
Zwaan, C.: 1962, *Bull. Astron. Inst. Netherlands* **16**, 225.

THE SOLAR CONTINUUM FROM 900 TO 130000 Å AND THE PHOTOSPHERIC TEMPERATURE MODEL

A. SAUVAL

Royal Belgian Observatory, Uccle-Brussels, Belgium

(Received 2 June, 1967)

Abstract. In order to obtain a better agreement between observed and computed values of the solar intensity, an improved temperature distribution is deduced for the range $0.02 < \tau_0 < 10$. The intensity observations here considered refer to the wavelength region between λ 1980 and λ 129500, and the center-limb variations generally go down to $\cos\theta = 0.1$. The improved model, given in Figure 4 and Table II, differs rather little from the Utrecht 1964 model, used here as a reference.

It appears necessary to introduce an empirical correction function to be applied to the continuous absorption coefficient. This function was derived for the spectral region between 2000 and 130000 Å; it is shown in Figure 5.

Furthermore, an extension of the model $(1.10^{-7} < \tau_0 < 2.10^{-2})$ is deduced (see Table III and Figure 8), which reasonably well represents the observations of the ultraviolet solar flux ($\lambda\lambda$ 900–1700 Å).

1. Observations of the Solar Intensity ($\lambda\lambda$ 1980–129500 Å)

The solar *limb darkening* $I_\lambda(\mu)/I_\lambda(1)$, with $\mu = \cos\theta$ ranging between 1 and 0.1 (0.02 for some wavelengths in the visible part of the solar spectrum), and the *intensity* at the centre of the sun's disk 'between the spectrum lines', $I_\lambda(1)$, were collected and summarized by ALLEN (1963) for wavelengths between 2000 and 120000 Å.

The following sources are considered for the absolute intensity in the disk centre:

DETWILER *et al.* (1961): $\lambda\lambda$ 900–2000 Å.
BONNET and BLAMONT (1967): $\lambda\lambda$ 2190–2885 Å.
CANAVAGGIA *et al.* (1950): $\lambda\lambda$ 3200–5000 Å.
PIERCE and WADDELL (1961): $\lambda\lambda$ 3200–25000 Å.
LABS and NECKEL (1967): $\lambda\lambda$ 3290–12500 Å.
MAKAROVA (1963): $\lambda\lambda$ 3400–8500 Å.
MURAŠEVA and SITNIK (1964): $\lambda\lambda$ 3400–10000 Å.
LABS and NECKEL (1962): $\lambda\lambda$ 4000–6600 Å.
LAMBERT and WILLSTROP (1965): $\lambda\lambda$ 4000–6500 Å.
PEYTURAUX (1952): $\lambda\lambda$ 6100–23000 Å.
PEYTURAUX (1961): $\lambda\lambda$ 6000–8600 Å.
LABS and NECKEL (1963): $\lambda\lambda$ 6400–12500 Å.
SAIEDY (1960): $\lambda\lambda$ 86300, 111000, 120200 Å.

From an intercomparison of the values of $I_\lambda(1)$ given by these authors, we find a maximum relative difference, $\Delta I_0 \equiv 2(M-m)/(M+m)$, (where M and m are respectively the maximum and minimum values ranging between 91% and some percents (see Table I, column 2). We note, that Labs and Neckel's values are the smallest in the spectral region λλ 4000–10000 Å.

TABLE I

Relative Differences at $\mu = 1$

λ (Å)	ΔI_0	$\Delta(T_i/T_m)$	$I_0 - I_c$ URP 64 Model	$I_0 - I_c$ Improved Model	$I_0 - I_c$ Impr. and κ_c Model
1980	91%	9%	−135%	−148%	−148%
2190	71	7	−127	−138	− 59
2660	18	15	−125	−127	+ 20
2880	40	5	− 70	− 75	+ 8
3200	29	4	− 25	− 30	+ 7
3400	39	5	− 8.6	− 12.5	+ 4
3500	38	5	− 5.4	− 9.2	+ 7
3700	38	5	− 25	− 30	− 6
4000	11	2	− 0.3	− 0.8	+ 4.6
4226	15	4	+ 4.1	− 0.4	+ 3
4500	22	5	+ 4.8	+ 0.6	+ 3.4
4674	25	6	+ 6.1	+ 2.4	+ 2.2
5000	14	5	+ 4.6	+ 0.2	− 2
5400	17	4.5	+ 3.9	+ 0.2	+ 0.2
5490	16	4	+ 3.3	− 0.2	− 0.2
5593	15	4	+ 3.9	+ 0.5	+ 0.4
6000	12	3	+ 3.2	− 0.3	− 3
6200	10	5	+ 2.5	− 0.9	− 3
6245	10	5	+ 2.5	− 0.6	+ 1.5
6563	10	2.5	+ 2.0	− 1	− 5
7000	12	3.5	+ 2.6	− 0.2	− 0.2
7850	12	3.5	− 1	− 2.4	− 1.5
8000	12	4	− 0.1	− 2.9	− 2.9
8500	11	4	− 0.1	− 2.7	− 2.7
9000	11	4	− 1.9	− 4.4	− 1.5
10000	10	4	− 2.3	− 4.7	− 2
10100	9	4	− 0.8	− 3.1	− 1.5
11000	5	2.2	− 2	− 3.9	− 1.5
12900	2	1	− 8	− 9.6	− 1
14900	2	1	− 10.3	− 11.6	+ 0.2
15000	1	0.5	− 11.8	− 13.3	+ 3.0
15400	3	1.5	− 13.2	− 14.5	+ 1
16500	6	3	− 15.2	− 15.8	+ 0.1
17800	11	5.5	− 11.6	− 12.6	+ 4
19870	13	8	− 14.2	− 15.2	+ 3
20000	14	9	− 18.2	− 19.2	+ 2
22400	11	7	− 9.2	− 9.9	+ 0.8
24000	10	5.5	− 8.5	− 9.3	+ 1.3
30000	3	2	− 3.1	− 3.8	− 2.1
40000	9	6	+ 2.1	+ 1.4	− 0.7
50000	16	11	+ 7.0	+ 6.2	− 1.5
86300	16	13	− 0.3	− 1.8	− 1
100000	18	16	− 3.7	− 5.3	− 0.3
111000	6	5.5	+ 0.8	− 0.8	− 1.5
115000	5	4.5	+ 1.4	− 0.1	− 0.5
120200	2	1.3	+ 4.0	+ 2.2	− 2
129500	6	5.5	+ 8	+ 6	− 2

Relative measurements of the solar limb darkening were made by several investigators:

BLAMONT and BONNET (1966): $\lambda\lambda$ 1980–2880 Å.
MITCHELL (1959): $\lambda\lambda$ 3200–24400 Å.
DAVID and ELSTE (1962): $\lambda\lambda$ 3810–24400 Å.
LAMLA and SCHEFFLER (1956): λ 4226 Å.
NECKEL (1958): λ 4674 Å.
ROGERSON (1959): λ 5440 Å.
GAUSTAD and ROGERSON (1961): λ 5490 Å.
MINNAERT *et al.* (1949): $\lambda\lambda$ 5593 and 6245 Å.
PEYTURAUX (1952): $\lambda\lambda$ 6100–23110 Å.
DUNN (1959): λ 6563 Å.
PIERCE (1954): $\lambda\lambda$ 7800–24400 Å.
KOŽEVNIKOV (1957): $\lambda\lambda$ 10700–39400 Å.
SAIEDY (1960): $\lambda\lambda$ 86300, 111000, 120200 and 129500 Å.

The relative differences between values of $I_\lambda(\mu)/I_\lambda(1)$ given in these various papers do not exceed a few percents.

2. Computed Intensity

The model of the solar photosphere considered as a reference in our computations is the 'Utrecht Reference Photosphere 1964' (HEINTZE, HUBENET, and DE JAGER, 1964). The variables of this model (T, P_g, P_e) were somewhat extrapolated by us at both sides of the given range of optical depths, such in order to have them for 80 values of τ_0 (optical depth at λ 5000) ranging between 10^{-9} and 20.

The following relations give the intensity $I_\lambda(\mu)$ and the mean depth of formation $\bar{\tau}_\lambda$ of the continuous radiation, respectively:

$$I_\lambda(\mu) = \int_0^\infty B_\lambda(\tau_\lambda)\cdot e^{-\frac{\tau_\lambda}{\mu}} \frac{d\tau_\lambda}{\mu}, \qquad (1)$$

$$\bar{\tau}_\lambda = \frac{\displaystyle\int_0^\infty \tau_\lambda B_\lambda(\tau_\lambda)\cdot e^{-\frac{\tau_\lambda}{\mu}} \frac{d\tau_\lambda}{\mu},}{I_\lambda(\mu)}, \qquad (2)$$

with $d\tau_\lambda = \kappa_\lambda/\kappa_0 \, d\tau_0$. $B_\lambda(\tau_\lambda)$ is the Planck function.

BODE (1965) gives the continuous absorption coefficient $\kappa_\lambda(T, \log P_e)$ in tabular form in the spectral region $\lambda\lambda$ 100–115000 Å; interpolation in these tables is required to obtain κ_λ for other λ-values (we applied linear interpolation near discontinuities and quadratic far from them). We must note that κ_λ is really the sum of the absorption coefficient and of the scattering coefficient. More recent results for the free-free absorption coefficient of H^-, computed by GELTMAN (1965), would conduct to somewhat larger values for κ_λ in the spectral range $\lambda\lambda$ 5000–25000 Å.

The computations of $I_\lambda(\mu)$ and $\overline{\tau_0}$ according to Equations (1) and (2) and 10 values of μ (from 1 to 0.01) were performed by numerical integration for some 60 wavelengths.

A. INTENSITY DISTRIBUTION

Figure 1 shows the computed variation of I_λ (average column) in terms of λ, for three values of μ (1, 0.3 and 0.1). We also determined I_λ (combination) from:

$$I_\lambda(\text{comb}) = \tfrac{1}{4}I_\lambda(\text{av}) + \tfrac{3}{8}I_\lambda(\text{hot}) + \tfrac{3}{8}I_\lambda(\text{cold}),$$

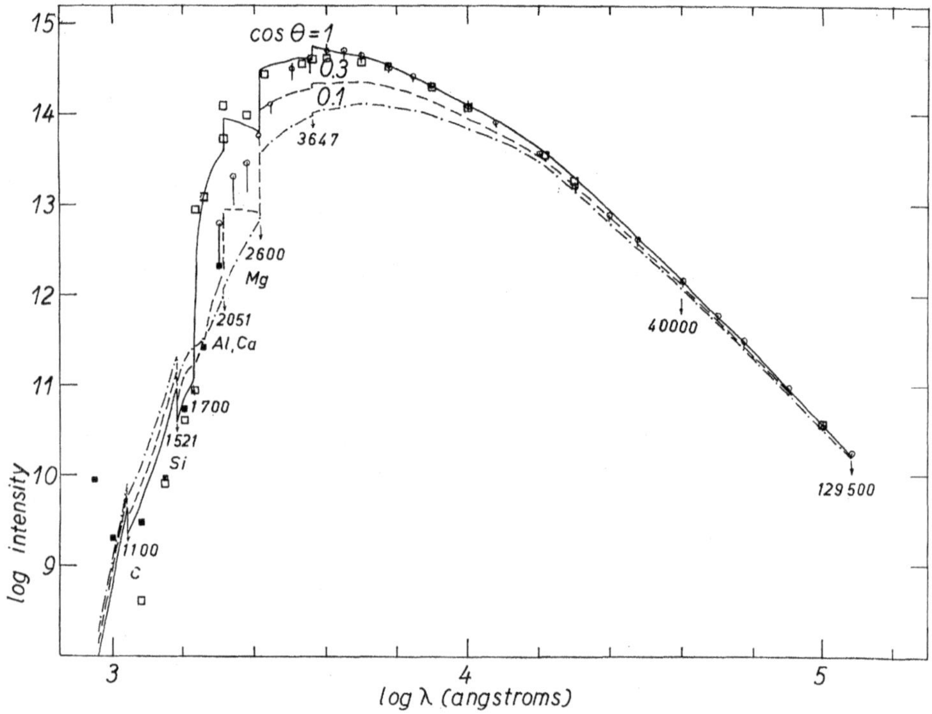

Fig. 1. Variation of the intensity of the solar radiation ($\lambda\lambda$ 911–129500 Å). Ordinate is $\log I_\lambda(\mu)$, with I_λ expressed in erg sterad^{-1} cm^{-2} s^{-1} ($\Delta\lambda = 1$ cm), and abscissa is $\log\lambda$, with λ in Å. The theoretical curves for three values of $\cos\theta$ are derived for the average column of the Utrecht solar model (computed from Equation (1)), open squares represent the theoretical values $I_\lambda(1)$ corresponding to the BCA model (computed by Gingerich, 1967). Some values of λ are indicated where the largest jumps occur. Circles represent Allen's compilation of observed values of the intensity at the centre of the sun's disk between the spectrum lines, full squares correspond to Detwiler *et al.*'s observations ($\lambda < 2000$ Å), and vertical lines give information on the accuracy of the observations (see Table I).

where the three numerical factors have been deduced from observations of lines of the C-multiplet near 10700 Å, by DE JAGER and NEVEN (1967). The relative difference between I_λ(av) and I_λ(comb) appears to decrease continuously towards long wavelengths: from 4.5% (λ 1800) to 0.05% (λ 129500).

Maxima in the intensity distribution occur near λ 3700 ($\mu=1$), λ 4000 ($\mu=0.3$), and λ 5200 ($\mu=0.1$). Discontinuities of κ_λ cause jumps in the intensity distribution: at $\mu=1$, the largest jumps occur at λ 2600 ($I_+/I_-=5.1$), λ 2051 (2.1), λ 3647 (1.3), λ 1100 (0.5) and λ 1521 (0.4).

A comparison of the three curves shows that the model predicts a conspicuous *limb brightening* for short wavelengths up to λ 1700, a transition region $\lambda\lambda$ 1700–1850 and a strong limb darkening for longer wavelengths onward. It is remarkable that the transition region is very short.

For the three columns of the model

$$I_\lambda(\text{cold}) \leqslant I_\lambda(\text{comb}) \leqslant I_\lambda(\text{av}) \leqslant I_\lambda(\text{hot}),$$

and the difference between the extreme values varies with λ: it is zero between λ 911 and λ 1700 and it decreases regularly from λ 1800 (80%) to λ 129500 (0.5%), in the case $\mu=1$.

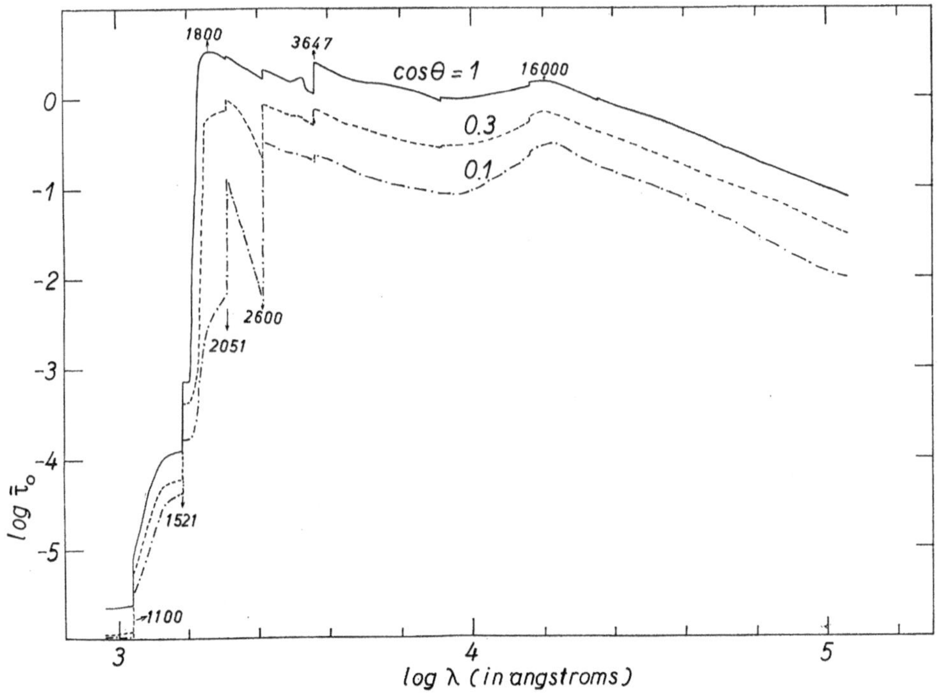

Fig. 2. Variation of the mean depth of formation of the continuous radiation ($\lambda\lambda$ 911–129500 Å). Ordinate is $\log \bar{\tau}_0$ (τ_0 is the optical depth at λ 5000) and abscissa is $\log \lambda$, with λ in Å. The theoretical curves for three values of $\cos\theta$ correspond to the average column of the Utrecht model (computed from Equation (2)). The largest jumps and the maxima in the mean depth distribution occur at the indicated wavelengths.

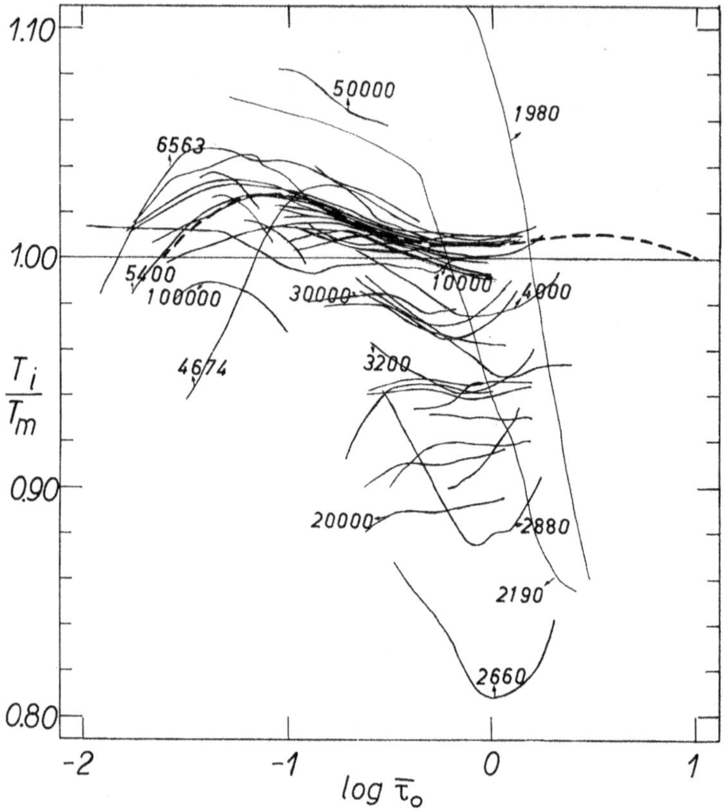
1.10
1.00
0.90
0.80
T_i/T_m
50000
1980
6563
5400
100000
30000
10000
4000
4674
3200
20000
2880
2190
2660
-2
-1
0
1
log $\bar{\tau}_0$

Fig. 3a

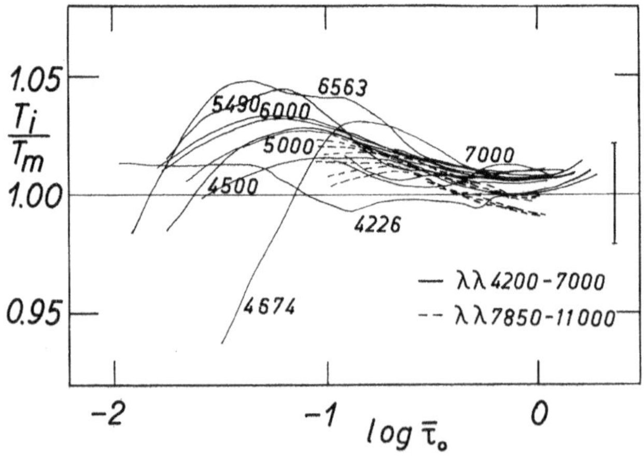
1.05
T_i/T_m
1.00
0.95
6563
5490
6000
5000
7000
4500
4226
4674
— λλ 4200-7000
-- λλ 7850-11000
-2
-1
0
log $\bar{\tau}_0$

Fig. 3b

In Figure 1, circles represent Allen's values for the disk centre; vertical lines give information on the accuracy of the observations. Table I, column 4 gives the difference between the observed (Allen) and computed intensities at $\mu=1$; the agreement is rather bad for short wavelengths ($\lambda<3700$), and there is also some disagreement in the range $\lambda\lambda$ 15000–24000.

B. MEAN DEPTH OF FORMATION

Figure 2 shows the variation of $\bar{\tau}_0$ (av) in terms of λ, for $\mu=1$, 0.3 and 0.1. Here too $\bar{\tau}_0$ (av) differs rather little from $\bar{\tau}_0$ (comb) (generally less than 1%).

There are three maxima in this function; for $\mu=1$, $\bar{\tau}_0$ (av) reaches the top values 3.4 (λ 1800), 2.4 (λ +3647) and 1.6 (λ 16000); this fact can easily be explained by the small values of the continuous absorption coefficient for these wavelengths.

The largest jumps occur at λ 1521 ($\bar{\tau}_{0+}/\bar{\tau}_{0-}=6.0$), λ 1100 (3.7), λ 3647 (2.1), λ 2600 (1.3), λ 8205 (1.1) and λ 14588 (1.1).

We have everywhere:

$$\bar{\tau}_0(\mu=0.1)<\bar{\tau}_0(\mu=1).$$

There is no simple relation between the values of $\bar{\tau}_0$ corresponding to the three columns of the model; the maximum difference between the extreme values varies with λ (from 15% to 0.1%).

3. Improved Model of the Photosphere ($0.02<\tau_0<10$)

A comparison of the observed (I_0) and computed (I_c) solar intensities for several wavelengths allows us to obtain in each point (λ, μ) an 'improved' temperature T_i, as follows.

We assume that for each (λ, μ):

$$\frac{B_\lambda(T_i)}{B_\lambda(T_m)}=\frac{I_0}{I_c}. \tag{3}$$

This relation defines T_i (T_m and I_c correspond to the same column of the Utrecht model); that means, we assume proportionality between the intensity and Planck function corresponding to observations and computations.

Fig. 3a and 3b. Temperature ratio as a function of the mean depth of formation corresponding to the average column of the Utrecht model. Ordinates are T_i/T_m (found from Equation (4)) and abscissae are $\log\bar{\tau}_0$ (given by Equation (2)) for the average column of the Utrecht model. Each solid curve corresponds to one λ-value between λ 1980 and λ 129500 Å (indicated for some of them). Figure 3b shows the results for $\lambda\lambda$ 4200 to 7000; these curves apparently show the same general character. The dashed curve is the 'mean' curve drawn through the observations referring to $\lambda\lambda$ 4200–7000 Å; it is extrapolated such to join the line $T_i/T_m=1$ at great depths. The vertical line in Figure 3b gives the mean accuracy (4%) of the temperature ratio.

Then, with Planck's law:

$$\left(\frac{T_i}{T_m}\right)_{\bar{\tau}_0} = \frac{c_2}{\lambda \cdot T_m \cdot \ln\left[1 + \frac{I_c}{I_0}\left(e^{\frac{c_2}{\lambda T_m}} - 1\right)\right]}. \tag{4}$$

Next we correlate the 'temperature ratio' (defined by Equation (4)) with $\bar{\tau}_0$ corresponding to $\bar{\tau}_\lambda$ found with Equation (2). Hence each temperature ratio corresponds to one value (λ, μ); so we obtain a cloud of N points, where $N = \Sigma_i (n\lambda)_i \cdot (n\mu)_i$. The computations were made separately, for the three columns and for 47 various wavelengths, from λ 1980 to λ 129500, with I_0 = Allen's values, and are preresented in Figure 3a (average column).

From Equation (4) it is clear that a larger value of I_0 corresponds to a larger temperature ratio, which depends further on the wavelength. Table I, column 3 gives the relative differences of the T_i/T_m values corresponding to various observed values I_0; at $\mu = 1$ the mean difference does not exceed about 4%.

Figure 3a shows the results for the average column; points belonging to the same wavelength are joined. We note a very large scattering $(0.81 < T_i/T_m < 1.08)$ in the region $0.01 < \tau_0 < 3$. However, points corresponding to the wavelength region $\lambda\lambda$ 4200–7000 are situated in a relatively narrow region 0.99–1.05 (see Figure 3b) and a rather close belt appears.

We next *assume* that the observations in the wavelength range 4200–7000 Å are the best, and that those in the other wavelength regions are influenced by a systematic effect. We assume that this effect occurs in the *absorption coefficients*, and in the next section we shall try to determine an empirical correction function to the absorption coefficient for wavelengths outside the region 4200–7000 Å.

On the basis of this assumption, we can draw a mean curve through the observations referring to $\lambda\lambda$ 4200–7000 Å (this is the thick broken line in Figure 3a), and so obtain an improved temperature distribution (Figure 4), which differs little (less than 3%) from the U.R.P. model distribution.

Assuming this new temperature distribution, and applying the following equations:

$$T = T(\tau_0)$$

$$\frac{\mathrm{d}P_g}{\mathrm{d}\tau_0} = \frac{g_{\text{eff}}}{\kappa_0}$$

$$P_e = P_e(P_g, T)$$

$$\kappa_0 = \kappa_0(P_e, T)$$

$$g_{\text{eff}} = \frac{1}{\rho}\frac{\mathrm{d}}{\mathrm{d}h}\left(P_g + \tfrac{1}{2}\rho v_t^2\right) \approx g_{\text{geom}}$$

we obtain by numerical integrations improved values for the functions $P_e(\tau_0)$ and $P_g(\tau_0)$ and thus get an *improved photospheric model* for the range $0.02 < \tau_0 < 10$ (see Table II).

Table I, column 5 gives the differences between the observed and newly computed central intensities and of course, we find a better agreement for $\lambda\lambda$ 4200–7000 Å.

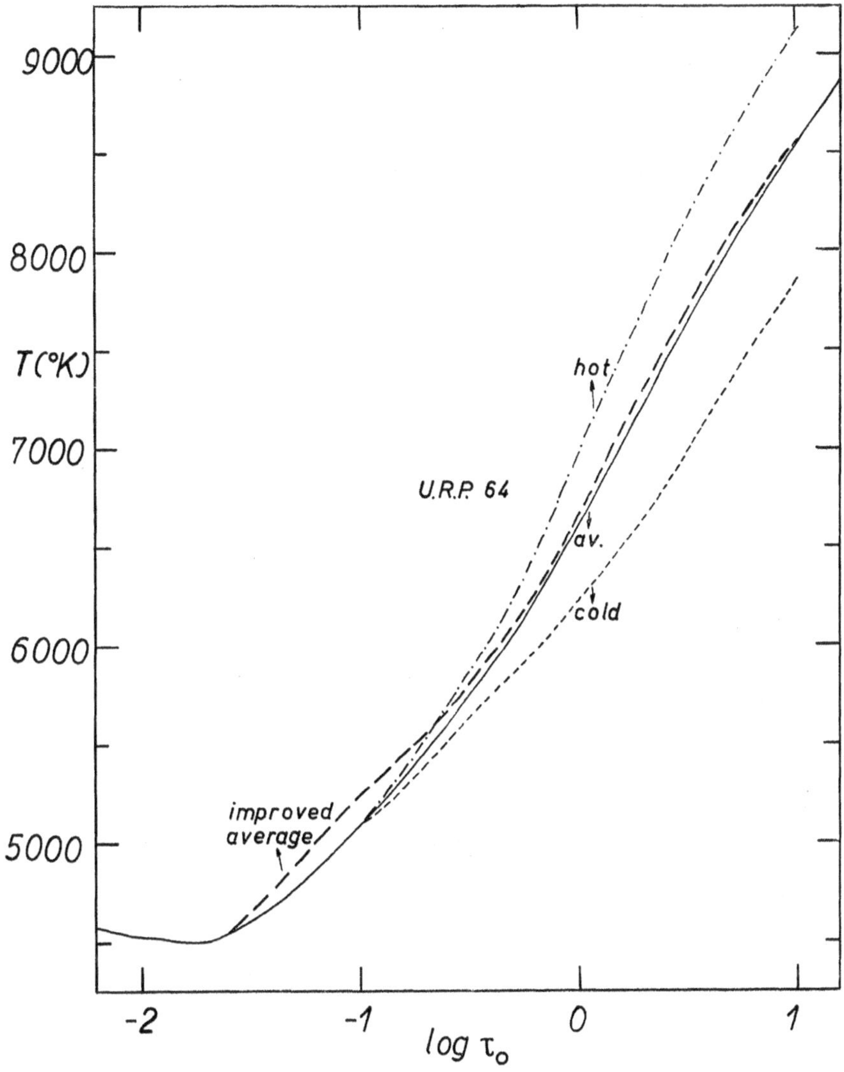

Fig. 4. Relation between the temperature and the optical depth. The temperature distributions for the three columns of the Utrecht solar model and the improved distribution for the average column ($0.02 < \tau_0 < 10$).

4. Correction Function to Continuous Absorption Coefficient

We next assume that the remaining disagreement between the observations and computations at $\lambda < 4000$ and $\lambda > 7000$ is entirely due to uncertainties in the absorption coefficient; we can thus find, for each wavelength, a correction factor κ_c/κ_m to be applied to κ_m, given in Bode's tables.

TABLE II

Improved Photospheric Model ($0.02 < \tau_0 < 10$)

H(km)	$\log P_g$	Average Column			Hot Column			Cold Column		
		τ_0	T(°K)	$\log P_e$	τ_0	T(°K)	$\log P_e$	τ_0	T(°K)	$\log P_e$
197	4.410	0.020	4500	−0.049						
157.7	4.576	0.040	4770	0.226						
116.2	4.737	0.080	5137	0.531						
104.5	4.787	0.100	5237	0.620						
66.8	4.929	0.200	5563	0.931	0.204	5638	0.989	0.202	5487	0.889
33.7	5.048	0.400	5963	1.332	0.437	6100	1.456	0.390	5810	1.216
7.2	5.131	0.800	6450	1.796	0.992	6750	2.031	0.698	6130	1.528
0	5.165	1.000	6663	1.982	1.283	7000	2.247	0.824	6275	1.663
−19.0	5.222	2.000	7313	2.499	2.937	7800	2.820	1.418	6750	2.080
−36.7	5.267	4.000	7913	2.913	6.660	8475	3.232	2.570	7275	2.495
−53.0	5.309	8.000	8400	3.214	12.820	9000	3.518	4.650	7725	2.818
−58.1	5.356	10.000	8550	3.310	15.930	9142	3.612	5.610	7857	2.929

The following relation defines the corrected absorption coefficient κ_0:

$$\frac{\kappa_c - \kappa_m}{\bar{\kappa}} = \Delta \ln \kappa = \frac{\mathrm{d} \ln \kappa_m}{\mathrm{d} \ln \tau_0} \frac{\mathrm{d} \ln \tau_0}{\mathrm{d} \ln T_i} \Delta \ln T' \tag{5}$$

where $(\mathrm{d} \ln \kappa_m)/(\mathrm{d} \ln \tau_0)$ is computed from the relation between the absorption coefficient corresponding to the improved model, and the optical depth $\kappa_m(\tau_0)$ (the values appear to range between 0.5 and 1.1), where $(\mathrm{d} \ln \tau_0)/(\mathrm{d} \ln T_i)$ is derived from the improved temperature distribution $T_i(\tau_0)$ (between 7 and 12), and where

$$\Delta \ln T' = \frac{T_i' - T_i}{\dfrac{T_i' + T_i}{2}} = 2\frac{(1 - T_i'/T_i)}{(1 + T_i'/T_i)} \quad (\text{negative if } T_i'/T_i > 1)$$

in which the temperature ratio is deduced from a relation similar to Equation (4), corresponding to the improved model (between −0.08 and +0.21).

For the average column and at each wavelength, we compute the correction factor κ_c/κ_m in some regions of the optical depth $\Delta\tau_0$, and the mean value thus obtained for each wavelength is considered as a first approximation of the correction function. With these values of κ_c/κ_m, we compute anew the temperature ratio T_i''/T_i, with an expression like Equation (4), for each wavelength, and we adjust this correction factor so that the mean value (taken over $\cos\theta$) of the temperature ratio for each λ-value approaches unity, as well as possible.

Figure 5 shows the relation between T_i''/T_i (computed from a relation similar to Equation (4), referring to the improved model with corrected absorption coefficients) and $\log \bar{\tau}_0''$ (deduced from an equation similar to Equation (2), corresponding to the improved model and with κ_c). Table I, column 6 gives the so obtained final agreement concerning the central intensity $I_\lambda(1)$. We must recall that the adopted value of κ_c was chosen in order to obtain a better agreement between observed and computed

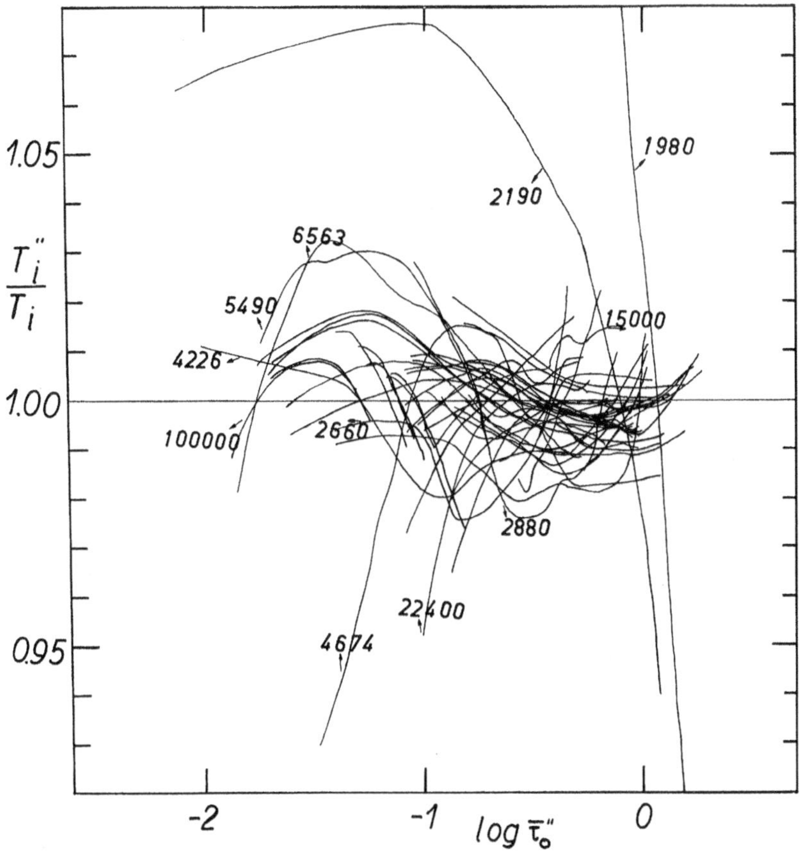

Fig. 5. Temperature ratio as a function of the mean depth of formation corresponding to the improved model with corrected absorption coefficients. The ordinates and abscissas are defined as in Figure 3. T_i''/T_i and $\log \bar{\tau}_0''$ are respectively deduced from relations similar to Equation (4) and to Equation (2), corresponding to the improved solar model with corrected absorption coefficients κ_c. The results refer to the spectral region λλ 1980–129500 Å.

values of the solar intensity *from centre to limb* (and not only at centre of sun's disk): this fact explains the apparent disagreement occurring in some values of the central intensity. It is gratifying to see that the greatest deviations from unity are now reduced to values $<3\%$ for most wavelengths, but for some in the U.V.

Figure 6 represents the adopted values of κ_c/κ_m as a function of λ. With our assumptions, we must note that the correction factor is *very different from unity* outside the spectral region λλ 4200–10000 Å.

The empirical function which describes the variation of the correction factor in terms of λ has the following characteristics; strong variation for short wavelengths (about 1 at λ 2000, 6.5 at λ 2660 with large uncertainties, and 1.2 at λ 4000); the function increases from λ 11000 (1.1) up to λ 20000 (4.2), then begins to decrease

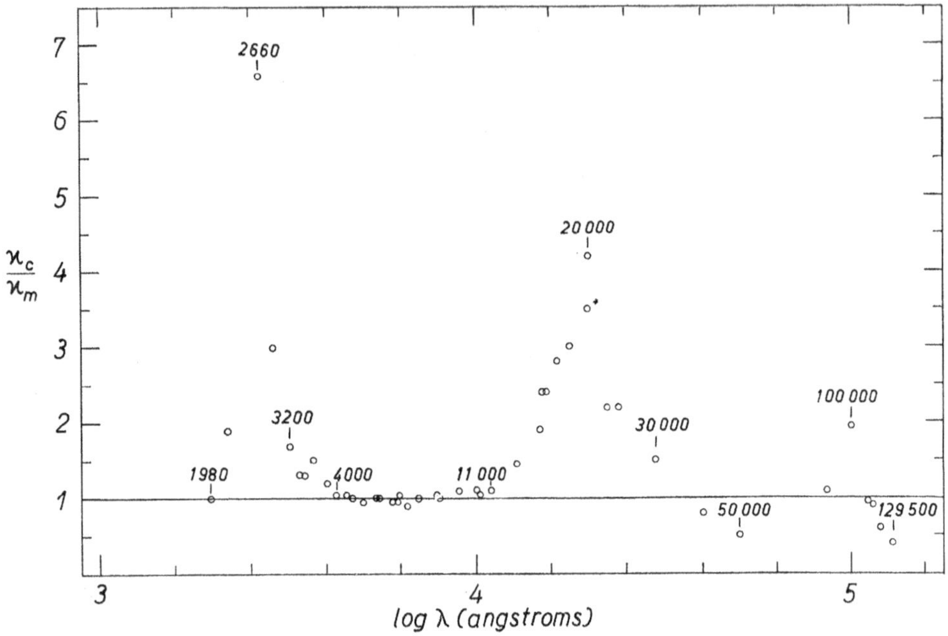

Fig. 6. Correction factor to the continuous absorption coefficient (λλ 2000–130000 Å). Circles represent the ratio κ_c/κ_m between the corrected and Bode's values of the absorption coefficients (computed from Equation (5)) as a function of λ (indicated for some values of λ).

until λ 50000 (0.5), reaches a maximum at λ 100000 (2) and again decreases down to λ 129500 (0.4).

Hence this result means that the continuous absorption coefficient, given in Bode's tables and used here, is too small in the spectral regions λλ 2000–4000, 11000–35000 and 85000–110000 Å (correction factor varying between 1.1 and 6); on the other hand, κ_λ seems to be too large in two narrow parts of the spectrum: λλ 35000–85000 and λλ 110000–130000 (κ_c/κ_m ranging between 0.4 and 1). If our hypothesis is correct, it will, hence, be necessary to explain why the absorption coefficient must be changed in this way: this may be a task for theoretical physics.

We may conclude that there is a very good agreement between observed and computed values of the solar intensity, from λλ 2000 up to 129500 Å, when using an *improved solar model*, not very different from the U.R.P. 64 model, with *absorption coefficients corrected* as indicated above.

5. Solar Flux from 900 to 2000 Å and a Model of the Solar Photosphere and Low Chromosphere

In order to test the temperature distribution *above the Utrecht minimum* ($\tau_0 < 0.02$), it is necessary to study the solar continuum in the extreme ultraviolet region: the radiation corresponding to 900–1700 Å originates from a region with a mean depth of formation ranging between 10^{-6} and 10^{-2} (see Figure 2).

Only the total radiation flux (and not the specific intensity) has been well recorded in this part of the solar spectrum down to 900 Å. Nevertheless, the true continuous flux at $\lambda < 1300$ Å is rather difficult to determine, owing to the presence of the Lyman α line and other emission lines. Figure 7 shows the wavelength dependence of $\log f_\lambda$, the solar flux at earth; circles represent ALLEN's (1963) summary (the dominating resonance lines are excluded from f_λ and the radiation from the other lines is smoothed) and the dashed curve corresponds to the observations given in graphical form by TOUSEY (1963).

The following relation gives the solar flux outside the earth atmosphere:

$$f_\lambda = 2\pi \left(\frac{R}{A}\right)^2 \int_0^\infty B_\lambda(\tau_\lambda) \cdot E_2(\tau_\lambda)\, \mathrm{d}\tau_\lambda,$$

where R and A are the solar radius and the astronomical unit respectively, $E_2(x)$ is the exponential integral function of the second order.

Figure 7 shows the theoretical curves of the solar flux computed for three recent solar models: the Utrecht Reference Photosphere (HEINTZE, HUBENET, and DE JAGER, 1964), MÜLLER and MUTSCHLECNER's (1964) model, and HOLWEGER's (1967) model (see Figure 8 for the temperature distributions). In order to compute the flux at short wavelengths, it has been found necessary somewhat to extrapolate the two last mentioned models: if we assume a uniform temperature near $\tau_0 = 0$ (4605 or 3900 °K), the corresponding flux at the smallest λ-values is Planckian:

$$f_\lambda = \pi \left(\frac{R}{A}\right)^2 \cdot B_\lambda(T).$$

We note that between 900 and 1700 Å none of the computed flux distributions agrees completely with the observations: Holweger's and Mutschlecner's models give far too small values (some orders of magnitude) of the flux (except at $\lambda > 1500$ Å); whereas that computed for the Utrecht model is slightly too large (except near 1000 Å). We next *assume* that the disagreement between the observed and computed values of the solar flux is only due to uncertainties in the *temperature distribution* and not to errors in the adopted continuous absorption coefficient (BODE, 1965), and we shall try to derive an empirical temperature distribution which better represents the observations.

For such empirical models, we first assume that the temperature model above the minimum ($\tau_{\min} = 0.02$) obeys the following law:

$$T(\tau_0) = T_{\min} + a(\log \tau_{\min} - \log \tau_0),$$

where various values between 0 and 200 °K are chosen for the parameter **a**, whereas below the minimum ($\tau_0 > 0.02$), we maintained the Utrecht temperature model.

Figure 7 shows the theoretical continuous spectral curves thus obtained for the

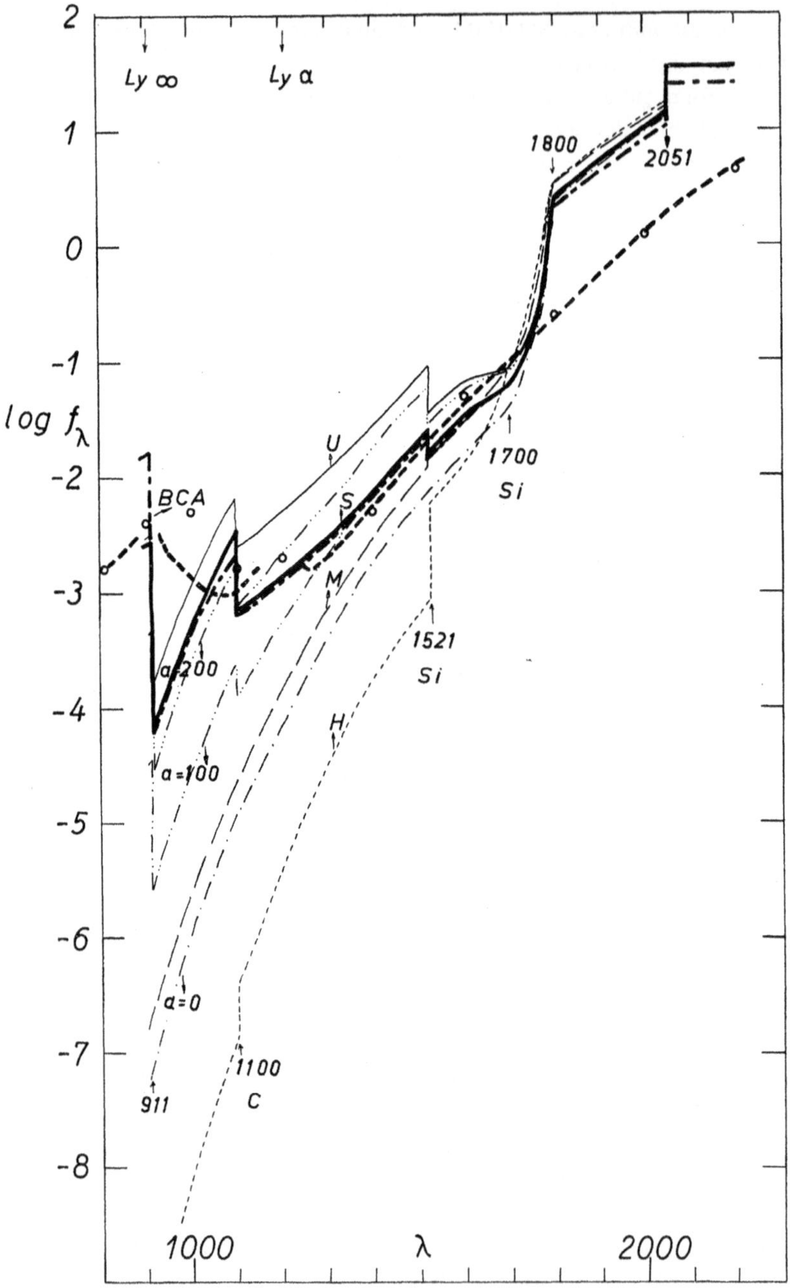

Fig. 7. Variation of the total flux of the solar radiation ($\lambda\lambda$ 900–2000 Å). The ordinate is $\log f_\lambda$ (flux at earth), with f_λ expressed in erg cm^{-2} s^{-1} Å^{-1}, and the abscissa is λ (in Å). Circles represent Allen's observed values and the thick broken line corresponds to Tousey's observations. The theoretical curves are labeled with: BCA = Bilderberg Continuum Atmosphere, H = Holweger, M = Müller-Mutschlecner, S = Sauval, U = Utrecht; curves $a = 0$, 100, and 200 correspond to three empirical models with parameter **a** respectively equal to the indicated values.

models with $\mathbf{a}=0$, 100 and 200°K. Clearly it is necessary to assume a broken shape of the $T(\log \tau_0)$ curve in the region $10^{-9}<\tau_0<0.02$, leading to a discontinuity in $\mathbf{a}$: near the minimum temperature ($2.10^{-5}<\tau_0<0.02$), 100°K is a good value for the parameter $\mathbf{a}$ and a steeper gradient (about 600°K) seems necessary for the region $\tau_0<2.10^{-5}$. Table III gives a model (S) which reasonably well represents the observations of the ultraviolet solar continuous spectral flux (see also Figures 7 and 8).

However we note that a model with a minimum temperature equal to 4600°K up to about $\tau_0=10^{-4}$ (Müller-Mutschlecner's model) would also fit rather well the observed flux at $\lambda>1500$ Å.

TABLE III

Model S ($1.10^{-7}<\tau_0<2.10^{-2}$)

H(km)	$\log P_g$	τ_0	T(°K)	$\log P_e$
1820	−1.847	1.10^{-7}	6250	−2.283
1640	−1.401	2.	6062	−2.046
1440	−0.771	4.	5875	−1.826
1210	0.023	8.10^{-7}	5685	−1.602
1140	0.299	1.10^{-6}	5625	−1.530
946	1.049	2.	5437	−1.369
809	1.604	4.	5250	−1.319
709	2.026	8.10^{-6}	5062	−1.343
679	2.153	1.10^{-5}	5000	−1.356
601	2.499	2.	4812	−1.389
545	2.835	4.	4770	−1.209
500	3.031	8.10^{-5}	4740	−1.084
486	3.107	1.10^{-4}	4730	−1.041
444	3.298	2.	4700	−0.906
404	3.480	4.	4670	−0.769
367	3.655	8.10^{-4}	4640	−0.632
354	3.712	1.10^{-3}	4630	−0.588
318	3.881	2.	4600	−0.456
282	4.049	4.	4570	−0.329
247	4.217	8.10^{-3}	4540	−0.206
237	4.274	1.10^{-2}	4530	−0.166
197	4.410	2.10^{-2}	4500	−0.049

Addendum

The solar model adopted at the Study Week on the Quiet Photosphere, Bilderberg, near Arnhem, 1967 (BCA = 'Bilderberg Continuum Atmosphere') is shown in Figure 8 (temperature distribution), in Figure 1 (variation of the solar intensity), and in Figure 7 (wavelength dependence of the solar flux).

We note, that the BCA solar model agrees better than the Utrecht model with the observations of the solar continuous radiation for short wavelengths ($\lambda<1700$ Å), in the near ultraviolet region ($2600<\lambda<3800$ Å), and in the infrared part of the solar spectrum ($\lambda>8000$ Å). In actual fact, the relative differences between the BCA and

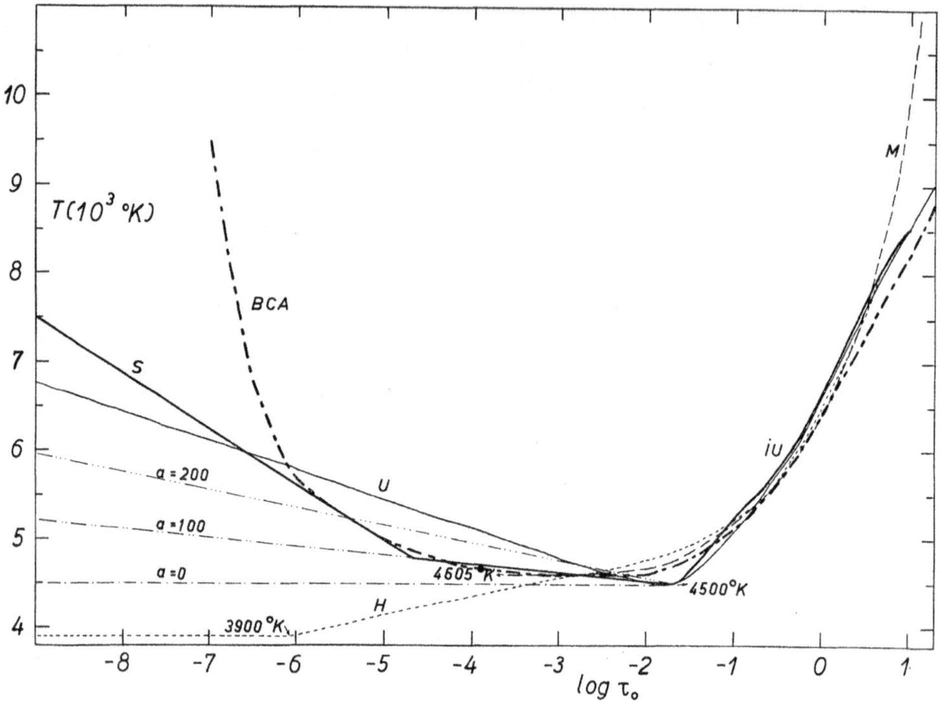

Fig. 8. Relation between the temperature and the optical depth. The figure shows the temperature distributions corresponding to some solar models: BCA = Bilderberg Continuum Atmosphere, H = Holweger; IU = Improved Utrecht ($0.02 < \tau_0 < 10$; average column); M = Müller-Mutschlecner; U = Utrecht; the distributions labeled with S = Sauval and $a = 0$, 100, 200 refer to the region: $10^{-9} < \tau_0 < 0.02$.

Utrecht values of the solar intensity at long wavelengths ($\lambda > 6000$ Å) are rather small (less than 7%).

Acknowledgements

We wish to thank Professor C. de Jager (Utrecht) and Dr. L. Neven (Uccle) for their continued advice and encouragement and for their helpful suggestions. It is also a pleasure to thank Dr. O. Gingerich for the computations of all our models. Our most cordial thanks to M. W. Nijs for his help in all the programming involved in the computations presented here.

References

Allen, C. W.: 1963, *Astrophysical Quantities*. Athlone Press, London.

Blamont, J. E. and Bonnet, R. M.: 1966, *Comptes Rendus Acad. Sci. Paris* **262 B,** 152.

Bode, G.: 1965, *Die kontinuierliche Absorption von Sternatmosphären in Abhängigkeit von Druck, Temperatur und Elementhäufigkeiten.* Kiel.

Bonnet, R. M. and Blamont, J. E.: 1967, preprint *Study-Week on the Quiet Photosphere*, Arnhem.

Canavaggia, R., Chalonge, D., Egger-Moreau, M., and Oziol-Peltey, H.: 1950, *Ann. Astrophys.* **13,** 355.

David, K. H. and Elste, G.: 1962, *Z. Astrophys.* **54,** 12.
De Jager, C. and Neven, L.: 1967, *Solar Phys.* **1,** 27.
Detwiler, C. R., Purcell, J. D., and Tousey, R.: 1961, *Astron. J.* **66,** 281.
Dunn, R. B.: 1959, *Astrophys. J.* **130,** 972.
Gaustad, J. E. and Rogerson, J. B. Jr.: 1961, *Astrophys. J.* **134,** 323.
Geltman, S.: 1965, *Astrophys. J.* **141,** 376.
Heintze, J. R. W., Hubenet, H., and De Jager, C.: 1964, *Bull. Astron. Inst. Neth.* **17,** 442.
Holweger, H.: 1967, *Z. Astrophys.* **65,** 365.
Koževnikov, N. I.: 1957, *Soviet Astron. – AJ* **1,** 856.
Labs, D. and Neckel, H.: 1962, *Z. Astrophys.* **55,** 269.
Labs, D. and Neckel, H.: 1963, *Z. Astrophys.* **57,** 283.
Labs, D. and Neckel, H.: 1967, *Z. Astrophys.* **65,** 133.
Lambert, D. L. and Willstrop, R. V.: 1965, *Observatory* **85,** 124.
Lamla, E. and Scheffler, H.: 1956, *Z. Astrophys.* **40,** 93.
Makarova, E. A.: 1963, *Observatory* **83,** 183.
Minnaert, M., Van den Hoven van Genderen, E., and Van Diggelen, J.: 1949, *Bull. Astron. Inst. Neth.* **11,** 55.
Mitchell, W. E.: 1959, *Astrophys. J.* **129,** 93.
Müller, E. A. and Mutschlecner, J. P.: 1964, *Astrophys. J., Suppl. Series* **85,** 1.
Muraševa, M. S. and Sitnik, G. F.: 1964, *Soviet Astron. – AJ* **7,** 623.
Neckel, H.: 1958, *Z. Astrophys.* **44,** 153.
Peyturaux, R.: 1952, *Ann. Astrophys.* **15,** 302.
Peyturaux, R.: 1961, *Comptes Rendus* **252,** 668.
Pierce, A. K.: 1954, *Astrophys. J.* **120,** 221.
Pierce, A. K. and Waddell, J. H.: 1961, *Mem. Roy. Astron. Soc.* **68,** 89.
Rogerson, J. B., Jr.: 1959, *Astrophys. J.* **130,** 985.
Saiedy, F.: 1960, *Monthly Notices Roy. Astron. Soc.* **121,** 483.
Tousey, R.: 1963, *Space Sci. Rev.* **2,** 3.

COMMENTS ON THE BILDERBERG CONTINUUM ATMOSPHERE*

GÜNTHER ELSTE

Dept. of Astronomy, The University of Michigan, Ann Arbor, Michigan, U.S.A.

(Received 24 October, 1967)

Abstract. The Bilderberg Continuum Atmosphere fails to reproduce the observed limb-darkening throughout the range of wavelengths 4500 Å $< \lambda <$ 25000 Å. The temperature-pressure diagram for the deep layers of this model is a curve which is flatter than the relation predicted from the mixing length theory. A modification of the Bilderberg Continuum Atmosphere that improves representation of the observations and theoretical results is proposed.

1. Introduction

The Bilderberg Continuum Atmosphere at the depth range of $0.1 \leqslant \tau_{5000} \leqslant 10$ has unfortunately been based on an average over several temperature distributions rather than on an optimum fitting of the observed limb-darkening over a large wavelength range. It is evident in Figure 1 that the adopted procedure leads to an unsatisfactory graduation of the best available measurements of limb-darkening for a range of wavelengths restricted to 4500 Å $< \lambda <$ 25000 Å.

2. Observational Basis

References to all known measurements of limb-darkening have been appended in a special section of this paper. These have been critically reviewed before the start of the present work. In order to compare observations made at different values of $\mu = \cos\theta$, a set of five limb distances was selected for which all limb-darkening observations were interpolated. It was found practical to interpolate the $\log [I_\lambda(\mu)/I_\lambda(1)]$ as a linear function of $\log \mu$. The data from the various observers were then plotted against wavelength. Whenever systematic differences were found, the observing techniques and possible sources of errors have been investigated. If the proper information was missing the observations were rejected. As a result one gains the impression that the extensive and careful photoelectric measurements by PIERCE (1954), obtained with the Snow telescope at Mt. Wilson, can be considered the best after proper correction for scattered light (DAVID and ELSTE, 1962). Up to about 15000 Å PEYTURAUX's (1952, 1955) observations agree with these within $\frac{1}{2}$ of 1%. This is quite good considering the lower spectral purity available, and the fact that nothing is known about the scattered light. Towards longer wavelengths Peyturaux's measurements lie systematically up to 1% below those of Pierce as well as PIERCE *et al.* (1950). All three of these observational sets are used in Figure 1.

* The work described here was supported in part by Contract N(onr)-1224(19) with the Office of Naval Research.

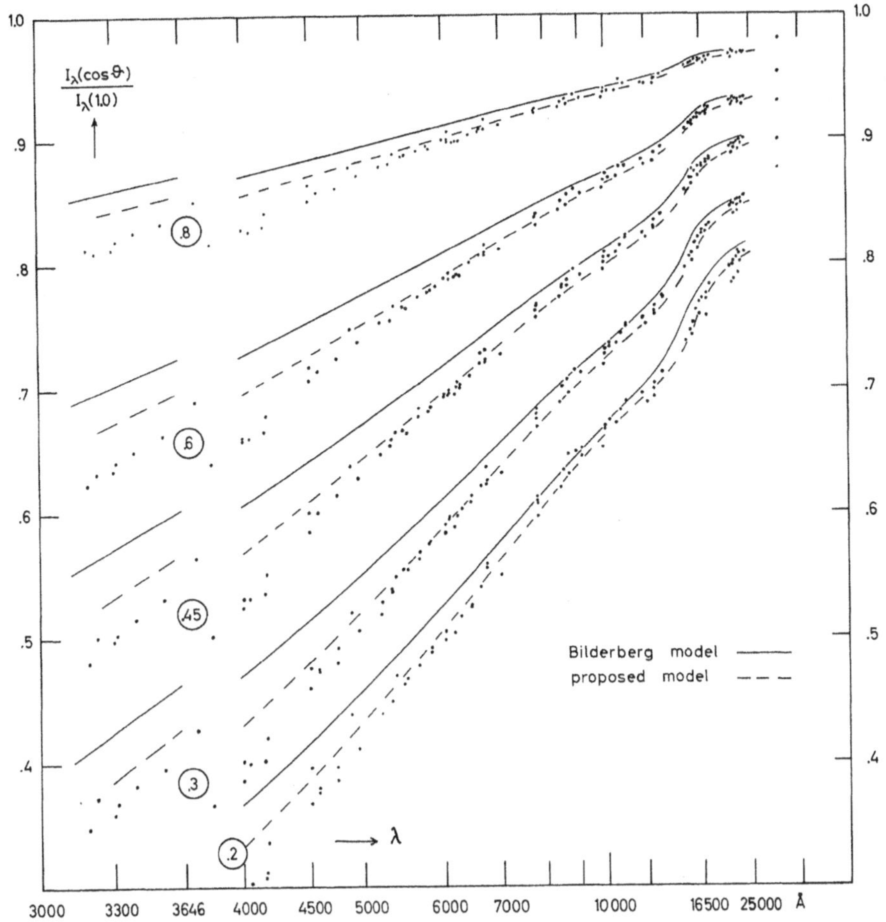

Fig. 1. Comparison between observed and predicted limb-darkening for $3000 < \lambda < 25000$ Å. The values of the parameter $\mu = \cos\theta$ are 0.8, 0.6, 0.45, 0.3, and 0.2, respectively.

3. Semi-Empirical Solar Models

Explaining the observed limb-darkening and energy distribution is not the only condition which we may impose on a model. We know from sufficient observational evidence that the hydrogen convection zone lies below the photosphere. Therefore it is reasonable to require for the deep layers that the model atmosphere should behave in the temperature-pressure diagram as the theory predicts for the transition region between photosphere and convection zone. Mrs. BÖHM-VITENSE (1958) has calculated such diagrams with the help of the mixing length theory for stars of different effective temperature and luminosity.

A model constructed on this basis was reported at the meeting of the 'Deutsche Astronomische Gesellschaft' at Weimar in 1960 (ELSTE, 1960). Instead of solving simultaneously for both the temperature model and the wavelength dependence of the optical depth τ_λ by employing an analytic method, the theoretically computed absorption coefficient was used to calculate the optical-depth scales. Since this procedure requires knowledge of the pressures, they first were determined by the integration of the hydrostatic equation, to be discussed later. The remaining steps follow standard procedures. Starting from the model published by GOLDBERG and PIERCE (1959) further improvements were obtained by repeated trials. Herein the contribution curves for the continuum emission, i.e., the integrands of the expressions for the specific intensity, served as indicators for the sensitivity range of the limb-darkening at each wavelength. As can be seen in Figure 2 for a few samples, the shapes of the contribution curves, their positions at the disk center, and their shifts with cos θ depend on the wavelength. Thus, by sounding out the atmosphere with probes of different widths this procedure allows to use the whole information content of the measurements with advantage.

For this first model, called No. 9, the energy distribution at the center of the solar disk has been taken from the measurements by LABS (1957) made at the Pic-du-Midi. MATTIG and SCHRÖTER (1961) have used this model in a successful explanation of the center-to-limb variation of the wings of the Na D-lines. It appears even more impor-

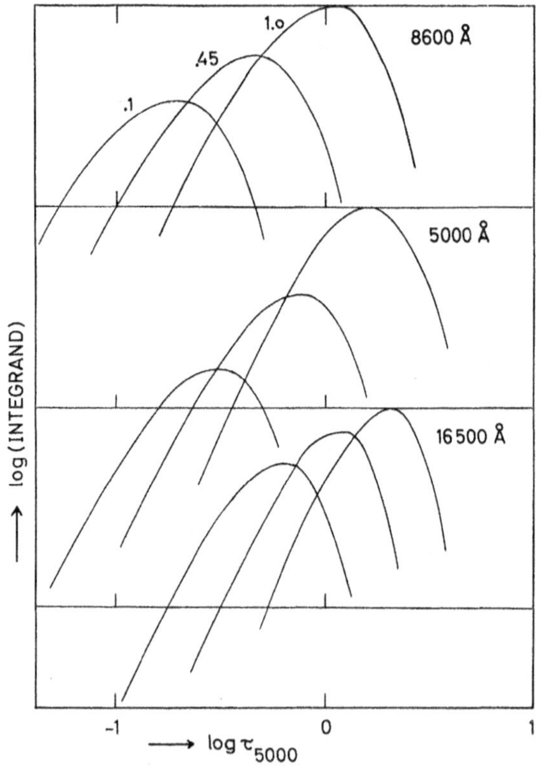

Fig. 2. Center-to-limb variation of the contribution to the continuum emission at the wavelengths 8600, 5000, and 16500 Å displayed as logarithms of the integrands as a function of log τ_{5000}. The three values of the parameter $\mu = \cos\theta$ are 1.0, 0.45, and 0.1.

tant to me, that the center-to-limb variation of the wings of the Balmer lines Hα–Hδ could be reproduced, as has been shown by DAVID (1961). The high excitation energy of the second quantum state of hydrogen prevents a contribution to the line absorption in the wings of these lines except from very deep layers around $\tau = 3$. Moreover, the contribution functions are rather sharp as compared with those for the continuum emission. Therefore these lines offer a possibility for checking the model in the transition region to the hydrogen convection zone, which otherwise is not accessible for continuum observations in the visible.

After completion of the new determination of the energy distribution at the disk center by LABS and NECKEL (1962, 1963, 1967) a small correction in $\Theta = 5040/T$ had to be applied, which then led to model 10 (ELSTE, 1967). Actually this difference is too small to affect the line spectrum appreciably. Mugglestone used it for the wings of the Na D-lines, WITHBROE (1967) for the profiles of the lines of the CH molecule, WITHBROE (1968) for the center-to-limb variation of the equivalent widths of C_2, CH, CN, CO, and MgH lines, and ELSTE (1967) for the investigation of the micro- and macroturbulence.

The models so far had been computed for a helium-to-hydrogen ratio of 0.16, and the abundances of the remaining elements according to GOLDBERG *et al.* (1960). But since (i) the calculations for the internal structure of the sun favor a lower value for the helium-to-hydrogen ratio of about 0.1, and (ii) the investigation of the ultraviolet spectrum points towards a temperature stratification of the outer photosphere as given in the Bilderberg Continuum Atmosphere, corresponding changes have been introduced in the model here proposed.

4. Discussion

Although not yet ideal the new model represents an improvement with respect to the Bilderberg model in both the limb-darkening and the behaviour in the temperature-pressure diagram (Figure 3) for the transition region between the photosphere and the hydrogen convection zone. The best reference curve for the latter is that obtained by KOHL (1967) for the chemical composition used by VARDYA (1964) with He/H= 0.125. It can be noticed that the Bilderberg model uses too low temperatures in the deepest layers, while the proposed model follows more closely in its depth dependence the curve by Kohl.

For the discussion of the limb-darkening we have plotted in Figure 1 as dots the observed $I_\lambda(\mu)/I_\lambda(1)$ for certain representative values of $\mu = \cos\theta$ as a function of $1/\lambda$. The relations predicted by the Bilderberg model and by the proposed model are drawn as full and broken lines, respectively. In addition to the values for the limb-darkening, attention should be directed to the almost constant gradient

$$\frac{\mathrm{d}\left[I_\lambda(\mu)/I_\lambda(1)\right]}{\mathrm{d}(1/\lambda)}$$

for a fixed value of μ. Evidently the Bilderberg model predicts too large values for the

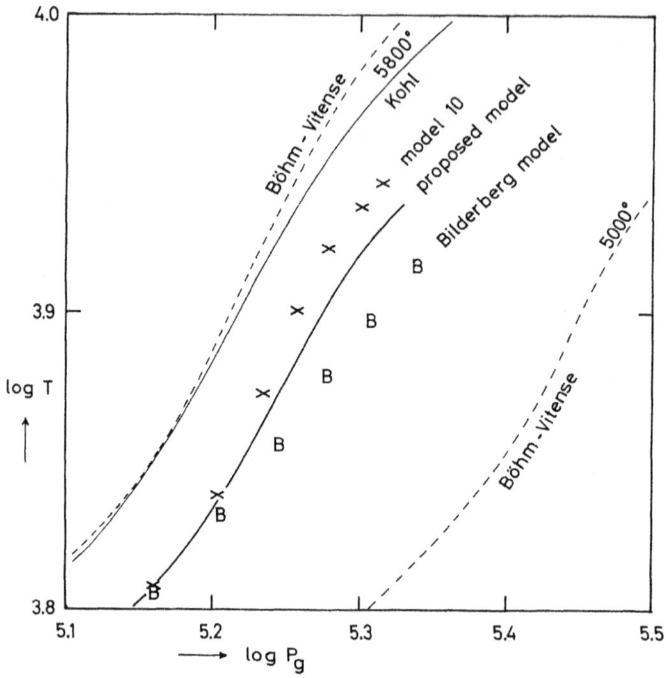

Fig. 3. Relation between temperature and pressure for the transition region to the hydrogen convection zone.

limb-darkening, and also gives a too small gradient, while the proposed model comes much closer to the observations in both respects. For the discussion of the infrared observations where the differences are becoming small, let us look at the situation for $\mu=0.6$ in more detail in Figure 4. At this limb distance the influence of the finite resolving power of the telescopes used by the different observers is not yet significant. At about 23000 Å the observed points lie between the lines for the two models, but at longer wavelengths both models predict too small intensity ratios. Judging from the quite good agreement between the three different observational sets, PIERCE *et al.* (1950), SAIEDY (1960), and LÉNA (1968), the observations should not be responsible for the discrepancy.

It may be mentioned, however, that the inhomogeneities could provide a key to understanding of at least part of the remaining differences. Some are hard to explain just by additional unidentified absorbers.

Let us finally compare the predicted energy distributions at the center of the solar disk with the observations. For this purpose the observations by LABS and NECKEL (1967) have been combined with the relative measurements by PIERCE (1954b) in the infrared and with the other absolute measurements by PEYTURAUX (1952) and by SAIEDY (1960). In order to make such a comparison meaningful in the badly blended violet and blue spectral region, the observed energies in the given intervals have been

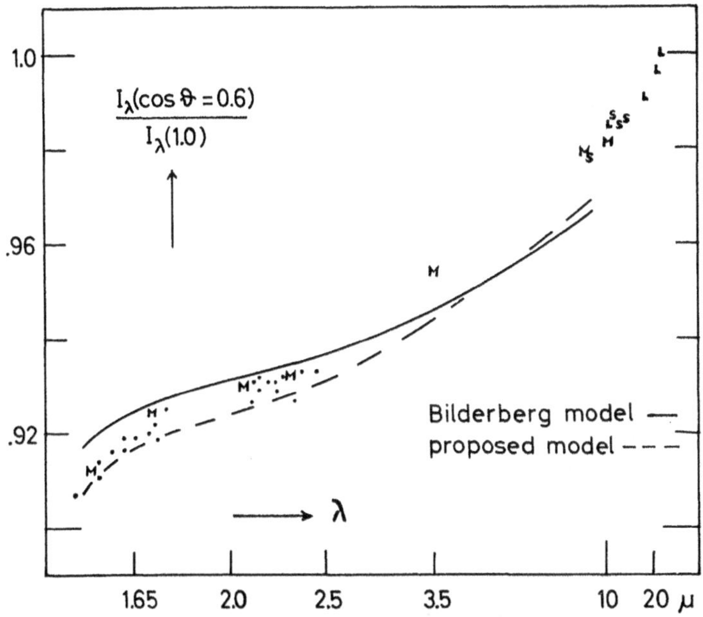

Fig. 4. Wavelength dependence of the limb-darkening in the infrared for $\cos\theta = 0.6$. Observations by PIERCE *et al.* (1950): *M*, SAIEDY (1959): *S*, LÉNA (1968): *L*; those by PIERCE (1954) and PEYTURAUX (1955) are plotted as dots.

transformed to the energies of an assumed continuum. This continuum was defined by the fitting of the wings of the strongest lines as they are predicted by the model 10. Only a few wavelength regions have been analyzed so far by employing special tracings of high resolution extended over such wavelength intervals. These tracings were obtained at the solar towers at Göttingen and at the McMath-Hulbert Observatory. The resulting logarithms of the energy of the assumed continuum at the disk center expressed in [erg cm^{-2} sec^{-1} $ster^{-1}$ for $\Delta\lambda = 1$ cm] are given in Table I.

TABLE I

λ	3298.1	3661.0	3775.2	4020.0	4211.2	4823.3
$\log E_\lambda$	14.476	14.551	14.605	14.667	14.661	14.616

For wavelength larger than 5000 Å it is sufficient to calculate the correction to the assumed continuum by using the sum over the equivalent widths of the lines in each interval. While in the wavelength region 4000 Å $< \lambda <$ 12000 Å the energy predicted by the proposed model agrees with the observations within $\frac{1}{2}$ of 1%, the comparison farther into the infrared is displayed in Figure 5. The agreement can still be considered satisfactory.

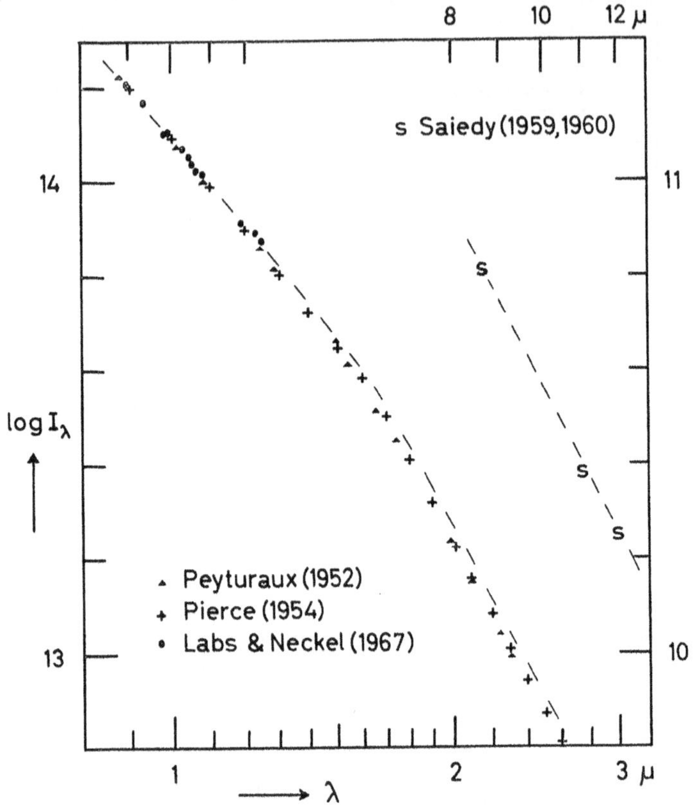

Fig. 5. Observed infrared energy distribution at the center of the solar disk compared with that predicted by the proposed model.

In order to allow the reader a quick orientation, the proposed temperature model is compared with model 10 and the Bilderberg model in Figure 6 in the form: $\Theta = 5040/T$ as a function of $\log \tau_{5000}$.

5. Pressures and Absorption Coefficients

As has been mentioned above, pressure calculations must precede the computations of the continuous absorption coefficients. Assuming hydrostatic equilibrium we followed WEIDEMANN's (1955) suggestion and integrated the expression

$$\sqrt{P_g}\, dP_g = g m_1 \Sigma(\varepsilon_i \mu_i) \sqrt{(P_g/P_e^2)}\, (\kappa_{5000}/P_e)^{-1}\, d\tau_{5000}, \tag{1}$$

in which the quantity under the square root on the right-hand side

$$(P_g/P_e^2) = \Sigma\varepsilon_i \left(1 + \frac{\Phi_i/P_e}{1 + \Phi_i/P_e}\right) \bigg/ \Sigma\varepsilon_i \frac{\Phi_i}{1 + \Phi_i/P_e}, \tag{2}$$

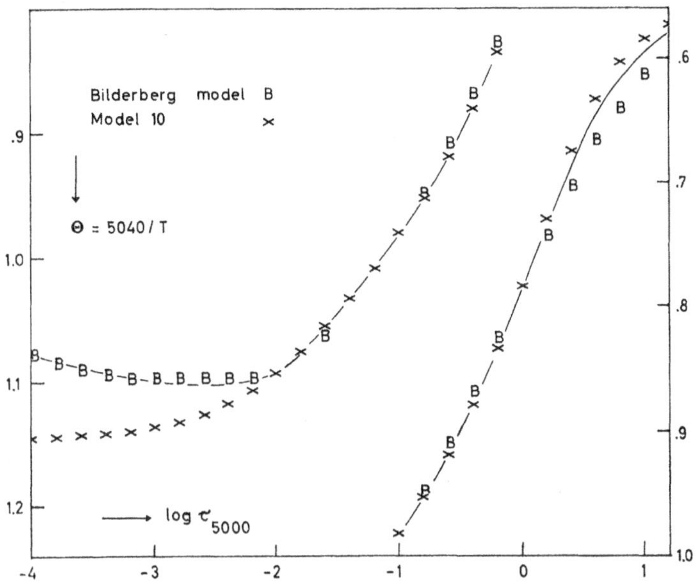

Fig. 6. Temperature stratification of the proposed model, model 10, and the Bilderberg model.

as well as κ_{5000}/P_e depends only slightly on the chosen value of the electron pressure, and thus enable a fast convergence of the iterative procedure.

In the above equations we have used the notation: P_g = gas pressure, P_e = electron pressure, g = surface gravity, m_1 = mass of unit atomic weight, ε_i = abundance of an element or group of elements with respect to hydrogen, μ_i = atomic weight of an element or group of elements, κ_{5000} = absorption coefficient per hydrogen particle, τ_{5000} = optical depth at the wavelength 5000 Å, $\Phi_i = (n_1/n_0) P_e$ = the Saha term for the ratio of ions to atoms, multiplied by P_e.

For the computations of the sums in Equations (1) and (2) the following elements were treated together in groups:

(H, N, O); (C, S, P); Si; (Fe, Cu, Co); Mg;
(Ni, Mn); (Cr, Ti, V); Ca; Na; K.

The integrations of Equation (1) as well as those of the optical depths have been calculated on log τ_{5000}-scale with the help of the Gauss-Encke quadrature formula (Elste, 1955, Eq. 37). For the starting integrals an exponential approximation was used for the integrands (Elste, 1957, Eq. A3b).

The data for the computation of the absorption coefficients were taken from the following sources:

H^- Ion:

Bound-Free	Geltman (1962), Doughty *et al.* (1966)
Free-Free	
for $\lambda > 12000$ Å	Doughty and Fraser (1966)
for $\lambda \leqslant 12000$ Å	John (1964)

H Neutral Atom:

Three nearest Bound-Free transitions and an integral over the remaining levels, as well as Free-Free transitions according to UNSÖLD (1955) with

Bound-Free Gaunt Factors

for $n \leqslant 3$ GINGERICH (1964)

for $n > 3$ MENZEL and PEKERIS (1935)

Free-Free Gaunt Factors: KARZAS and LATTER (1961) approximated

for $\chi_\lambda \Theta \geqslant 2.0$ by $g_{FF} - 1 = \Theta^{-1/4}\, 0.107$

for $\chi_\lambda \Theta < 2.0$ by $g_{FF} - 1 = \Theta^{-1/4} [0.082 + 0.0397 (\chi_\lambda \Theta)^{-2/3}]$

H_2^+ Molecules: BATES (1952), BOGGESS (1959)

Rayleigh Scattering: DALGARNO and WILLIAMS (1962), GINGERICH (1964)

Metals: These are treated in hydrogenic approximation with screening factors adjusted in such a way that the resulting cross sections agree with laboratory measurements or theoretical values (employing the quantum defect method) whenever known (summarized by BODE, 1965).

TABLE II

Proposed Solar Model

$\log \tau_0$	Θ	$\log P_e$	$\log P_g$	$\log \kappa / P_e$
−4.0	1.077	−1.113	3.040	−24.934
−3.8	1.084	− .995	3.198	−24.938
−3.6	1.089	− .910	3.310	−24.938
−3.4	1.094	− .823	3.423	−24.937
−3.2	1.097	− .731	3.536	−24.938
−3.0	1.099	− .639	3.648	−24.941
−2.8	1.102	− .550	3.759	−24.940
−2.6	1.102	− .457	3.870	−24.945
−2.4	1.102	− .365	3.982	−24.948
−2.2	1.099	− .269	4.093	−24.957
−2.0	1.092	− .166	4.205	−24.971
−1.8	1.076	− .046	4.316	−25.001
−1.6	1.056	.081	4.427	−25.037
−1.4	1.034	.213	4.537	−25.077
−1.2	1.009	.355	4.647	−25.122
−1.0	.982	.506	4.754	−25.171
− .8	.952	.675	4.858	−25.225
− .6	.918	.876	4.956	−25.286
− .4	.879	1.123	5.043	−25.357
− .2	.834	1.432	5.115	−25.441
.0	.784	1.796	5.171	−25.538
.2	.733	2.181	5.213	−25.638
.4	.686	2.543	5.245	−25.730
.6	.646	2.856	5.272	−25.804
.8	.616	3.094	5.296	−25.853
1.0	.595	3.266	5.320	−25.884
1.2	.582	3.372	5.336	−25.901

6. Proposed Model

The present model is given in steps of $\Delta \log\tau_{5000}=0.2$ for the range $-4.0 \leqslant \log\tau_{5000} \leqslant +1.2$. For the higher layers, $\log\tau_{5000} < -4.0$, the Bilderberg Continuum Atmosphere should be followed. Subsequent columns in Table II give the logarithm of the optical depth at 5000 Å, $\Theta=5040/T$, the logarithms of both the electron pressure, and the gas pressure in [dyn/cm^{-2}], the logarithm of the continuous absorption coefficient at 5000 Å per unit electron pressure per hydrogen particle.

Acknowledgement

The author wishes to express his gratitude to the late Professor Dr. P. ten Bruggencate, Director of the Sternwarte Göttingen, for his supporting interest during the early stage of the development of a semi-empirical solar model, as well as to the 'Aerodynamische Versuchsanstalt', the 'Max-Planck-Gesellschaft', and the University at Göttingen for granting the extensive use of the IBM-650 computer. I also wish to thank Professor Dr. O.C. Mohler for his continued interest and to Dr. R.C.F. Bartels, Director of The University of Michigan Computing Center, for permission to use the IBM-7090 computer.

References

A. CONCERNING LIMB-DARKENING OBSERVATIONS

Abbot, C. G., Aldrich, L. B., and Fowle, F. E.: 1913, *Ann. Smithson. Astrophys. Obs.* **3.**

Abbot, C. G., Aldrich, L. B., and Fowle, F. E.: 1922, *Ann. Smithson. Astrophys. Obs.* **4.**

Canavaggia, R. and Chalonge, D.: 1946, 'Recherches sur le spectre continu du soleil, II: Nouvelles données d'observation', *Ann. Astrophys.* **9,** 143.

David, K. H. and Elste, G.: 1962, 'Der Einfluss von Streulicht auf die Photometrie der Sonnenoberfläche', *Z. Astrophys.* **54,** 12.*

Krat, T. W.: 1948, *Poulkowo Publ.* **17,** no. 137.

Léna, P. J.: 1968, 'Observations of the Center-to-limb Variation of the Solar Brightness in the Far Infrared (10 to 25 μ)', *Solar Phys.* **3,** 28.

Mitchell, Jr., W. E.: 1959, 'The Center-limb Variation of the Intensities of Selected Solar Lines', *Astrophys. J.* **129,** 93.*

Moll, W. J. H., Burger, H. C., and Van der Bilt, J.: 1925, 'The Distribution of the Energy over the Sun's Disk', *Bull. Astron. Inst. Neth.* **3,** 83.

Peyturaux, R.: 1952, 'Contribution **a** l'étude du fond continu du spectre solaire dans le proche infrarouge', *Ann. Astrophys.* **15,** 302.

Peyturaux, R.: 1955, 'Étude du fond continu du spectre solaire, IV: L'assombrissement centre-bord du soleil entre 3190 et 23130 Å', *Ann. Astrophys.* **18,** 34.

Pierce, A. K.: 1954a, 'Solar Limb-Darkening in the Region λλ 7793–24388', *Astrophys. J.* **120,** 221.*

Pierce, A. K. and Waddell, J. H.: 1961, 'Analysis of Limb-Darkening Observations', *Mem. Roy. Astron. Soc.* **68,** 89.*

Pierce, A. K., McMath, R. R., Goldberg, L., and Mohler, O.: 1950, 'Observations of Solar Limb-Darkening between 0.5 and 10.2 μ', *Astrophys. J.* **112,** 289.

Raudenbusch, H.: 1938, 'Messungen der Intensitätsverteilung auf der Sonnenscheibe in verschiedenen Wellenlängen', *Astr. Nachr.* **266,** 301.

* These investigations are based on the original observations by Pierce (1954a).

SAIEDY, F.: 1960, 'Solar Intensity and Limb-Darkening between 8.6 and 13 μ', *Monthly Notices Roy. Astron. Soc.* **121,** 483.

SCHMIDT, Th.: 1961, 'Photometrie der ultravioletten Kupferresonanzlinien bei 3248 und 3274 Å auf der Sonne', *Z. Astrophys.* **53,** 273.

B. GENERAL REFERENCES

BATES, D. R.: 1952, 'Absorption of Radiation by an Atmosphere of H, H^+ and H_2^+-Semiclassical Treatment', *Monthly Notices Roy. Astron. Soc.* **112,** 40.

BODE, G.: 1965, *Die kontinuierliche Absorption von Sternatmosphären in Abhängigkeit von Druck, Temperatur und Elementhäufigkeiten.* Institut für Theoretische Physik und Sternwarte der Univ. Kiel.

BOGGESS, III, A.: 1959, 'The Emission of H_2^+', *Astrophys. J.* **129,** 432.

BÖHM-VITENSE, E.: 1958, 'Über die Wasserstoffkonvektionszone in Sternen verschiedener Effektivtemperaturen und Leuchtkräfte', *Z. Astrophys.* **46,** 108.

DALGARNO, A. and WILLIAMS, D. A.: 1962, 'Rayleigh Scattering by Molecular Hydrogen', *Astrophys. J.* **136,** 690.

DAVID, K. H.: 1961, 'Die Mitte-Rand-Variation der Balmerlinien $H\alpha$–$H\delta$ auf der Sonnenscheibe', *Z. Astrophys.* **53,** 37.

DOUGHTY, N. A. and FRASER, P. A.: 1966, 'The Free-Free Absorption Coefficient of the Negative Hydrogen Ion', *Monthly Notices Roy. Astron. Soc.* **132,** 267.

DOUGHTY, N. A., FRASER, P. A., and MCEACHRAN, R. P.: 1966, 'The Bound-Free Absorption Coefficient of the Negative Hydrogen Ion', *Monthly Notices Roy. Astron. Soc.* **132,** 255.

ELSTE, G.: 1955, 'Die Mitte-Rand-Variation schwacher bis mittelstarker Fraunhoferlinien auf der Sonnenscheibe', *Z. Astrophys.* **37,** 184.

ELSTE, G.: 1957, 'Nomographic Method for Integrating a Function given on a Logarithmic Scale', *Astrophys. J. Suppl. Ser.* **3,** 31.

ELSTE, G.: 1960, 'Ein neues Photosphärenmodell, abgeleitet aus Randverdunklungsmessungen unter Berücksichtigung der Wellenlängenabhängigkeit der kontinuierlichen Absorption,' *Physikalische Verhandlungen* **11,** Nr. 11, p. 8; *Mitt. astr. Ges.* **14.**

ELSTE, G.: 1967, 'The Distinction between Micro- and Macroturbulence in the Solar Photosphere using Line Profiles', *Astrophys. J.* **148,** 857.

GELTMAN, S.: 1962, 'The Bound-Free Absorption Coefficient of the Hydrogen Negative Ion,' *Astrophys. J.* **136,** 935.

GINGERICH, O.: 1964, Review of opacity calculations, *First Harvard-Smithsonian Conference on Stellar Atm., Cambridge, Mass.*, 1964, in *Smithson. Inst. Astrophys. Obs., spec. Rep.*, no. **167,** p. 17.

GOLDBERG, L. and PIERCE, A. K.: 1959, 'The Photosphere of the Sun', in *Handbuch der Physik*, LII: *Das Sonnensystem* (Ed. by S. Flügge). Springer-Verlag, Berlin-Göttingen-Heidelberg.

GOLDBERG, L., MÜLLER, E. A., and ALLER, L. H.: 1960, 'The Abundances of the Elements in the Solar Atmosphere', *Astrophys. J. Suppl. Ser.* **5,** 1.

JOHN, T. L.: 1964, 'The Free-Free Transitions of the Negative Hydrogen Ion in the Exchange Approximation,' *Monthly Notices Roy. Astron. Soc.* **128,** 93.

KARZAS, W. J. and LATTER, R.: 1961, 'Electron Radiative Transitions in a Coulomb Field', *Astrophys. J. Suppl. Ser.* **6,** 167.

KOHL, K.: 1967, 'Das Eindringen von Strömungen in die stabile Schicht unterhalb der solaren Konvektionszone', *Z. Astrophys.* **64,** 472.

LABS, D.: 1957, 'Die Intensität des kontinuierlichen Spektrums der Sonnenmitte im Wellenlängenbereich $3300 \leqslant \lambda \leqslant 6900$ Å', *Z. Astrophys.* **44,** 37.

LABS, D. and NECKEL, H.: 1962, 'Die absolute Strahlungsintensität der Sonnenmitte im Spektralbereich $4010 \leqslant \lambda \leqslant 6569$ Å', *Z. Astrophys.* **55,** 269.

LABS, D. and NECKEL, H.: 1963, 'Die absolute Strahlungsintensität der Sonnenmitte im Spektralbereich $6389 \leqslant \lambda \leqslant 12480$ Å', *Z. Astrophys.* **57,** 283.

LABS, D. and NECKEL, H.: 1967, 'Die absolute Strahlungsintensität der Mitte der Sonnenscheibe im Spektralbereich $3288 \leqslant \lambda \leqslant 12480$ Å', *Z. Astrophys.* **65,** 133.

MATTIG, W. and SCHRÖTER, E. H.: 1961, 'Die Mitte-Rand-Variation des Kontinuums für $\lambda = 5893$ Å und der Flügel der Na D-Linien im Sonnenspektrum', *Z. Astrophys.* **52,** 195.

MENZEL, H. D. and PEKERIS, C. L.: 1935, 'Absorption Coefficients and Hydrogen Line Intensities', *Monthly Notices Roy. Astron. Soc.* **96,** 77.

MUGGLESTONE, D.: 1964, Colloquium at The University of Michigan.
PIERCE, A. K.: 1954b, 'Relative Solar Energy Distribution in the Spectral Region 10000–25000 Å', *Astrophys. J.* **119**, 312.
SAIEDY, F. and GOODY, R. M.: 1959, 'The Solar Emission Intensity at 11 μ', *Monthly Notices Roy. Astron. Soc.* **119**, 213.
UNSÖLD, A.: 1955, *Physik der Sternatmosphären.* 2nd ed., Springer-Verlag, Berlin-Göttingen-Heidelberg.
VARDYA, M. S.: 1964, 'Atmospheric Rosseland Mean Opacities', *Astrophys. J. Suppl. Ser.* **8**, 277.
WEIDEMANN, V.: 1955, 'Metallhäufigkeiten, Druckschichtung und Stossdämpfung in der Sonnenatmosphäre', *Z. Astrophys.* **36**, 101.
WITHBROE, J. L.: 1967, 'An Analysis of CH in the Solar Atmosphere', *Astrophys. J.* **147**, 1117.
WITHBROE, J. L.: 1968, 'The Center-Limb Behavior of Solar Molecular Lines', *Solar Phys.* **3**, 146.

A SIMPLIFIED MODEL SOLAR ATMOSPHERE

D. L. LAMBERT

Dept. of Astrophysics, University Observatory, Oxford

(Received 19 June, 1967)

Abstract. A simplified representation of the temperature distribution in the solar photosphere is proposed: $\theta(\tau_0) = \theta_0 - \theta_1 \log \tau_0$. An expression is derived for the emergent continuous spectrum from the simple model. The limitations and applications of the simple model are discussed.

1. Introduction

With an electronic computer the emergent continuous spectrum may be rapidly and accurately calculated for a given model solar atmosphere. By a process of trial and error a model providing the optimum fit to any observations can be readily obtained. The possibility of machine computation renders redundant a variety of approximate methods for deriving a model atmosphere from observations. However, in the course of a detailed investigation of the solar continuous spectrum, an approximate representation for the photospheric temperature distribution was discovered, which enabled a fuller understanding of certain aspects of the machine calculations to be obtained.

Two model atmospheres (HEINTZE *et al.*, 1964; LAMBERT, 1965) are displayed in Figure 1. By inspection it is seen that to a certain approximation the temperature distribution may be written in the form

$$\theta(\tau_0) = \theta_0 - \theta_1 \log \tau_0 \tag{1}$$

where $\theta = 5040/T$, and τ_0 is the optical depth at the standard wavelength $\lambda_0 = 5000$ Å. In Figure 1 the model described by $\theta_0 = 0.770$ and $\theta_1 = 0.222$ is drawn. It may be noted that the Utrecht Reference Photosphere (URP) is apparently well represented by Equation (1) throughout the interval $-1.5 \leqslant \log \tau_0 \leqslant +0.5$. It is these layers which are responsible for the continuous spectrum for $\lambda \lesssim 3$ microns and at disk positions $\cos\phi \gtrsim 0.2$. The applications of Equation (1) are explored in this paper.

2. The Predicted Continuous Spectrum

With the standard assumptions and notation, the emergent continuous intensity is given by the equation

$$I_\lambda(0, \phi) = \int_0^\infty B_\lambda(\theta) \exp(-\tau_\lambda \sec\phi) \sec\phi \, d\tau_\lambda . \tag{2}$$

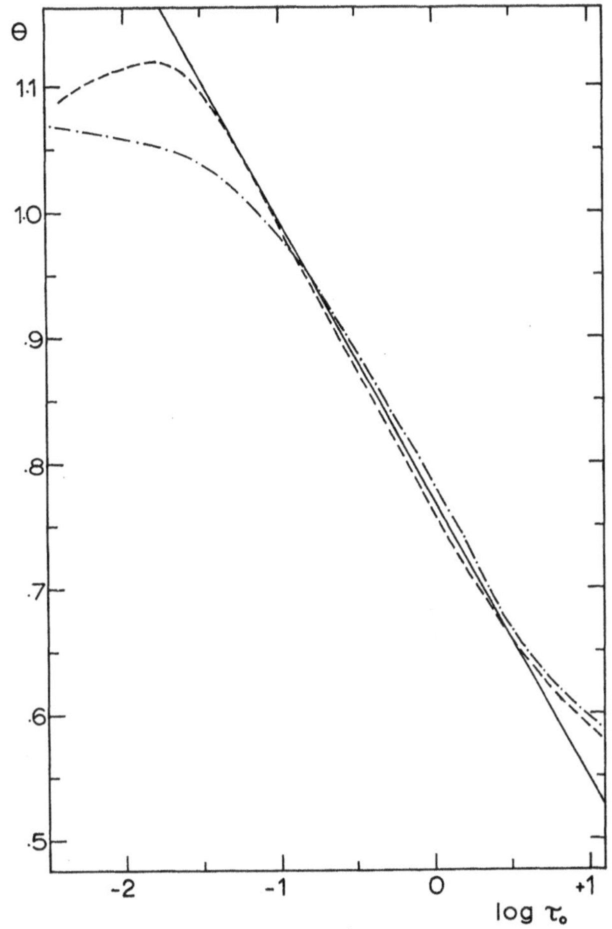

Fig. 1. The temperature distribution in the photosphere: $\theta(\tau_0)$ vs $\log \tau_0$. The URP and Lambert models are given by the dashed and the dot-dashed lines respectively. The solid line represents the simplified model given by Equation (1) with $\theta_0 = 0.770$ and $\theta_1 = 0.222$.

The evaluation of this integral for a temperature distribution given by Equation (1) proved successful with the following generalizations of the Planck function and the continuous absorption coefficients.

The Planck function may be replaced by the following direct generalization of Wien's Law:

$$B_\lambda(\theta) = \frac{2hc^2}{\lambda^5}\, b_\lambda\, 10^{-r\theta/\lambda_W} \tag{3}$$

where $r = hc/5040.16\kappa = 1.240$, if λ is in microns. For the appropriate choice of the constants b_λ and λ_W (see Table I) Equation (3) represents the Planck function to an accuracy of better than 1% over the required range of temperature. Of course, when

Wien's Law is valid $b_\lambda = 1$ and $\lambda_W = \lambda$. Combining Equations (1) and (3) gives

$$B_\lambda(\tau_0) = \frac{2hc^2}{\lambda^5} b_\lambda^* \tau_0^{S/\lambda_W} \tag{4}$$

where $b_\lambda^* = b_\lambda\, 10^{-r\theta_0/\lambda_W}$ and $S = r\theta_1$.

The assumption that the negative hydrogen (H^-) ion is the only contributor to the continuous opacity is valid for $\lambda \gtrsim 4000$ Å. With this assumption, the ratio κ_λ/κ_0 is

TABLE I

The generalization of Wien's Law

λ (microns)	b_λ	λ_W^{-1}
0.6	0.907	1.635
0.8	0.820	1.190
1.0	0.746	0.917
1.2	0.690	0.734
1.4	0.650	0.606
1.6	0.620	0.512
1.8	0.598	0.440
2.0	0.581	0.385
2.2	0.569	0.338
2.4	0.557	0.303

TABLE II

An analytical representation of the continuous absorption coefficients for the H^- ion calculated by DOUGHTY and FRASER (1966)

λ (microns)	p_λ	q_λ
3.038	29.4	1.52
2.278	16.7	1.52
1.823	10.9	1.52
1.519	5.12	1.12
1.302	2.28	0.45
1.139	1.80	0.21
1.013	1.61	0.09
0.911	1.51	0.04
0.759	1.36	0.00

independent of the gas and electron pressures and a function only of temperature. Examination of tables for the H^- absorption coefficient shows that the representation

$$\frac{\kappa_\lambda}{\kappa_0} = p_\lambda\, 10^{-q_\lambda \theta} \tag{5}$$

is accurate to a few per cent. For $\lambda \leqslant 8000$ Å, the bound-free absorption is dominant and $q_\lambda = 0$. Values for p_λ and q_λ at selected wavelengths are given in Table II for a recent set of absorption coefficient calculations (Doughty and Fraser, 1966). From Equations (1) and (5) the optical depth τ_λ is given as

$$\tau_\lambda = \int \frac{\kappa_\lambda}{\kappa_0} \, d\tau_0 = p_\lambda^* \frac{\tau_0^{1+q_\lambda\theta_1}}{1 + q_\lambda\theta_1} \tag{6}$$

where $p_\lambda^* = p_\lambda \, 10^{-q_\lambda\theta_0}$.

Substitution of Equations (4) and (6) in (2) gives the following expression for the emergent intensity:

$$I_\lambda(0, \phi) = \frac{2hc^2}{\lambda^5} \, b_\lambda^* \left(\frac{1 + q_\lambda\theta_1}{p_\lambda^*}\right)^{t_\lambda/\lambda_W} \Gamma\left(1 + \frac{t_\lambda}{\lambda_W}\right) (\cos\phi)^{t_\lambda/\lambda_W} \tag{7}$$

where $t_\lambda = S/(1 + q_\lambda\theta_1)$ and $\Gamma(m)$ is the Gamma function defined as

$$\Gamma(m) = \int_0^\infty x^{m-1} \, e^{-x} \, dx .$$

The intensity ratio, which is referred to as the limb darkening, is given by the following simple equation

$$i_\lambda(\phi) = \frac{I_\lambda(0, \phi)}{I_\lambda(0, 0)} = (\cos\phi)^{t_\lambda/\lambda_W} \tag{8}$$

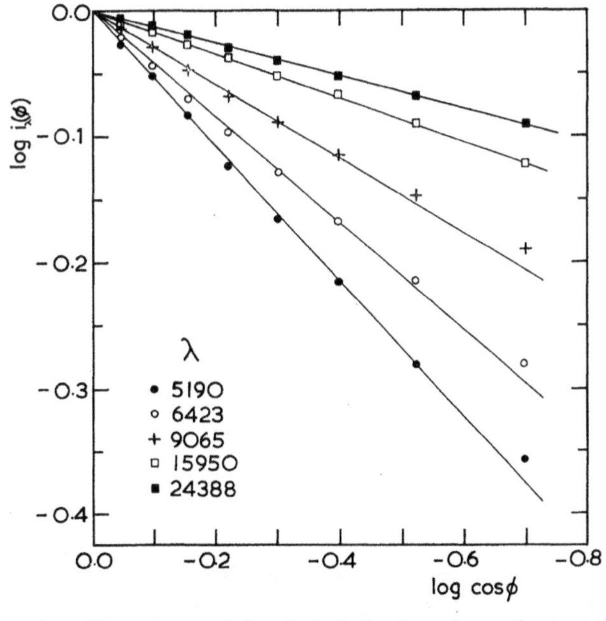

Fig. 2. A plot of $\log i_\lambda(\phi)$ vs $\log\cos\phi$ for limb-darkening observations at five representative wavelengths.

3. Limb-Darkening Observations

The present discussion will be based on the limb-darkening observations obtained by Pierce and corrected for scattered light by DAVID and ELSTE (1962). Results at five representative wavelengths are shown in Figure 2, which is a plot of $\log i_\lambda(\phi)$ vs. log $\cos\phi$. Clearly, the observations satisfy Equation (8) with the exception that at $\cos\phi = 0.2$ small departures (about 2%) are present, which increase towards the limb. This property of limb-darkening observations was first pointed out by HERTZSPRUNG (1937), but the interpretation in terms of a temperature distribution does not appear to have been recognized.

The limitations of Equation (1) should be outlined. In Figure 3 plots of $\log i_\lambda(\phi)$ vs $1/\lambda_W$ are presented for $\cos\phi = 0.4$ and 0.6. The straight line drawn through the observations at $\lambda_{W1}^{-} = 1.1$ assumes $q_\lambda = 0$ and corresponds to a model with $\theta_1 = 0.222$.

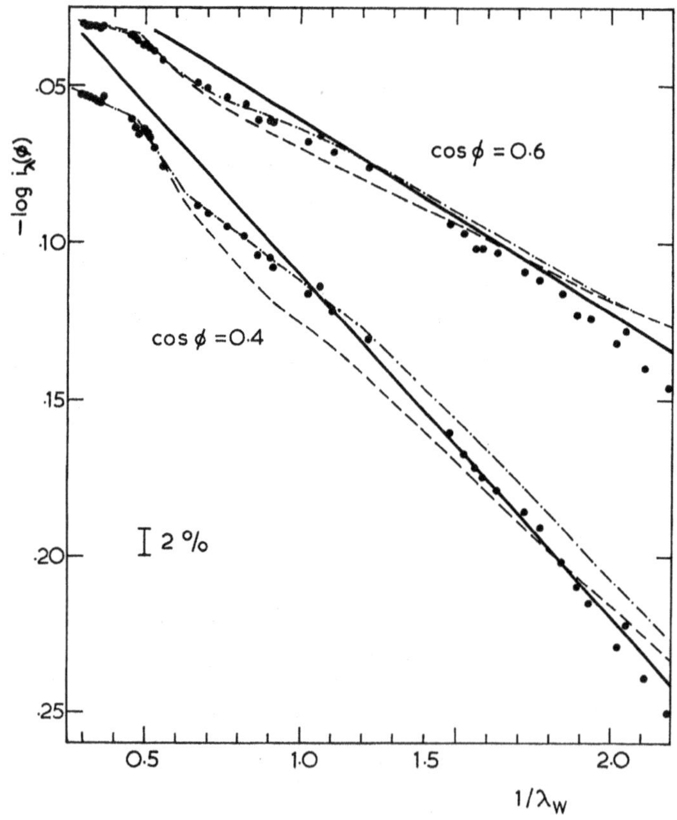

Fig. 3. A comparison of limb-darkening observations and predictions. Plots of $\log i_\lambda(\phi)$ vs $1/\lambda_W$ are shown for the two disk positions $\cos\phi = 0.4$ and 0.6. The observations (DAVID and ELSTE, 1962) are given by the filled circles. The predictions based on the URP and Lambert models are given by the dashed and the dot-dashed line respectively. The solid line gives the predictions for the simplified model with $\theta_0 = 0.770$ and $\theta_1 = 0.222$.

This model predicts the limb darkening at these wavelengths to an accuracy of about 2%. Its limitations are revealed by a consideration of the observations longward of about 1 micron. The continuous absorption coefficient is defined by p_λ and q_λ. The limb darkening is predicted to be independent of p_λ. Since $q_\lambda > 0$ the limb intensities are predicted to be greater than the corresponding values for $q_\lambda = 0$. The reverse is shown to be the case (see Figure 3). This discrepancy suggests that a more detailed model is required.

The predicted intensities for the URP and Lambert model are plotted in Figure 3. The differences between these predictions and those for the simple model $\theta_1 = 0.222$ correspond to small temperature differences. For example, the URP predictions for $\lambda \simeq 0.9$ microns ($\lambda_W^{-1} \simeq 1.0$) are about 2 to 3% less than simple model predictions. This is largely attributable to an increased slope in the URP model between about $\log \tau_0 = 0$ and -1.5 (see Figure 1).

4. Applications of the Simplified Model

It is clear that the simple model cannot be expected to predict sufficiently accurately the limb darkening over a wide range of wavelengths. However, it may be expected to prove sufficiently reliable for an assessment of differential corrections arising from errors in the absorption coefficients.

Machine calculations demonstrated that the predicted limb-darkening for $\lambda < 2.5$ microns was insensitive to the assumed set of theoretical H^- absorption coefficients. The original calculations by Chandrasekhar and Breen (1946) and the later ones by Ohmura (1964) differ by approximately 50%. Yet, the limb-darkening predictions differ by less than 0.5%. This insensitivity is predicted by the simple model. According to Equation (8) the limb-darkening is independent of a straight increase (or change in p_λ in Equation (5)) in the absorption coefficient, and insensitive to changes in the temperature dependence (or q_λ).

Clearly, the limb-darkening observations at these wavelengths are an unsatisfactory test of the continuous absorption-coefficient calculations. These calculations can be checked only by comparison with the absolute intensity observations (see Equation (7)). The simple model provides accurate estimates for the effect of absorption coefficient changes on the absolute intensities.

The various theoretical calculations for κ_λ/κ_0 differ in the p_λ value with only small differences in q_λ. From Equation (7) it/is seen that

$$\Delta \log I_\lambda(0, 0) = -\frac{t_\lambda}{\lambda_W} \Delta \log p_\lambda .$$

A change of p_λ by 50% corresponds to intensity changes of 11 and 5% at 1 and 2 microns respectively. For the absolute intensity observations by Labs and Neckel (1967) an accuracy of about 2.5% is claimed. These measurements extend to about 1.2 microns. The accurate relative measurements by Pierce (1954) can be used to extend the absolute measurements to about 2.5 microns with an overall uncertainty of about 3%. The estimates based on the simple model indicate that with these measurements the accuracy of present calculations of the H^- absorption coefficient might be assessed.

Although the simple model cannot predict sufficiently accurately the observed continuous spectrum, it can provide accurate estimates for certain differential corrections. It is for this latter purpose that the model is presented. It is hoped that it may also be of use in problems associated with the Fraunhofer lines.

References

Chandrasekhar, S. and Breen, F. H.: 1946, *Astrophys. J.* **104**, 430.
David, K. H. and Elste, G.: 1962, *Z. Astrophys.* **54**, 12.
Doughty, N. A. and Fraser, P. A.: 1966, *Monthly Notices Roy. Astron. Soc.* **132**, 267.
Heintze, J. R. W., Hubenet, H., and De Jager, C.: 1964, *Bull. Astron. Inst. Neth.* **17**, 442.
Hertzsprung, G.: 1937, *Bull. Astron. Inst. Neth.* **8**, 140.
Labs, D. and Neckel, H.: 1967, *Z. Astrophys.* **65**, 133.
Lambert, D. L.: 1965, Thesis, Oxford University.
Ohmura, T.: 1964, *Astrophys. J.* **140**, 282.
Pierce, A. K.: 1954, *Astrophys. J.* **119**, 312.

CENTER-TO-LIMB ANALYSIS OF THE SOLAR OXYGEN LINES

EDITH A. MÜLLER
Observatoire de Genève, Switzerland
and
BODO BASCHEK and HARTMUT HOLWEGER
Inst. für Theor. Phys. und Sternw., University of Kiel, Germany

(Received 2 June, 1967)

Abstract. Several lines of neutral oxygen observed at various positions on the solar disk were used to study the influence of (1) the temperature distribution, (2) the velocity field, and (3) the damping on the line profiles and the abundance of oxygen in the photosphere. Theoretical profiles were calculated on the basis of four different model atmospheres in LTE. It was found that the model proposed by HOLWEGER (1967) best reproduced the center-to-limb observations of the lines studied. The weighted mean of the oxygen abundance turned out to be $\log \varepsilon_O = 8.83$ on the basis of $\log \varepsilon_H = 12.00$.

1. Introduction

In recent years several new photospheric models have been proposed which differ both by the temperature and the turbulence distributions. The first question that arises is which of the models will best reproduce the observed line shapes and, hence, may be used for the determination of the chemical composition of the photosphere. Secondly one may ask whether or not the differences in the proposed models seriously affect the abundance of the elements in the photosphere. It is clear that answers to these questions can only be given after having analysed many lines formed in quite different atmospheric depths.

As a start we have chosen the lines of neutral oxygen, because in the solar spectrum both low and high-excitation O I lines are observed. The high-excitation lines (around 10 eV) are permitted transitions. Some of them are of medium strength, the others are very faint. The low-excitation lines are forbidden transitions. They are very faint and are O or 2 eV lines. Thus, although not many O I lines occur in the solar spectrum, they originate in quite different photospheric layers. Their analysis may give some insight as to the validity of some of the photospheric models recently proposed.

2. Observations

For our analysis we have selected the medium strong infrared triplet at λ 7774 and λ 8446 Å, two faint permitted lines near λ 6157 Å, and three faint forbidden lines at $\lambda\lambda$ 6364, 6300, and 5577 Å. The photoelectric spectrograms were kindly secured for us by Dr. R. Teske with the vacuum spectrograph of the McMath-Hulbert Observatory. The spectrograph was used in the single pass arrangement. The observations were made at the following positions on the solar disk: $\mu = 1.0$, 0.7, 0.5, 0.4, and 0.3.

The local continuum was drawn in the usual manner through the highest intensity level in the neighbourhood of the lines in question. The dispersion on the tracings is about 60 mm/Å for the stronger lines and about 130 mm/Å for the fainter lines. The observed profiles were corrected for stray light, using a stray-light correction of $s=2.5\%$ estimated by MOHLER (1963). The stray-light correction was applied in the following way: Let I_λ' be the observed intensity in percent of the continuum. Then the true intensity in percent of the continuum will be given by:

$$I_\lambda = \frac{I_\lambda' - s}{100 - s}.$$

The reduction of the observed lines was done by means of a two-dimensional measuring engine. A mean profile was then derived from the two to four observations of each line at any given position on the solar disk. Since no appreciable systematic asymmetry was detected in the profiles of the λ 7774 triplet, the violet and the red side of each line was averaged. The profiles of the faint lines are not as well defined as those of the stronger lines. Tracing noise, uncertainties in the location of the continuum, and disturbances by neighbouring lines make it difficult to obtain a reliable profile of a very faint line. Out of the five faint O I lines we selected the two best ones for which the profiles at $\mu=1.0$ and 0.3 appeared to be reasonably well established. They are the λ 6158 and the λ 6300 lines. The latter line was observed by SWINGS (1966) with the Liège spectrograph at the Jungfraujoch Scientific Station. Its profile observed at the center of the solar disk agrees very well with the corresponding profile reduced from the McMath-Hulbert tracings.

The equivalent widths were measured both with a planimeter and by integration of the profiles registered with the two-dimensional measuring engine. Both results are in excellent agreement. The random scatter of the equivalent widths due to different tracings is of the order of 2% for the λ 7774 triplet lines and about 5% for the λ 8446 triplet lines. Near the limb ($\mu=0.3$) the scatter is only slightly larger. The random scatter of the equivalent widths of the fainter lines is larger, due to the fact that a small displacement of the local continuum appreciably affects the measured equivalent width. The scatter between individual measurements of the faint lines is of the order of 10% and, in some instances, may reach 20%. SWINGS' (1966) compilation gives an idea of the scatter in equivalent widths of the forbidden [O I] lines measured by different authors.

In Table I the observed equivalent widths and central depths at the five selected positions on the solar disk have been collected. The values followed by a double point (:) are uncertain, due to the relatively large random scatter of the individual observations. The central depths are given only for those lines for which profiles were analysed. The lines of the triplet transition at λ 8446 have very similar wavelengths and, hence, form a blend in which only two absorption lines are clearly visible. These two lines have been listed in Table I with their corresponding central depths. The equivalent width was measured for the entire blend. It is interesting to note that the two faint permitted lines display a center-limb behaviour similar to the stronger O I

TABLE I

Observed equivalent widths and central depths of the oxygen lines at various positions on the solar disk (stray light correction was applied)

	W_λ(mÅ)					r_0(%)				
λ	$\mu=1.0$	0.7	0.5	0.4	0.3	$\mu=1.0$	0.7	0.5	0.4	0.3
7771.95	88.0	79.5	70.0	65.0	59.5	33.1	29.9	26.9	25.8	23.9
7774.18	71.0	63.5	56.5	52.5	48.0	29.9	26.5	23.7	22.0	20.4
7775.39	54.0	47.5	41.0	37.5	33.5	24.8	21.5	18.6	18.3	15.5
8446.38*	147	133	120	113.5	106	30.3	26.6	24.0	23.1	22.1
8446.76*						16.8	13.7	11.2	10.3	9.0
6156.80	4.9:	4.1:	3.7:	3.5:	3.4:					
6158.17	5.0	3.0	2.3	2.0	2.0	3.1				1.8
6363.81	1.8:	1.9:	2.2:	2.5:	3.1:					
6300.31	4.3	4.7	5.2	5.5	5.7	3.8				4.5
5577.34	3.2:	3.7:	4.0:	4.1:	4.6:					

* See discussion of this blend in Section 4.D.

lines. This is to be expected, since they are also high-excitation lines. Their equivalent widths decrease from center to limb. In contrast, the equivalent widths of the low-excitation forbidden lines increase from center to limb. For all lines the variation of the equivalent width across the solar disk follows a remarkably smooth curve, as is shown in Figure 1 for the strong lines and in Figure 2 for the faint lines.

Recently De Jager and Neven (1967) published center-to-limb profiles of the O I triplet at λ 7774, observed with the Liège spectrograph at the Jungfraujoch Scientific Station. Their profiles (averaged over the violet and the red side) are in good agreement with our observations, except for the wings, which they observe to be appreciably deeper. Consequently, their equivalent widths are greater than ours by about 20% at the disk's center. For comparison with our measurements we include in Figure 1 the center-to-limb variation of the equivalent widths as measured by De Jager and Neven (dashed lines). Their results do not follow a smooth curve.

Our observed profiles are presented in Figures 3–12 by a solid line.

3. Theory and Numerical Computations

The theoretical line profiles and the abundances were derived for the following atmospheric models: Mutschlecner's model (see Müller and Mutschlecner, 1964); the Utrecht Reference Model (Heintze, Hubenet, and De Jager, 1964); Heintze's (1965) model; and the Holweger (1967) model. Essentially, these four models differ by (a) the value of the temperature minimum $T_{\min}$, (b) the extent in optical depth $\bar{\tau}_{\min}$ at which the temperature minimum occurs, and (c) the variation of the microturbulence with depth. These three characteristics of the atmospheric models employed are presented in columns 2, 3, and 4 of Table II respectively. Only the average column of the Utrecht Reference Model (URM) was considered, i.e. the atmospheric inhomogeneities as described by the three columns in the URM were neglected.

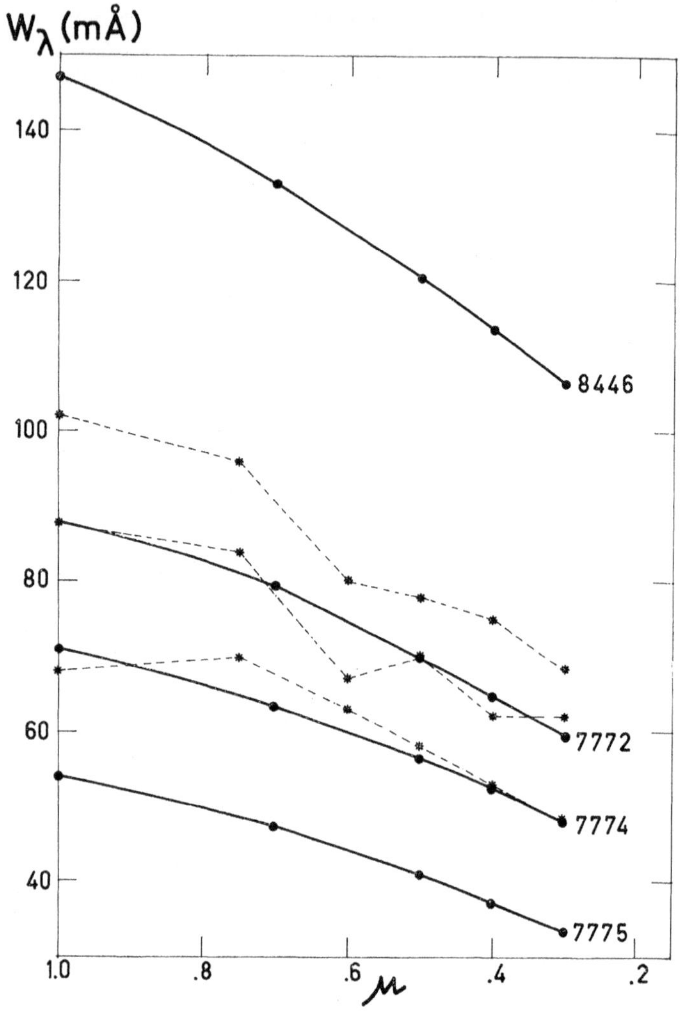

Fig. 1. Center-to-limb variation of the observed equivalent widths of the O I triplet near λ 7774 Å and of the O I blend at λ 8446 Å. Solid lines: our measurements; dashed lines: observations by De Jager and Neven.

In Holweger's model the anisotropic, depth-dependent turbulence is given for $\mu = 1.0$ and 0.3. Our observations of the oxygen lines refer to three additional positions on the solar disk, i.e. $\mu = 0.7$, 0.5, and 0.4. It was, therefore, necessary to compute the turbulent velocities for the three additional values of μ. The turbulence ξ_μ for any given $\mu = \cos\theta$ may be defined by the empirical relation:

$$\xi_\mu^2 = \frac{\xi_t^2 \xi_r^2}{\xi_r^2 \sin^2\theta + \xi_t^2 \cos^2\theta}$$

where ξ_t and ξ_r are the tangential and the radial components of the turbulence field

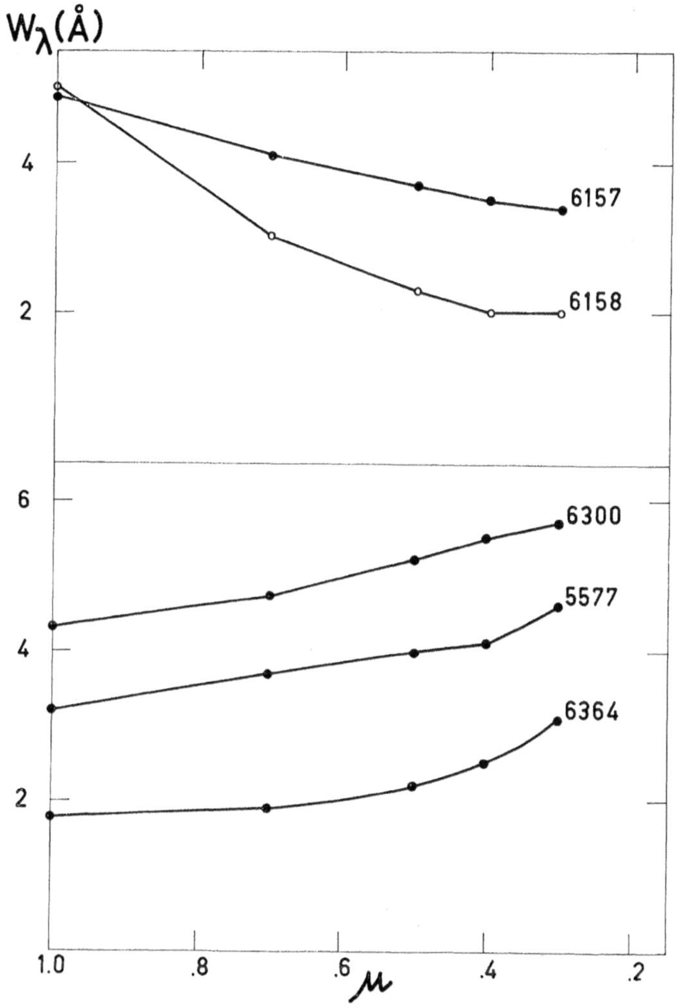

Fig. 2. Center-to-limb variation of the observed equivalent widths of the faint O I lines. Top: permitted lines; bottom: forbidden lines.

TABLE II

Characteristics of the atmospheric models

Model	T_{min}(°K)	$\log \bar{\tau}_{min}$	Microturbulence
Mutschlecner (Mu.)	4600	$\leqslant -2.7$	isotropic, constant
Utrecht Reference Model (URM)			
average column	4500	-1.8	isotropic, increas. with height
Heintze (He.)	4550	-1.15	isotropic, increas. with height
Holweger (Ho.)	3900	$\leqslant -6.7$	anisotrop., decreas. with height

respectively. Table III gives the resulting values of $\xi_{\mu=0.7}$, $\xi_{\mu=0.5}$ and $\xi_{\mu=0.4}$ as a function of depth derived from Holweger's $\xi_{\mu=1.0}=\xi_r$ and $\xi_{\mu=0.3}$ which, for completeness, are also listed in the table. For small depths, $\bar{\tau}\leqslant 0.00005$, and for great depths, $\bar{\tau}\geqslant 4.0$, the ξ_μ remain constant and, hence, are not explicitly given in Table III.

TABLE III

The anisotropic, depth dependent turbulence corresponding to Holweger's model at various positions on the solar disk

$\bar{\tau}$	$\xi_{1.0}$	$\xi_{0.7}$	$\xi_{0.5}$	$\xi_{0.4}$	$\xi_{0.3}$
.00005	1.00	1.00	1.00	1.00	1.00
.0001	1.00	1.00	1.00	1.00	1.00
.00016	1.00	1.05	1.08	1.09	1.10
.00058	1.00	1.16	1.26	1.31	1.35
.001	1.00	1.19	1.33	1.39	1.45
.0025	1.00	1.26	1.47	1.59	1.70
.004	1.00	1.28	1.52	1.66	1.80
.01	1.15	1.47	1.74	1.90	2.05
.016	1.25	1.58	1.85	2.00	2.15
.058	1.60	1.96	2.23	2.37	2.50
.10	1.70	2.06	2.34	2.48	2.60
.25	1.95	2.33	2.60	2.73	2.85
.40	2.10	2.47	2.73	2.85	2.95
.70	2.20	2.56	2.80	2.91	3.00
1.0	2.30	2.62	2.83	2.92	3.00
1.5	2.40	2.69	2.86	2.94	3.00
2.0	2.50	2.75	2.89	2.95	3.00
4.0	2.50	2.75	2.89	2.95	3.00

The required values for the continuous absorption coefficient at different depths and wavelengths were taken from BODE's (1965) tables.

For the profile computations the following basic assumptions were made:

(1) the atmosphere is homogeneous and plane-parallel;

(2) the lines are formed under conditions of local thermodynamic equilibrium (LTE);

(3) the macroturbulence is neglected.

We made use of the fine analysis program of BASCHEK, HOLWEGER, and TRAVING (1966), by which a direct integration of intensities is performed in order to obtain the line profile. The abundances were derived by an iterative procedure fitting the computed equivalent widths to the observed ones. Details of the physical and numerical methods are described by Baschek *et al.* We carried out the numerical computations at Kiel (X1) and Hamburg (TR4) in ALGOL and at Genève (CDC 6600) in the corresponding FORTRAN translation by Peytremann (see PEYTREMANN *et al.*, 1967).

The computations are based on the line data collected in Table IV. Columns 1, 2, and 3 list the wavelengths of the oxygen lines analysed, the corresponding multiplets, and the excitation potentials of the lower level respectively, as given by MOORE (1959).

TABLE IV

Line data for the profile calculations

λ(Å)	mult.	χ_1(eV)	log*gf*	log C_6
7771.96	1	9.11	+ 0.33	−30.87
7774.18		9.11	+ 0.18	−30.87
7775.40		9.11	− 0.04	−30.87
8446.38	4	9.48	+ 0.17	−30.81
8446.76		9.48	− 0.05	−30.81
8446.26		9.48	− 0.52	−30.81
6156.78	10	10.69	− 0.45	−29.64
6158.19		10.69	− 0.30	−29.64
6363.88	(1F)	0.02	−10.18	−32.43
6300.23		0.00	− 9.68	−32.43
5577.35	(3F)	1.96	− 8.22	−32.13

The wavelengths of the λ 8446 triplet were derived from the energy levels published by Moore (1949). The log*gf* values (column 4) of the permitted lines were obtained with the Coulomb approximation and LS coupling (see Goldberg, Müller, and Aller, 1960), and for the forbidden lines the log*gf* values of Garstang (1951) were used. The last column of Table IV lists the logarithm of the interaction constants C_6, due to van der Waals broadening. The values of C_6 were derived with Unsöld's (1955) approximation:

$$C_6 = 1.61 \times 10^{-32}\left[\left(\frac{13.6}{\chi - \chi_u}\right)^2 - \left(\frac{13.6}{\chi - \chi_l}\right)^2\right]$$

where χ is the ionisation potential of O I and χ_u and χ_l represent the excitation potentials of the upper and lower levels respectively. It should be noted that here the interaction constants are defined by the relation

$$\Delta\omega = C_i/r^i$$

$\Delta\omega$ being the angular frequency shift and r the relative distance of the two colliding particles. The broadening due to electron collisions is small compared to the broadening by van der Waals forces. Nevertheless, the damping due to collisions with electrons was taken into account in the computations of the strong lines. The interaction constants, C_4, employed are those given by Hunger (1960), i.e. $\log C_4 = -14.46$ for multiplet 1 and $\log C_4 = -14.35$ for multiplet 4.

In order to compare directly the observed and the theoretical profiles, we have folded the latter with Elste's (1964) provisional instrumental profile of the McMath-Hulbert vacuum spectrograph.

4. Results and Discussions

We now are ready to compare the observed lines with the theoretical profiles based on the four selected models. We shall proceed in discussing first the infrared triplet near λ 7774 Å, then the five faint lines, followed by a presentation of the abundance results, and finally discuss the blend at λ 8446 Å.

A. THE INFRARED TRIPLET NEAR λ 7774 Å

1. *The Temperature Distribution*

In Figures 3 to 7 the observed averaged profiles of the $\lambda\lambda$ 7772, 7774, and 7775 Å lines at the five positions on the solar disk ($\mu=1.0$, 0.7, 0.5, 0.4 and 0.3) are shown, as well as the theoretical profiles computed with the models described in Table II. The observed profiles are drawn with a solid line. The profiles predicted by the Holweger model are shown by dots connected with a dashed line. The open circles, the straight crosses (+) and the tilted crosses (×) represent the profiles calculated with the models Mutschlecner, Heintze, and the URM respectively.

The Heintze model predicts an emission core which becomes more and more

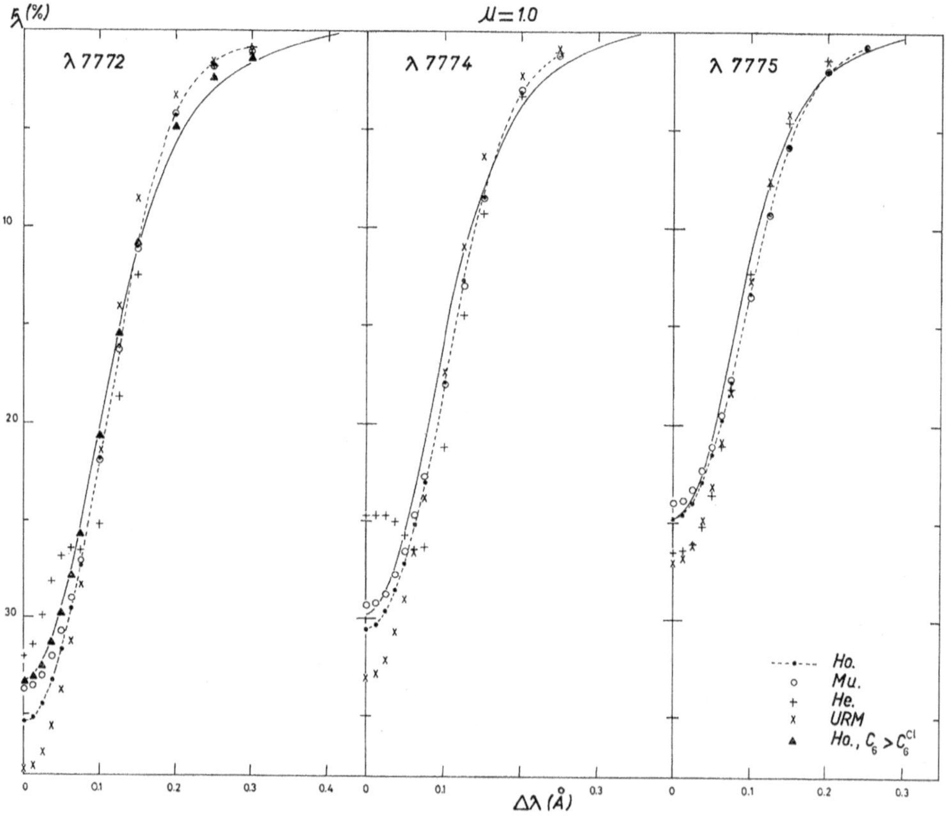

Fig. 3. Averaged line profiles of the O I $\lambda\lambda$ 7772, 7774 and 7775 Å lines at $\mu=1.0$. Solid line: observed; the theoretical profiles are: dots and dashed line: model Holweger; open circles: model Mutschlecner; straight crosses: model Heintze; tilted crosses: Utrecht Reference Model; filled triangles: model Holweger with increased van der Waals damping broadening.

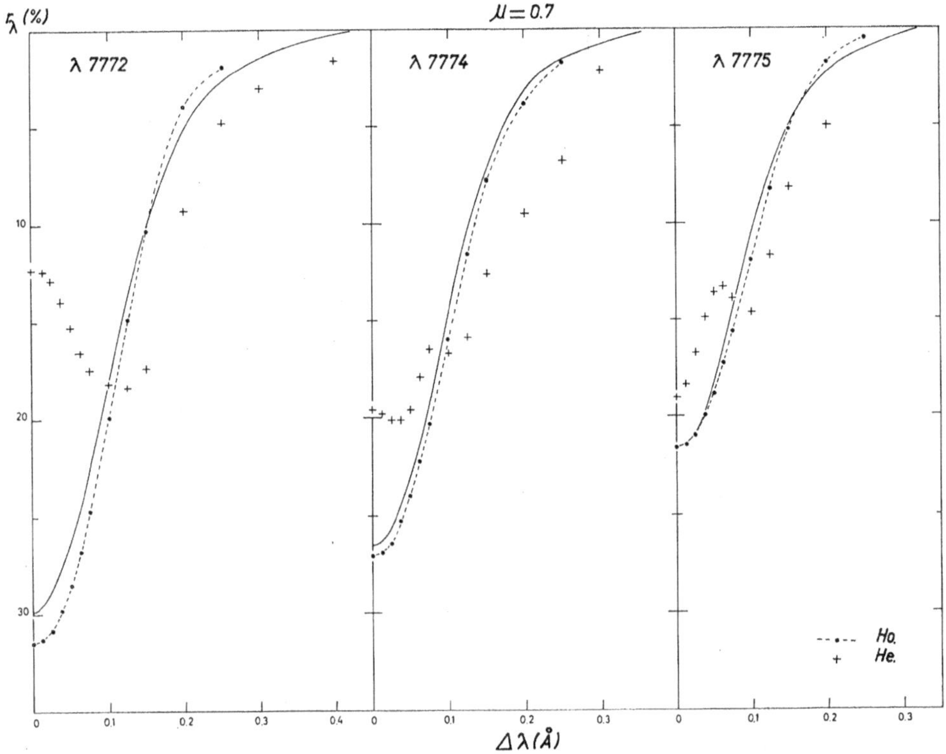

Fig. 4. Averaged line profiles of the O I λλ 7772/4/5 Å lines at $\mu = 0.7$. For symbols see Figure 3.

pronounced as one goes towards the limb. This clearly demonstrates the influence of the sharp temperature minimum proposed in Heintze's model at the depths that contribute most to the cores of these lines. The central emission is due to the rapidly increasing temperature in the higher layers. For illustration the profiles based on the Heintze model are included in Figures 3 and 4 corresponding to $\mu = 1.0$ and 0.7.

In contrast to Heintze's model the three other models on the whole reproduce the shape of the observed profiles reasonably well. This is due to the fact that all three models exhibit a similar temperature distribution in the most important regions for the formation of these lines. Some differences are noticed, however, between the profiles computed with the three models. For all three lines and at all values of μ the URM predicts central depths which are too low compared to the observations. Consequently, the profiles based on the URM appear too steep. The main cause of this behaviour is the small value of the turbulence given in the URM for the deeper layers.

Both the Holweger and the Mutschlecner models reproduce the observed profiles equally well at $\mu = 1.0$. Towards the limb the profiles obtained with the Holweger model fit the observations just as well as at the disk's center, whereas the central depths computed with Mutschlecner's model become too deep. This result can again be

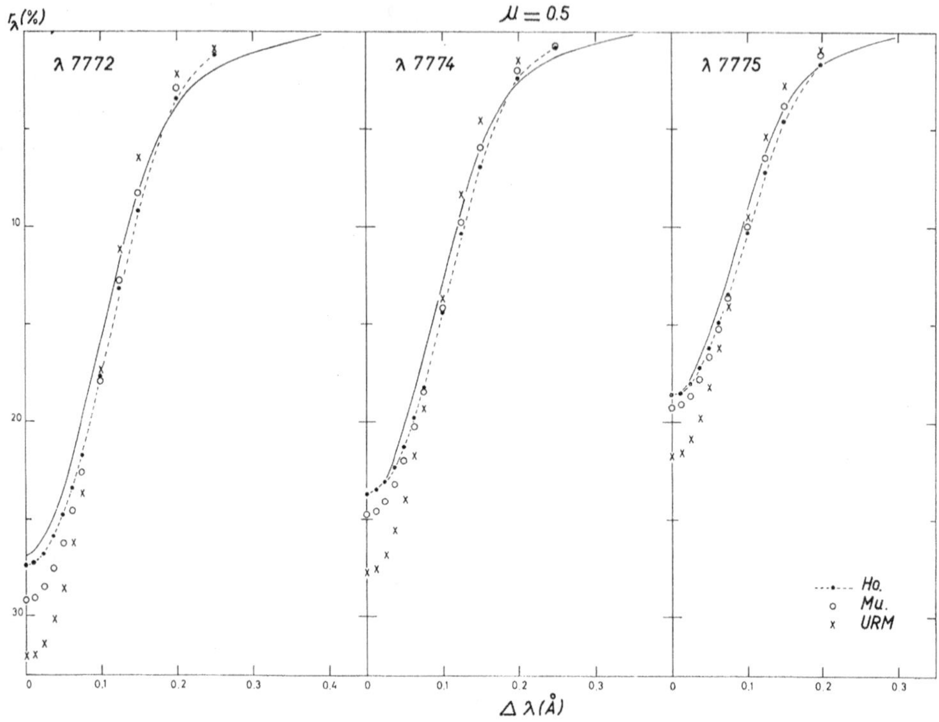

Fig. 5. Averaged line profiles of the O I $\lambda\lambda$ 7772/4/5 Å lines at $\mu = 0.5$. For symbols see Figure 3.

attributed mainly to an inadequate turbulence: compared to Holweger's turbulence field Mutschlecner's isotropic turbulence is smaller near the limb.

2. *The Turbulence Field*

In order to gain insight into the question of the microturbulence variation with depth, we combined the temperature and pressure distributions proposed by HOLWEGER (1967) with different microturbulence distributions, such as (A) an anisotropic but depth-independent turbulence, or (B) an isotropic turbulence which increases with height.

In case (A) the value of the depth-independent turbulence was chosen such that it corresponds to the value of Holweger's anisotropic and depth-dependent turbulence field at the approximate depth $\bar{\tau}^*$ of the line-forming region. This depth may be defined as the $\bar{\tau}$ corresponding to $\tau_{1+c}/\mu = 1$. The profile computations based on Holweger's model showed that for the triplet at λ 7774 seen on the disk's center the depth of line formation is $\bar{\tau}^* = 0.20$. At this depth the Holweger model gives a radial turbulence velocity of about $\xi_{\mu=0.3} = 1.9$ km/sec. Correspondingly, for $\mu = 0.3$ the turbulence velocity is about $\xi_{\mu=0.3} = 2.4$ km/sec. For case (B) we chose the isotropic turbulence increasing with height as proposed by the URM.

Thus we computed the profiles of the lines at $\lambda\lambda$ 7772, 7774, and 7775 Å for the

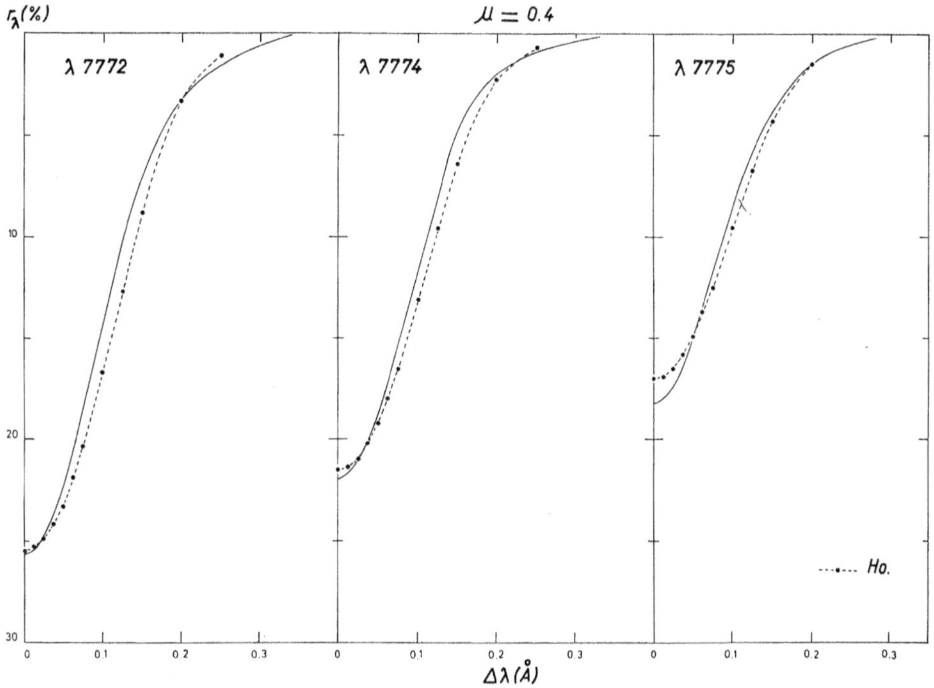

Fig. 6. Averaged line profiles of the O I λλ 7772/4/5 Å lines at $\mu = 0.4$. For symbols see Figure 3.

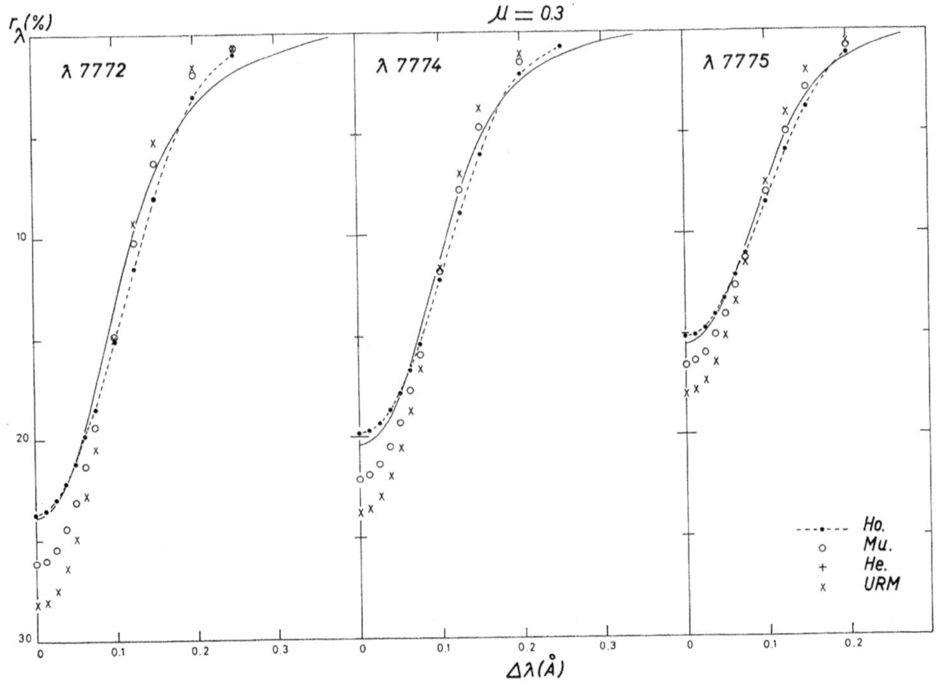

Fig. 7. Averaged line profiles of the O I λλ 7772/4/5 Å lines at $\mu = 0.3$. For symbols see Figure 3.

two positions $\mu=1.0$ and 0.3. The results are presented in Figures 8 and 9 for $\mu=1.0$ and 0.3 respectively, where again the full lines reproduce the observed profiles. The open circles are the results of case (A), i.e. $\xi_{\mu=1.0}=1.9$ km/sec and $\xi_{\mu=0.3}=2.4$ km/sec. The crosses correspond to the results obtained with the URM turbulence distribution. For comparison we included in Figures 8 and 9 the theoretical profiles derived with Holweger's turbulence field (dots connected by a dashed line). The figures clearly show that if appropriate values of the depth-independent but anisotropic turbulence are chosen, the theoretical profiles of these lines reproduce the observed ones just as good as Holweger's turbulence field. The calculations yield an abundance difference of only 6% at the center and near the limb, the depth-independent turbulence requiring the larger value. In case (B) the turbulence given in the URM is relatively small, its value being about 1.1 km/sec for $\bar{\tau}^*=0.20$. As a result the predicted profiles are too steep and the central depths too low. The abundance based on the URM turbulence distribution is up to 35% greater than the value found with Holweger's turbulence field.

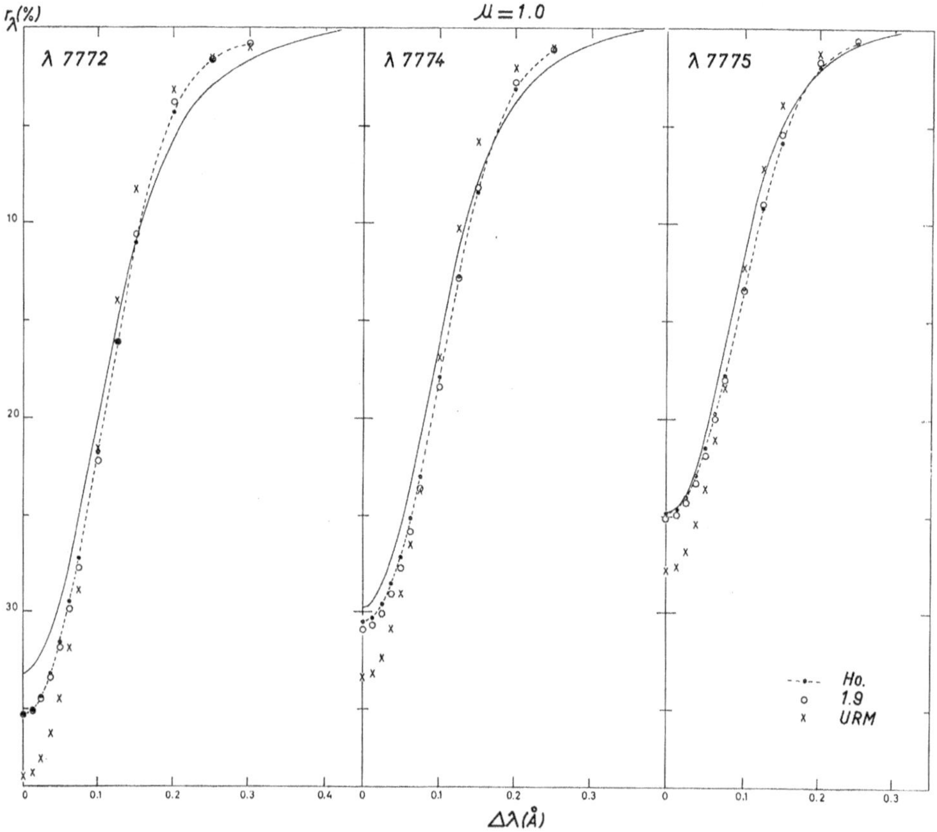

Fig. 8. Averaged line profiles of the O I λλ 7772/4/5 Å lines at $\mu=1.0$. Solid line: observed; dots and dashed line: temperature and turbulence distribution of the Holweger model; open circles: temperature distribution of the Holweger model, depth independent turbulence, $\xi_{\text{rad.}}=1.9$ km/sec; tilted crosses: turbulence distribution of the Utrecht Reference Model.

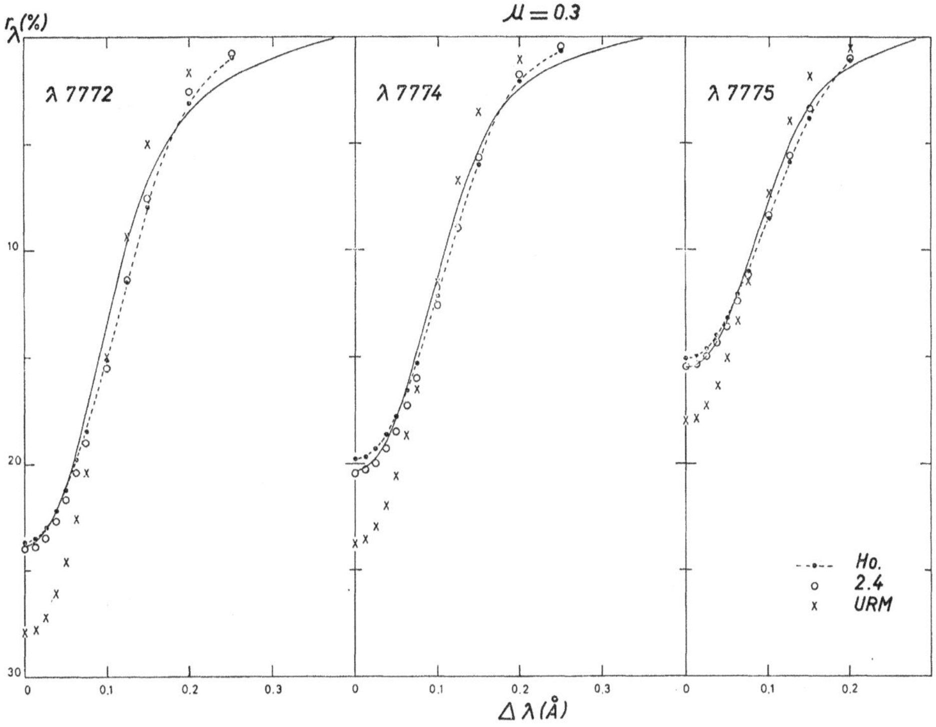

Fig. 9. Averaged line profiles of the O I λλ 7772/4/5 Å lines at $\mu=0.3$. Symbols same as in Figure 8, except for the open circles: temperature distribution of the Holweger model, depth independent turbulence $\xi_{0.3}=2.4$ km/sec.

The calculations show that a turbulence which either remains constant or decreases with height, requires an anisotropic turbulence field, in which the radial component is smaller than the horizontal component. If, however, the turbulence field is assumed to be isotropic, then the two different values of the turbulence – which are necessary to reproduce the profiles at the disk's center and near the limb – indicate that over a depth range of about $\bar{\tau}^*=0.20$ to about 0.06 the turbulence must increase from 1.9 to about 2.4 km/sec. The variation of the turbulence with depth cannot be decided from calculations of lines of one multiplet only. Other lines formed in various different depths of the atmosphere must be analysed in order to obtain an indication as to the way the turbulence varies with depth.

3. *The Damping*

As can be seen from Figures 3 to 9 all models predict profiles which are too flat in the outer wings compared to the observations. Accordingly, the theoretical halfwidths are larger than the observed ones, inasmuch as the computations are normalised to the observed equivalent widths. The flat theoretical wings suggest that, possibly, an inadequate damping factor was employed. Since for these lines the damping is mainly due to van der Waals broadening, we computed a few profiles increasing the interaction constant C_6. We arbitrarily chose $\log C_6 = -30$ instead of $\log C_6 = -30.88$ obtained with Unsöld's approximation and found an excellent agreement between the calculated and the observed profiles. For illustration we show, in Figure 3, by filled triangles the profile of λ 7772 at $\mu = 1.0$, calculated with the higher value of C_6.

Due to the increase of the van der Waals broadening the resulting abundance becomes smaller. The amount by which the abundance is lowered obviously depends on the strength of the line. When increasing C_6 by a factor of 7.6 the abundances obtained from the analysis of the λ 7772, 7774, and 7775 Å lines at $\mu = 1.0$ are smaller by 15, 10, and 5% respectively than the corresponding values found with $\log C_6 = -30.88$.

B. THE FAINT LINES

As mentioned earlier, we have selected the better of the two faint permitted lines and the best of the three forbidden lines for the comparison of observed and theoretical profiles.

The observed shape of the faint permitted line at λ 6158 Å is shown in Figure 10 (solid line). The upper and lower portions of the figure correspond to the center and to the limb of the solar disk respectively. The theoretical profiles based on the models of Holweger, Mutschlecner or Utrecht are marked by the same symbols as previously

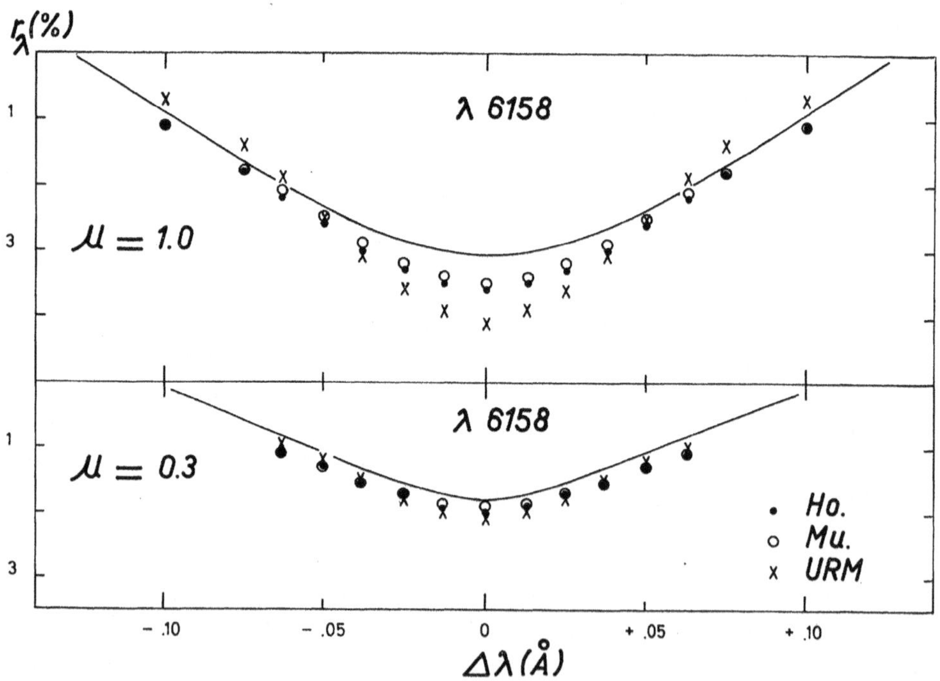

Fig. 10. Permitted O I line profile at λ 6158 Å. Top: corresponds to $\mu = 1.0$, bottom: to $\mu = 0.3$. For symbols see Figure 3.

used in Figures 3 to 7. The center-limb behaviour of the λ 6158 line confirms what was found for the λ 7774 triplet. At $\mu=1.0$ the Holweger and the Mutschlecner models predict the observed line shape equally well. The URM requires a deeper central depth and, consequently, a steeper profile, thus reflecting the small value of the turbulence proposed by this model for the deeper layers. Towards the limb the line weakens appreciably. For such a very faint line of high excitation the theoretical profiles calculated with all three models coincide at $\mu=0.3$ and are in accord with the observations.

The faint forbidden line at λ 6300 Å, being a zero Volt line, originates in higher layers of the photosphere. Its center-limb behaviour is shown in Figure 11, the upper part corresponding to $\mu=1.0$, the lower to $\mu=0.3$. Within the uncertainty of the observations the models of Holweger, Mutschlecner or Utrecht reproduce the observed line shapes very well, both at the center and near the limb of the sun. We have included in Figure 11 the results obtained with the Heintze model. Whereas the profile predicted by the Heintze model agrees with the others at the disk's center, it becomes deeper and, hence, steeper at $\mu=0.3$, as a consequence of the different temperature distribution proposed by this model in the relevant layers.

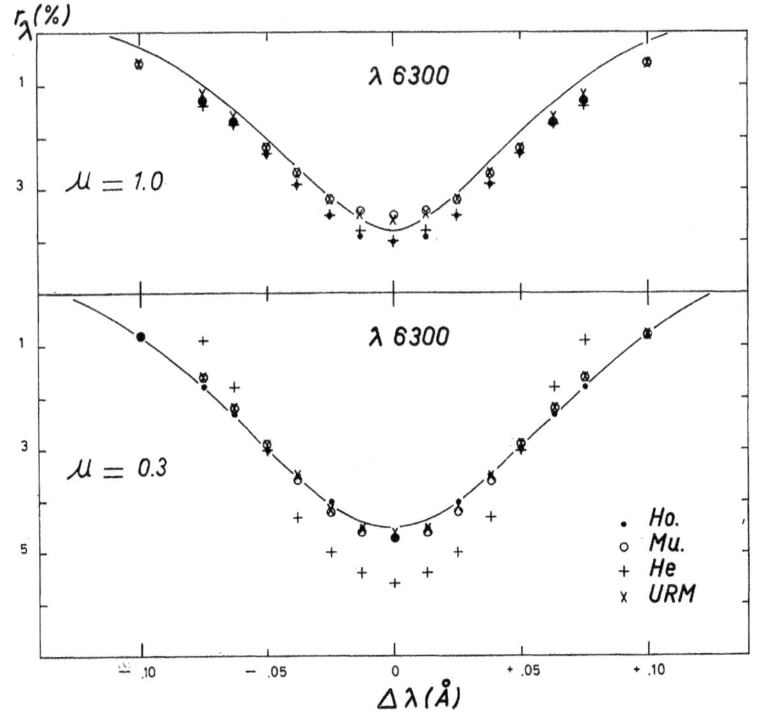

Fig. 11. Forbidden O I line profile at λ 6300 Å. Top: corresponds to $\mu=1.0$, bottom: to $\mu=0.3$. For symbols see Figure 3.

C. THE OXYGEN ABUNDANCE

The oxygen abundance was derived by iteration from the observed equivalent widths of the three medium strong lines at λ 7774 Å and the five faint lines at all five selected positions on the solar disk. In Tables V, VI, and VII the individual results deduced from the models of Holweger, Mutschlecner, and Utrecht respectively, have been collected. They are normalised to the hydrogen abundance $\log \varepsilon_H = 12.00$. The values followed by a double point (:) are uncertain results, due to the larger random scatter of the measured equivalent widths. The last column of each table lists the mean of $\log \varepsilon$ of the abundances found for each line at the five positions across the solar disk. The fourth row of each table gives for any of the five μ-values the mean of the $\log \varepsilon$ obtained for the three medium strong lines individually.

Let us now take a closer look at Tables V to VII. Firstly, we see that for the

TABLE V

Oxygen Abundance Results (Holweger Model)

λ	$\log \varepsilon$					$\overline{\log \varepsilon}$
	$\mu = 1.0$	0.7	0.5	0.4	0.3	$(c-1)$
7772	8.78	8.80	8.80	8.88	8.98	8.85
7774	8.72	8.73	8.75	8.80	8.91	8.78
7775	8.67	8.69	8.69	8.75	8.83	8.73
$\overline{\log \varepsilon}$ (7772/4/5)	8.72	8.74	8.75	8.81	8.91	8.79
6157	8.90:	9.00:	9.09:	9.22:	9.35:	9.11:
6158	8.76	8.69	8.77	8.89	9.03	8.83
6364 (*f.*)	8.81:	8.73:	8.76:	8.72:	8.81:	8.77:
6300 (*f.*)	8.81	8.72	8.71	8.69	8.70	8.73
5577 (*f.*)	9.10:	9.11:	9.12:	9.04:	9.10:	9.09:

TABLE VI

Oxygen Abundance Results (Mutschlecner Model)

λ	$\log \varepsilon$			$\overline{\log \varepsilon}$
	$\mu = 1.0$	0.5	0.3	$(c-1)$
7772	8.85	9.00	9.10	8.98
7774	8.78	8.91	9.04	8.91
7775	8.73	8.84	8.95	8.84
$\overline{\log \varepsilon}$ (7772/4/5)	8.79	8.92	9.03	8.91
6157	8.91:	9.19:	9.41:	9.17:
6158	8.77	8.87	9.09	8.91
6364 (*f.*)	8.77:	8.75:	8.77:	8.76:
6300 (*f.*)	8.75	8.71	8.68	8.71
5577 (*f.*)	9.10:	9.12:	9.14:	9.12:

TABLE VII

Oxygen Abundance Results (Utrecht Reference Model)

λ	$\log \varepsilon$			$\overline{\log \varepsilon}$
	$\mu = 1.0$	0.5	0.3	$(c-1)$
7772	8.92	8.93	8.98	8.94
7774	8.86	8.88	8.93	8.89
7775	8.81	8.81	8.85	8.82
$\overline{\log \varepsilon}$ (7772/4/5)	8.86	8.87	8.92	8.88
6157	8.93:	9.12:	9.36:	9.14:
6158	8.78	8.79	8.98	8.85
6354 (*f.*)	8.79:	8.65:	8.72:	8.72:
6300 (*f.*)	8.77	8.63	8.62	8.67
5577 (*f.*)	9.19:	9.06:	9.09:	9.11:

λ 7774 triplet all the models yield the largest abundance for the strongest line (λ 7772), the smallest value for the faintest line (λ 7775). This is true at all positions on the solar disk. The difference is similar for all three models and amounts to about 30% at $\mu = 1.0$ and about 40% at $\mu = 0.3$.

Secondly, we notice a systematic increase of the abundance from center to limb for all permitted lines, the effect being more pronounced for the very faint lines. Among the lines of medium strength the strongest line (λ 7772) tends to have the greatest center-limb variation. For this line the Holweger and the Mutschlecner model yield a center-limb abundance difference which is about double the value of the abundance difference found between the strongest and the weakest of the triplet lines at $\mu = 1.0$. With the URM the center-limb difference of the triplet lines is smaller than the abundance difference between the strongest and the weakest of the three lines. The forbidden lines behave differently. The abundances derived from these lines either remain unchanged or tend to decrease from center to limb. For the best line, λ 6300, the Holweger model yields a center-limb abundance decrease of about 30%. The other models give smaller variations.

It should be pointed out that the abundance differences from center to limb or from stronger to fainter lines are all relatively small and remain within a factor of 2 (except for the uncertain results of the λ 6157 line). The effects may be caused by various factors, such as uncertainties in the turbulence field (see Section 4.A.2), inadequate damping (see Section 4.A.3), temperature inhomogeneities, and observational inaccuracies.

The intercomparison of the last column in the Tables V to VII indicates that for the permitted lines the Holweger model yields the smallest, the Mutschlecner model the largest abundances. The greatest difference amounts to about 35%. In the case of the forbidden lines the Holweger model tends to give the largest, the URM the smallest abundances, but the differences are only 15% at the most. These effects clearly reflect the differences in the temperature distribution and the turbulence field proposed by the three models. In the deeper atmospheric layers, where all three models have a similar temperature distribution, it is mainly the turbulence field which produces the abundance differences. Thus, the URM with its small turbulence in the deeper layers yields the largest abundance for the λ 7774 triplet at $\mu = 1.0$. In higher layers both the temperature and the turbulence distributions differ in the three models and, therefore, both play a role in the abundance results obtained from the forbidden lines and from the permitted lines observed at $\mu = 0.3$.

Finally, we derive an average abundance for each of the three models. We use the results derived from all lines and all positions on the solar disk. The results are weighted such that the uncertain values are given only half the weight. Thus, the weighted mean of the oxygen abundance is found to be 8.83, 8.90, and 8.87 (on the basis of $\log \varepsilon_H = 12.00$) for the models of Holweger, Mutschlecner or Utrecht respectively. It is interesting to note that the three results differ by less than 20%, although the three models propose a different physical structure of the photosphere. Since the Holweger model best reproduces the observed profiles of the strong and the faint O I lines both at the center and near the limb, we consider

$$\log \varepsilon_O = 8.83$$

the most reliable result of the photospheric oxygen abundance. The oxygen abundance determined by Goldberg, Müller, and Aller (1960), corrected by Zwaan (1962) for an inadequate electron pressure, is $\log \varepsilon_O = 8.86$, in close agreement with the results obtained here.

D. The Blend at λ 8446 Å

The blend at λ 8446 Å is composed of three lines corresponding to the triplet transition $3s\ {}^3S^\circ - 3p\ {}^3P$. As mentioned earlier, two of the lines, i.e. λ 8446.26 and λ 8446.38, lie very close together and are not resolved in the spectrograms. The third line, λ 8446.76, lies far enough on the redward side to be resolved but not far enough to be entirely separated from the main component of the triplet. The observed blend is shown in Figure 12 by the solid line. For an abundance determination this blend is not suitable. Instead, we may employ the mean abundance derived from the other neutral oxygen lines to compute the profile of the blend.

We restricted ourselves to the Holweger model and, accordingly introduced the oxygen abundance $\log \varepsilon_O = 8.83$ into the calculations. As a result it was found that the central intensity ratio of the two main components was not in accord with the observations. The calculations yielded 1.2 for the λ 8446.38/8446.76 central intensity ratio, whereas the observations indicate a ratio of 1.8. This discrepancy cannot be accounted for by an effect of the model. Likewise it cannot be due to incorrect relative f-values, since these are obtained by means of the J-file sum rule which is independent of coupling. Furthermore, Richter and Haag (1967) made a high-frequency discharge experiment and determined the intensity ratio in question. Their measurements confirmed the theoretical f-value ratio. Therefore, the discrepancy can only be

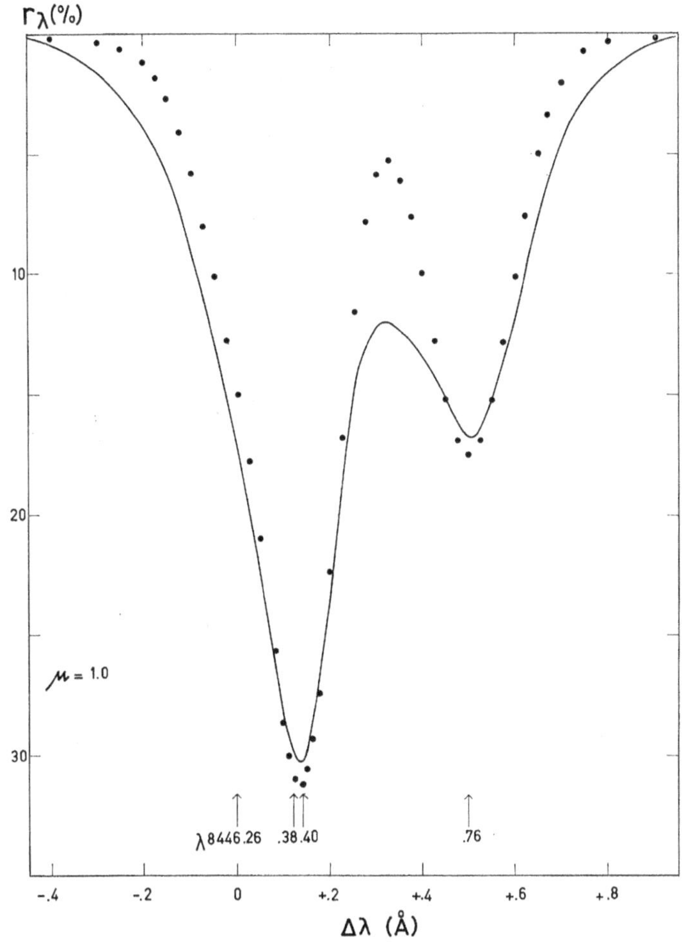

Fig. 12. Blend at λ 8446 Å for $\mu = 1.0$. Solid line: observed. Dots: theoretical profile calculated on the basis of Holweger's model for a blend composed of the three O I lines and the Fe I line at λ 8446.40 Å. The arrows mark the wavelengths of the four components.

explained by the existence of further components of other elements contributing to the observed blend. According to the recently published Solar Spectrum Wavelengths (Moore, Minnaert, and Houtgast, 1966) the following two Fe I lines might be expected to occur in the blend: λ 8446.42 of multiplet 1272 and λ 8446.56 of multiplet 1267. On high-dispersion laboratory spectra of neutral iron Richter and Haag (1967) confirmed the presence of an Fe I line of multiplet 1272, which has a wavelength of λ 8446.40. The predicted Fe I line of multiplet 1267 is not detectable. The *f*-value of the λ 8446.40 line was measured on these spectra by Miss Haag. Reduced to the absolute scale given by Corliss and Warner (1964) it becomes $\log gf = -0.27$. The lower excitation potential of the line is $\chi_l = 4.97$ eV. The iron abundance derived from Holweger's model for a number of Fe I lines is 6.76 on the basis of the Corliss and

Warner f-values. With this value for the iron abundance the Holweger model predicts an equivalent width of 40 mÅ for the Fe I line at λ 8446.40 Å if it appeared unblended in the solar spectrum.

Knowing all the relevant data concerning the Fe I line at λ 8447.40 Å it is possible to compute the blend composed of the three O I lines and the one Fe I line. The theoretical profile (folded with the instrumental profile) based on the Holweger model is shown in Figure 12 by the dots. The two main absorption features now have central intensities, which are in accord with the observations. The calculated λ 8446.38/8446.76 central intensity ratio is 1.8, which is just what is observed. The wings of the theoretical profile, however, do not reproduce the observed blend correctly. The theoretical profile is too steep and, thus, the deviations from the observed blend become more pronounced further out in the wings and in the intermediate region between the two main absorption features. Part of these differences may be accounted for by additional faint lines occurring in the blend. Also, an inadequate damping and the temperature inhomogeneities may play a role. The blend is to be investigated further.

5. Conclusions

The analysis of the lines of neutral oxygen revealed that Holweger's temperature distribution, together with his proposed anistropic turbulence which decreases with height, best reproduces the observations. The profiles predicted by the Mutschlecner model are in good agreement with the observed line shapes at $\mu = 1.0$, but reveal some differences with the limb observations on account of the isotropic turbulence assumed in this model. The URM produces profiles of the high-excitation lines which are too deep and too steep due to the low turbulence proposed for the greater photospheric depths. The sharp temperature minimum of the Heintze model occurring relatively deep in the photosphere produces line profiles with central emission cores. From the oxygen lines only, no definite conclusion can be drawn as to the increase or decrease of the turbulence with height. For this, further analyses are necessary of lines of other elements formed in different layers of the photosphere. The oxygen abundance based on the Holweger model turns out to be $\log \varepsilon_O = 8.83$ relative to the hydrogen abundance which is set to be $\log \varepsilon_H = 12.00$. The abundance derived from the other models differ by less than 20% from this value. Our result is in very good agreement with the abundance determined by GOLDBERG, MÜLLER, and ALLER (1960) after applying ZWAAN's (1962) correction which yields $\log \varepsilon_O = 8.86$.

Acknowledgements

We wish to express our appreciation to Dr. R. G. Teske for securing the observations for us. Our thanks are due to the Kiel Computing Center, to Professor G. Traving of the Hamburg University, and to the CERN Computing Center for the use of the high-speed computors. We are most grateful to Professor J. Richter and to Miss T. Haag of the 'Institut für Experimentalphysik' at Kiel for having kindly furnished to us in advance of publication the necessary laboratory data of the O I and the Fe I spectra around λ 8446 Å. We acknowledge the assistance of MM. Gehlich and Messerschmidt (Hamburg Observatory) and of MM. P. Bochsler, M. A. Erbetta, and E. Peytremann (Geneva Observatory) in part of the reductions and computations. One of us (E.A.M.) particularly wishes to thank Professor A. Unsöld for providing her the opportunity and facilities for working at his Institute during several short periods of time.

References

Baschek, B., Holweger, H., and Traving, G.: 1966, *Abhandl. Hamburger Sternw.* **8**, No. 1.
Bode, G.: 1965, *Veröff. Inst. Theor. Phys. Univ. Kiel.*
Corliss, C. H. and Warner, B.: 1964, *Astrophys. J. Suppl.* **8**, 395.
De Jager, C. and Neven, L.: 1967, *Bull. Astron. Inst. Neth. Suppl. Ser.* **1**, No. 8.
Elste, G.: 1964, private communication.
Garstang, R. H.: 1951, *Mon. Not. Roy. Astron. Soc.* **111**, 115.
Goldberg, L., Müller, E. A., and Aller, L. H.: 1960, *Astrophys. J. Suppl.* **5**, 1.
Heintze, J. R. W.: 1965, *Rech. Astron. Obs. Utrecht* **17**, 1.
Heintze, J. R. W., Hubenet, H., and De Jager, C.: 1964, *Bull. Astr. Inst. Neth.* **17**, 442.
Holweger, H.: 1967, *Z. Astrophys.* **65**, 365.
Hunger, K.: 1960, *Z. Astrophys.* **49**, 129.
Mohler, O. C.: 1963, private communication.
Moore, C. E.: 1949, *Atomic Energy Levels*, Vol. I, N.B.S. Circular 467.
Moore, C. E.: 1959, *N.B.S. Technical Note No. 36.*
Moore, C. E., Minnaert, M. G. J., and Houtgast, J.: 1966, *Second Revision of Rowland's Preliminary Table of Solar Spectrum Wave-lengths.* N.B.S. Monograph 61.
Müller, E. A. and Mutschlecner, J. P.: 1964, *Astrophys. J. Suppl.* **9**, 1.
Peytremann, E., Baschek, B., Holweger, H., and Traving, G.: 1967, *Transact. I.A.U.* XIII B.
Richter, J. and Haag, T.: 1967, private communication.
Swings, J.-P.: 1966, *Ann. Astrophys.* **29**, 371.
Unsöld, A.: 1955, *Physik der Sternatmosphären*, 2nd ed., Springer Verlag, Berlin-Göttingen-Heidelberg.
Zwaan, C.: 1962, *Bull. Astron. Inst. Neth.* **16**, 225.

THE CENTER-LIMB BEHAVIOR OF SOLAR MOLECULAR LINES

GEORGE L. WITHBROE
Harvard College Observatory, Cambridge, Mass., U.S.A.

(Received 13 June, 1967)

Abstract. The center-limb behavior of C_2, CH, CN, CO, and MgH lines have been analyzed using five photospheric models. A three-stream model developed by G. Elste gave the most satisfactory results, providing evidence for the existence of inhomogeneities in the photospheric layers $-2.5 \leqslant \log \tau_{5000} \leqslant -0.5$.

1. Introduction

In the past decade different investigators have developed a variety of photospheric models that explain one or more of the observable features of the solar spectrum. These models often exhibit marked differences in their upper layers, at optical depths $\log \tau_{5000} \lesssim -1.0$ (e.g. Figure 1). Since molecular lines are formed at optical depths between $\log \tau_{5000} = -1.0$ and $\log \tau_{5000} = -2.0$, they are useful for studying the atmospheric layers where existing models are most uncertain. In this paper we shall compare the observed center-limb behavior of lines of five molecules with the corresponding

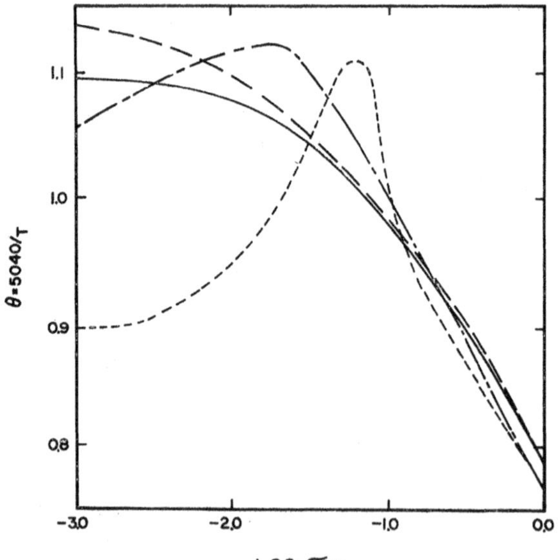

Fig. 1. The depth dependence of the electron temperature in several photospheric models. Long dash line: Elste's Model 10; short dash line: Heintze's model; solid line: Mutschlecner's model; dash-dot line: Utrecht Reference Model.

behavior predicted by several different models. This should give us some indication as to the best model for the upper layers of the photosphere.

2. Observations

The molecules that we shall use are CO, CN, CH, C_2, and MgH. The CO observations are from NEWKIRK (1957), the CN observations from COWLEY (1964), and the CH observations from WITHBROE (1967). The equivalent widths of the C_2 and MgH lines, measured with a planimeter by the author from the *McMath-Hulbert Atlas of the Solar Spectrum* (unpublished), are given in Table I and Table II.

TABLE I

Equivalent Widths of the C_2 Lines (mÅ)

λ	$W_\lambda(1.0)$	$W_\lambda(0.5)$	$W_\lambda(0.3)$	λ	$W_\lambda(1.0)$	$W_\lambda(0.5)$	$W_\lambda(0.3)$
5063.177	10.50	15.90	18.55	5107.872	8.86	12.50	16.60
5063.301	4.94	6.36	8.0	5109.294	3.83	5.85	6.84
5066.732	12.30	15.74	17.95	5111.833	8.74	11.03	13.70
5066.863	6.67	8.22	9.12	5119.364	6.02	8.96	9.54
5073.453	11.27	15.68	16.90	5120.617 } .702 }	9.95	15.52	17.83
5073.578	5.20	6.80	7.85	5135.561 } .672 }	13.80	22.75	24.65
5084.688	6.47	8.78	10.42				
5086.234	13.80	18.35	19.02	5136.256	8.56	11.30	12.35
5089.236	14.08	19.50	21.20	5136.426	4.43	6.61	6.48
5089.350	6.70	10.07	10.74	5138.094	3.55	4.93	4.41
5092.292	15.50	22.20	24.40	5140.363	2.59	4.24	4.90
5094.095	10.00	12.27	16.41	5142.092	3.20	4.58	4.95
5095.194	11.40	15.75	18.30	5143.307	4.35	6.06	6.77
5098.133	18.90	23.70	25.20				
5102.422	11.23	15.28	17.70				

TABLE II

Equivalent Widths of the MgH Lines (mÅ)

λ	$W_\lambda(1.0)$	$W_\lambda(0.5)$	$W_\lambda(0.3)$	λ	$W_\lambda(1.0)$	$W_\lambda(0.5)$	$W_\lambda(0.3)$
5061.526	2.60	3.62	4.35	5157.006	3.40	5.00	6.45
5063.519	2.30	3.56	4.87	5159.790	2.66	4.22	5.58
5082.903	2.54	4.00	3.42	5165.958	2.70	4.22	3.73
5106.206	2.94	3.84	4.68				
5106.860	3.50	5.65	5.82	5175.419	4.07	5.15	5.60
				5178.503	4.62	6.48	7.48
5121.237	1.93	3.09	3.37	5180.593	3.96	4.52	6.54
5125.850	5.82	5.25	5.72				

3. Photospheric Models

In interpreting the observations we shall use five recent photospheric models. The first of these, Elste's Model 10 (ELSTE, 1964, unpublished), was derived from an analysis of limb-darkening observations of the spectral continuum. The second model, that of MUTSCHLECNER (1963), is very similar to Model 10 and has been used to explain the center-limb behavior of lines of the iron group of elements (MÜLLER and MUTSCHLECNER, 1964). Both Elste and Mutschlecner proposed classical models in which the electron temperature decreases monotonically with decreasing optical depth. The third model, the Utrecht Reference Model (HEINTZE, HUBENET, and DE JAGER, 1964), has a temperature minimum located at $\tau_{5000}=0.02$. In the upper layers of the photosphere, $\tau_{5000}<0.1$, the Utrecht model is homogeneous while in the lower layers it is inhomogeneously multicolumnar. The fourth model, HEINTZE's (1965), was developed from the Utrecht Reference Model and an analysis of limb-darkening observations made at the extreme limb during an eclipse. Like the Utrecht Reference Model, it contains a temperature minimum. In this case, it is located at

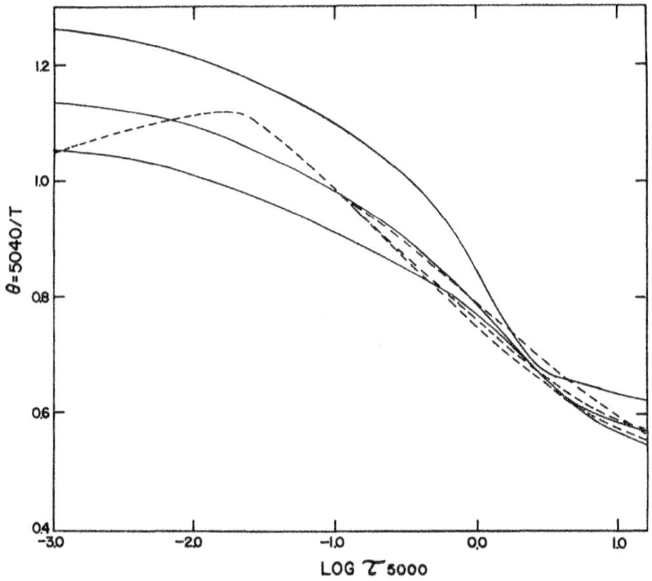

Fig. 2. The depth dependence of the electron temperature in the inhomogeneous models. Solid lines: Elste's three-stream model; dashed lines: Utrecht Reference Model.

$\tau_{5000}=0.07$. The final model is a preliminary inhomogeneous three-stream model developed by G. Elste from Model 10 and EDMONDS' (1962) measurements of the center-limb behavior of the photospheric granular contrast. The temperature variations of the different models are given in Figures 1 and 2.

4. Theory

The center-limb behavior of the equivalent widths, W_λ, of the molecular lines was computed by the method of weighting functions. The appropriate formulae are given in an earlier paper (WITHBROE, 1967), hereafter called Paper I. In Paper I we pointed out that CO formation seriously depletes the population of free carbon atoms in the upper layers of the photosphere. The computations discussed below allow for this. Unless otherwise stated we have used an isotropic microturbulence, which is independent of depth and has a magnitude of 1.4 km/sec, and have also assumed local thermodynamic equilibrium (LTE). This appears to be a fairly good first approximation for molecular lines (e.g. NEWKIRK, 1957; COWLEY, 1964; WITHBROE, 1967).

The theoretical equivalent widths for the three-stream model were evaluated from the equation,

$$W_\lambda = \frac{\sum_{i=1}^{3} A_i \cdot I_i(\mu) \cdot W_i(\mu)}{\sum_{i=1}^{3} A_i \cdot I_i(\mu)}, \tag{1}$$

where A_i is the area of stream i, I_i is the intensity of the continuum in stream i, $\mu = \cos\Theta$, Θ is the angle of radiation flow measured from the outward normal to the atmospheric surface, and W_i is the equivalent width of a line formed in stream i. The relative areas of the hot, medium, and cool streams are respectively 1:2:1. Equation (1) is applicable only as long as the three streams can be considered independently. Near the limb we cannot make this assumption, since the observer's line-of-sight will pass through more than one column. That is, molecules in one column may absorb energy from the continuum formed in a neighboring column. If we assume that the columns have diameters comparable with the size of the solar granulation, about 700 km, then Equation (1) should be valid for $\mu \gtrsim 0.3$.

5. The Excitation Temperature of C_2 and MgH

NEWKIRK (1957), COWLEY (1964), and WITHBROE (1967) found that the excitation temperatures of CO, CN, and CH can be satisfactorily represented by the local electron temperatures in photospheric models similar to Model 10 or Mutschlecner's model. In order to see whether this is also the case for C_2 and MgH, I computed theoretical equivalent widths for lines of these two molecules, using Model 10.

The C_2 and MgH lines are sufficiently weak that their equivalent widths are proportional to the number of molecules excited to the lower level of the transition responsible for the line. That is

$$W_\lambda = C f_J \, e^{-(E_v + E_r)/kT}, \tag{2}$$

where C is a constant, f_J is the rotational contribution to the oscillator strength, E_v is the vibrational energy of the lower level of the transition, E_r is the rotational energy of the lower level, and T is a mean temperature in the region of line formation. The vibrational energy is the same for all lines in a given vibrational band and $E_r = hcBJ(J+1)$ where B is a constant for all lines in a given band and J is the rotational quantum number. Therefore it follows from Equation (2) that

$$\log W_\lambda/f_J = \text{constant} - 0.43429448 \frac{hcB}{kT} J(J+1). \tag{3}$$

In Figure 3 the observed values of $\log W_\lambda/f_J$ for the C_2 lines are plotted as a function of $J(J+1)$. Also plotted is the relation between $\log W_\lambda/f_J$ and $J(J+1)$

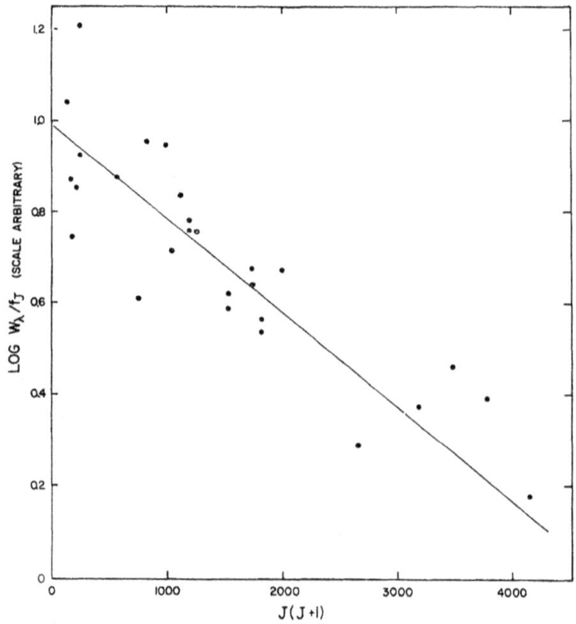

Fig. 3. A graph giving $\log W_\lambda/f_J$ as a function of $J(J+1)$ for C_2. Points: observations; line: theory.

predicted by Model 10. The theoretical line fits the observations very well, suggesting that the excitation temperature of the C_2 lines can be adequately represented by the photospheric local electron temperature. Figure 4 presents similar data for MgH. The agreement between the observations and the theoretical line is not as good as it is for C_2. However, the MgH equivalent widths are so small, the order of 3 to 4 mÅ, that they may be in error by as much as 50%. Therefore, within the limits imposed by observational error, the theoretical line fits the observations.

Thus, these results and those mentioned earlier provide evidence that the excitation temperatures of solar molecular lines do not differ significantly from the local electron temperature in the region of line formation.

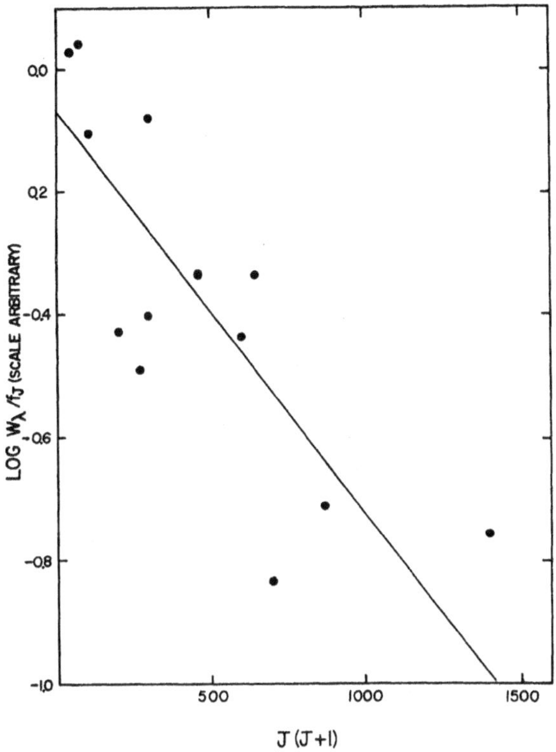

Fig. 4. A graph giving log W_λ / f_J as a function of $J(J+1)$ for MgH. Points: observations; line: theory.

6. The Center-Limb Behavior of Molecular Lines

A. C_2

The observed center-limb behavior of the ratio $W_\lambda(\mu)/W_\lambda(1.0)$ for the C_2 lines is given in Figure 5. Within the observational scatter the center-limb behavior of these lines is independent of line strength. Consequently all of them could be grouped together and placed in a common diagram. In Figure 5 we have also plotted lines to show the results of computations made with the different photospheric models. As is apparent, Elste's three-stream model reproduces the observed center-limb behavior of the C_2 lines better than any of the other models. Elste's Model 10 and Mutschlecner's model both predict values of $W_\lambda(\mu)/W_\lambda(1.0)$ that are somewhat larger than those observed (3% at $\mu=0.5$, 11% at $\mu=0.3$). The two models with temperature minima, Heintze's model and the Utrecht Reference Model, predict even larger values of $W_\lambda(\mu)/W_\lambda(1.0)$, particularly at $\mu=0.3$.

Only the hot and cool columns were used to compute the results illustrated from the Utrecht Reference Model. If the medium or average column is used alone as a homogeneous model, or if all three columns are used as a three-stream model, the computed values of $W_\lambda(\mu)/W_\lambda(1.0)$ are larger than the values computed with the two-column model. The same behavior holds for the other molecules discussed below.

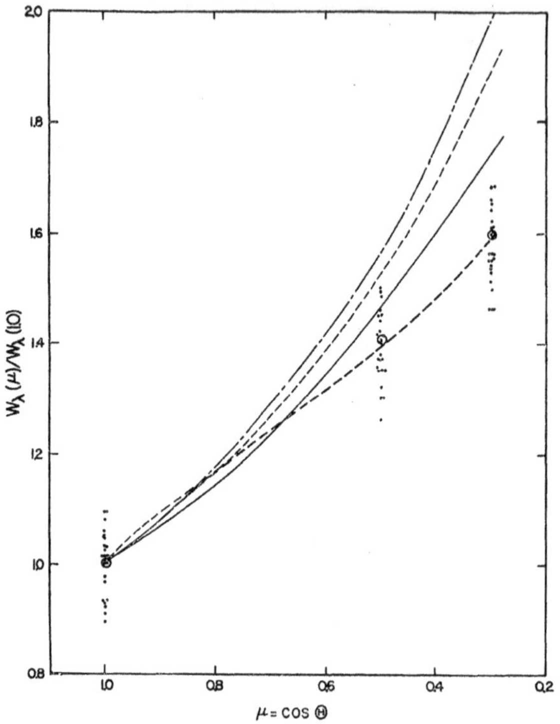

Fig. 5. A comparison between the observed center-limb variation of the C_2 equivalent widths and the theoretical variation predicted by the different photospheric models. Points: observations; circles: means of observations; dash-dot line: predicted by the Utrecht Reference Model; upper dashed line: Heintze's model; solid line: Model 10 and Mutschlecner's model; lower dashed line: Elste's three-stream model.

The fact that we can explain the center-limb behavior of the C_2 lines with the three-stream model is of interest. LA BORDE (1961) was unable to account for the behavior of these lines and suggested several reasons for the difficulty. It now appears that the primary reason for La Borde's problem was inadequate observations.

B. CO

The observed center-limb behavior of the infrared rotation-vibration lines of CO is given in Figure 6, which shows that the center-limb behavior predicted by the different models for CO is similar to that predicted for C_2. The Utrecht Reference Model, Elste's Model 10, and the models of Mutschlecner and Heintze all predict values of $W_\lambda(\mu)/W_\lambda(1.0)$ that are systematically too large. The values of $W_\lambda(\mu)/W_\lambda(1.0)$ computed with the three-stream model are in good agreement with the observed values.

Note that although there are fairly large differences in the temperature variations in the upper layers of the different models, they all predict values of $W_\lambda(\mu)/W_\lambda(1.0)$ that are comparable in magnitude. The reason for this is that the number of CO molecules formed in the upper layers of the photosphere is limited by the number

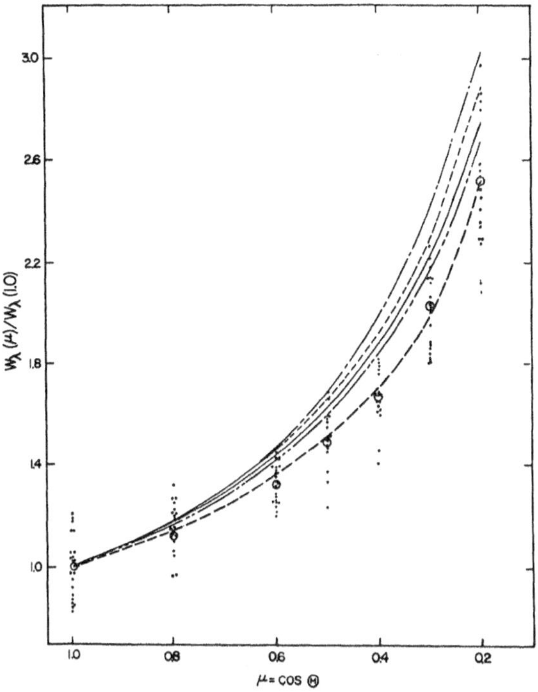

Fig. 6. Same as Figure 5, but for CO. Points: observations; circles: means of observations: dash-dot line: Utrecht Reference Model; short dash line: Heintze's model; solid line: Model 10; dash-double-dot line: Mutschlecner's model; dashed line: Elste's three-stream model.

of carbon atoms in these layers. If a large fraction of the free carbon atoms in an atmospheric layer have been combined into CO molecules, the layer becomes relatively insensitive to changes in temperature. In the uppermost layers of the models considered the temperatures and pressures are such that this does occur. Otherwise, the differences in the values of $W_\lambda(\mu)/W_\lambda(1.0)$ computed for the various models would be much larger.

C. CH

The equivalent widths of most of the observed CH lines are large enough that their center-limb behavior is a function of line strength. A convenient way of illustrating this is to plot $W_\lambda(\mu)/W_\lambda(1.0)$ as a function of $W_\lambda(1.0)$. This has been done for two limb positions in Figures 7 and 8.

Several of the CH lines used in constructing these figures are doublets with appreciable separations between the two components that form them. The splitting of a spectral line into a doublet reduces the central intensity of the line and therefore reduces the amount of saturation in the line. This in turn affects the ratio $R = W_\lambda(\mu)/W_\lambda(1.0)$, since R is smaller for a saturated line than for an unsaturated line. In order to allow for the difference between the behavior of $R(W_\lambda(1.0))$ in doublets and in singlets, we computed curves of $W_\lambda(\mu)/W_\lambda(1.0)$ as a function of $W_\lambda(1.0)$ for

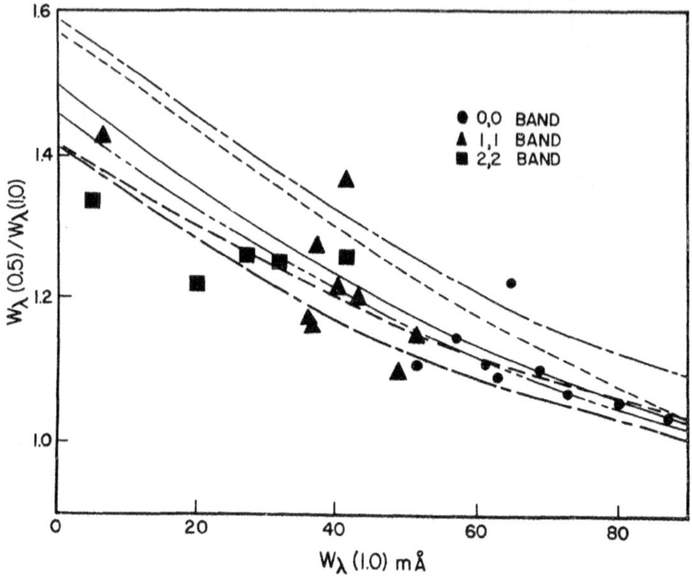

Fig. 7. A comparison between the observed values of $W_\lambda(\mu)/W_\lambda(1.0)$ for the CH lines at $\mu = 0.5$ and the values predicted by different models. Upper dash-dot line: Utrecht Reference Model; upper dashed line: Heintze's model; solid line: Model 10; dash-double-dot line: Mutschlecner's model; lower dashed line: three-stream model and anisotropic microturbulence; lower dash-dot line: three-stream model and isotropic microturbulence.

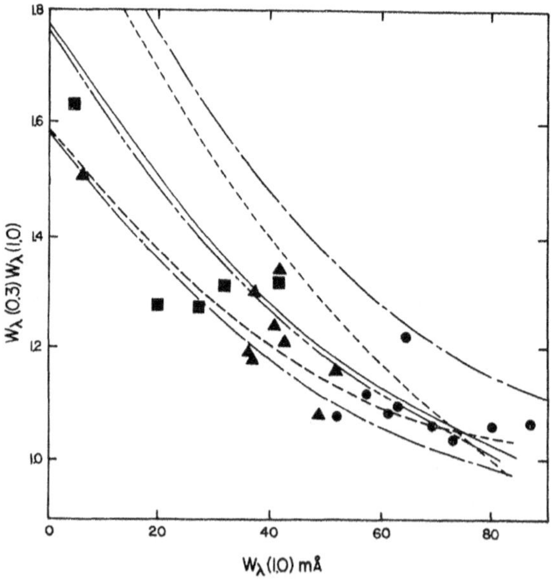

Fig. 8. Same as Figure 7, but for $\mu = 0.3$.

singlets and for each of the doublets given in Paper I. A comparison between the two sets of curves showed that each doublet of strength $W_\lambda(1.0)$ behaves like a singlet of strength $W'_\lambda(1.0)$. For the majority of the lines, the separation between the two components of the doublet is so small that $W'_\lambda(1.0) \approx W_\lambda(1.0)$. In Figures 7 and 8 are plotted the observed values of $W_\lambda(\mu)/W_\lambda(1.0)$ for the doublets as a function of $W'_\lambda(1.0)$.

In these figures, we have also drawn as curves the behavior of $W_\lambda(\mu)/W_\lambda(1.0)$ as predicted by different photospheric models under the assumption of LTE. Mutschlecner's model, Model 10, and the three-stream model explain the observations fairly well (also see Paper I). The Utrecht Reference Model and Heintze's model predict values of $W_\lambda(\mu)/W_\lambda(1.0)$ that are too large. Paper I showed that the CH observations can be accounted for with the three-stream model more closely if the assumed microturbulence model is changed to one which has a radial velocity of 1.0 km/sec and a tangential velocity of 1.5 km/sec. The curves computed with this turbulence model are also given in Figures 7 and 8.

D. CN

Figure 9 shows a plot of the CN observations similar to that for CH. All of the models tested predict values of $W_\lambda(\mu)/W_\lambda(1.0)$ that tend to be larger than the observed

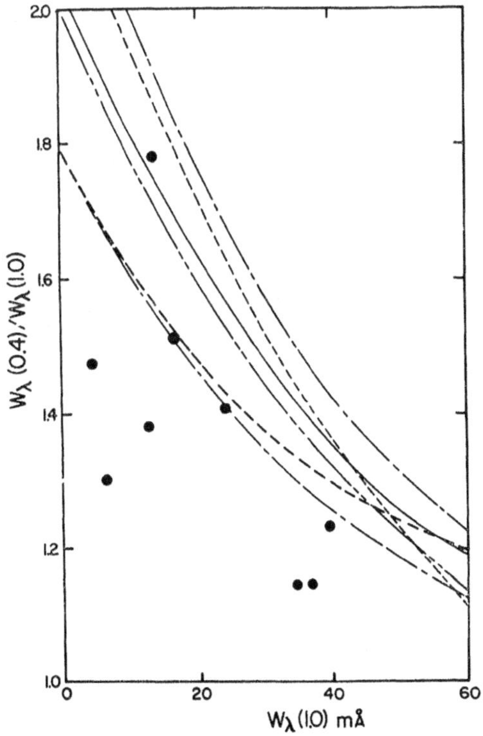

Fig. 9. Same as Figure 7, but for CN at $\mu = 0.4$.

values. This may or may not be significant, because of the unreliability of the CN equivalent widths. Accurate equivalent widths are difficult to obtain for these lines, since they fall in a region of the solar spectrum ($\lambda < 3840$ Å) that is very crowded with spectral lines. The two curves computed with the three-stream model are in better agreement with the observations than the other curves. The curve computed with this model and the isotropic microturbulence gives the best agreement.

E. MGH

The MgH lines are so weak that their center-limb variation can be considered to be independent of line strength. The observed values of the ratio $W_\lambda(\mu)/W_\lambda(1.0)$ for the MgH lines are plotted in Figure 10. In this figure we have also plotted lines to show

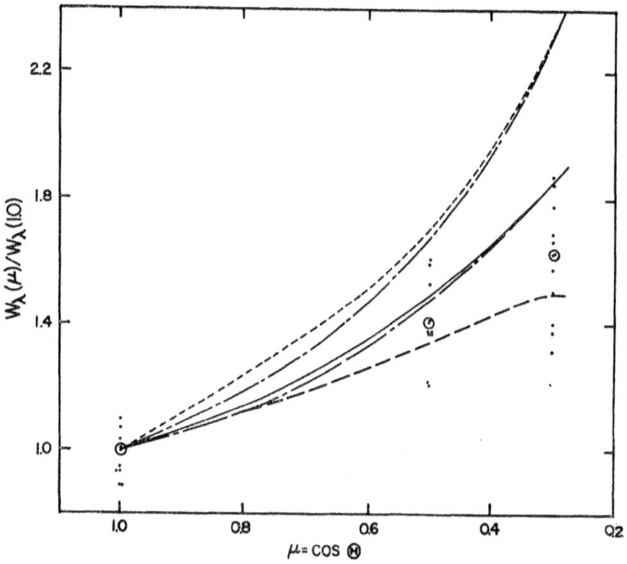

Fig. 10. A comparison between the observed variation of the MgH equivalent widths and the theoretical variations predicted by different models. Points: observations; circles: means of observed points; short dash line: Heintze's model; upper dash-dot line: Utrecht Reference Model; solid line: Model 10; lower dash-dot line: Mutschlecner's model; long dash line: three-stream model.

the results of computations made with the different photospheric models. The observations tend to fall between the curve predicted by Elste's three-stream model and the curves predicted by Model 10 and Mutschlecner's model. As has been the case with the carbon bearing molecules, the Utrecht Reference Model and Heintze's model give unsatisfactory results.

7. The Depth of Formation of Molecular Lines

It is of interest to know in which layers the lines of the different molecules are formed. Information on the depth of formation of spectral lines may be obtained from their contribution functions, which tell how much each layer contributes to the lines (see Paper I). Figure 11 shows a plot of the contribution functions for the centers of typical lines of C_2, CH, CN, CO and MgH. In general, molecular lines are formed in the atmospheric layers at optical depths between $\log\tau_{5000} = -2.5$ and $\log\tau_{5000} = -0.5$.

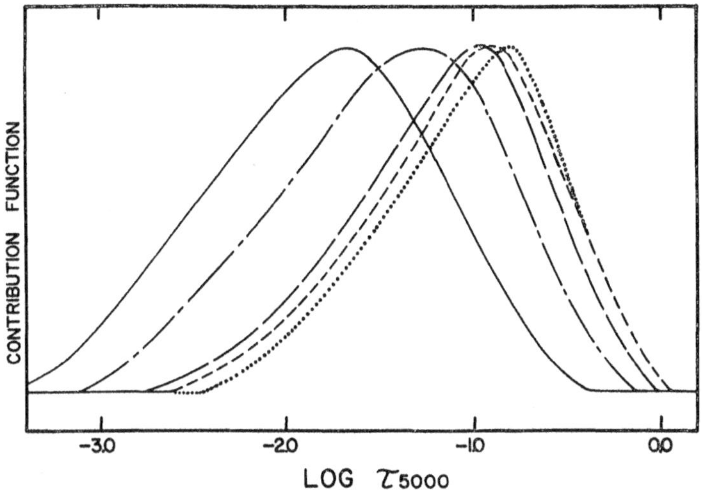

Fig. 11. Contribution functions for the centers of typical molecular lines for Model 10. Dotted line: MgH; short dash line: CH; long dash line: C_2; dash-dot line: CN; solid line: CO.

8. Discussion of the Results

The results presented in Figures 5 through 10 show that the different photospheric models have varying degrees of success in accounting for the center-limb behavior of the molecular lines. The three-stream model gives the most satisfactory results for C_2 and CO and accounts fairly well for the MgH observations. It also explains the CH observations, particularly when the anisotropic microturbulence is used. (The C_2, CO, and MgH lines are so weak that their center-limb variation is almost identical for the isotropic and anisotropic microturbulence models.) The CN observations are not explained very well by any of the photospheric models; however, as suggested above, this may result from the inaccuracy of the equivalent widths. New observations of the CN lines would be useful. In general the homogeneous models and the Utrecht Reference Model are not as successful as the three-stream model in accounting for the observed center-limb variation of molecular lines. These models, particularly the Utrecht Reference Model and Heintze's model, tend to predict equivalent widths that are too large at the limb. Elste's preliminary three-stream model better explains the observations of the molecular lines because, in contrast to the others, it contains fairly large inhomogeneities in temperature in the molecular line-forming layers.

Of the different homogeneous models tested, Mutschlecner's model and Elste's Model 10 appear to be the most satisfactory with Mutschlecner's yielding slightly better results. Both the Utrecht Reference Model and Heintze's model appear to be unsatisfactory for the upper layers of the photosphere. The possibility cannot be ruled out that the molecular lines can be explained with these models and departures from LTE. However, since the observations of molecular lines are fairly well explained by the three-stream model under the assumption of LTE, it does not appear to be necessary to introduce the difficulties of departures from LTE when analysing the behavior of molecular lines.

9. Conclusions

These results present evidence for the existence of significant inhomogeneities in the upper layers of the photosphere where molecular lines are formed, $-2.5 \leqslant \log \tau_{5000} \leqslant -0.5$. If the temperature minimum is located in these layers, as several investigators have suggested (Pagel, 1956, 1961; De Jager, 1963; Heintze, 1965), then our results suggest that inhomogeneities may be important in the region of the temperature minimum. Mutschlecner's model and/or Model 10 seems to represent the average physical conditions in the upper photosphere better than Heintze's model or the Utrecht model. The assumption of LTE appears to be adequate for explaining the behavior of molecular lines in the solar photosphere.

Acknowledgements

I wish to express my appreciation to Dr. G. Elste for the use of two of his unpublished photospheric models. I also would like to thank Dr. Leo Goldberg for his comments on this work. This work was supported by grant NsG-438 and contract NASw-184 with the National Aeronautics and Space Administration.

References

Cowley, C. R.: 1964, *Astrophys. J.* **139,** 1344.
De Jager, C.: 1963, *Bull. Astron. Inst. Netherl.* **17,** 209.
Edmonds, F. N. Jr.: 1962, *Astrophys. J. Suppl.* **6,** 357.
Heintze, J. R. W.: 1965, *Rech. Obs. Utrecht* **17,** 1.
Heintze, J. R. W., Hubenet, H., and De Jager, C.: 1964, *Bull. Astron. Inst. Netherl.* **17,** 442.
La Borde, G.: 1961, *Ann. d'Astrophys.* **24,** 81.
Müller, E. A. and Mutschlecner, J. P.: 1964, *Astrophys. J. Suppl.* **9,** 1.
Mutschlecner, J. P.: 1963, Thesis, University of Michigan.
Newkirk, G.: 1957, *Astrophys. J.* **125,** 571.
Pagel, B. E. J.: 1956, *Mon. Not. Roy. Astron. Soc.* **116,** 608.
Pagel, B. E. J.: 1961, *Astrophys. J.* **133,** 924.
Withbroe, G. L.: 1967, *Astrophys. J.* **147,** 1117.

SOURCE FUNCTIONS IN THE CORES OF INFRARED FRAUNHOFER LINES

C. DE JAGER
University Observatory and Space Research Laboratory, Utrecht, The Netherlands
and
L. NEVEN
Royal Belgian Observatory, Uccle-Brussels, Belgium

(Received 14 June, 1967)

If the profiles of all Fraunhofer lines were formed according to the mechanism of pure absorption (the L.T.E.-hypothesis) the source functions $S_\lambda(\tau)$ deduced empirically from the central intensities of these lines should all be identical, and equal to the blackbody function $B_\lambda(\tau)$. In order to examine this hypothesis the source functions $S_\lambda(\tau)$ were deduced empirically according to a method applied earlier by us (De Jager and Neven, 1967a) for some twenty infrared lines for which accurate centre-limb observations were available (De Jager and Neven, 1967b). The lines belonged to five different multiplets.

An example of such a determination is given in Figure 1 for the Si I multiplet $4p^3D$–$4d^3F^0$. Here the solid line gives the temperature-optical depth variation as assumed in the Utrecht Reference Photosphere 1964; the broken line gives this function for the Bilderberg Continuum Atmosphere. The source-function intensities have also been transformed into temperature values (these, then, have the meaning of *excitation* temperatures). The various symbols refer to results found from centre-limb observations of different lines from the *same* multiplet.

We have also computed the possible errors in the derived $T_{ex}(\tau_0)$ values. In order not to overload the figure these errors have been given for a few points only.

The figure shows that these errors appear to be large for 'large' optical depths ($\tau_0 > 0.1$) and small in the higher levels. The results for the Si I multiplet show that within the limit of errors the source function is identical either to the URP or to the BCA Planck functions for $\log \tau_0 \gtrsim -2$; it deviates in less deep layers.

Roughly the same results were derived for the other multiplets which we examined:

Si I	$4s^3P^0$–$4p^3P$	(Figure 2);
O I	3^5S^0 –3^5P	(Figure 3);
C I	$3s^3P^0$–$3p^3D$	(Figure 4);
Sr II	4^2D –5^2P^0	(Figure 5).

A summary of the results for these five multiplets is given in Figure 6. Altogether one may conclude:

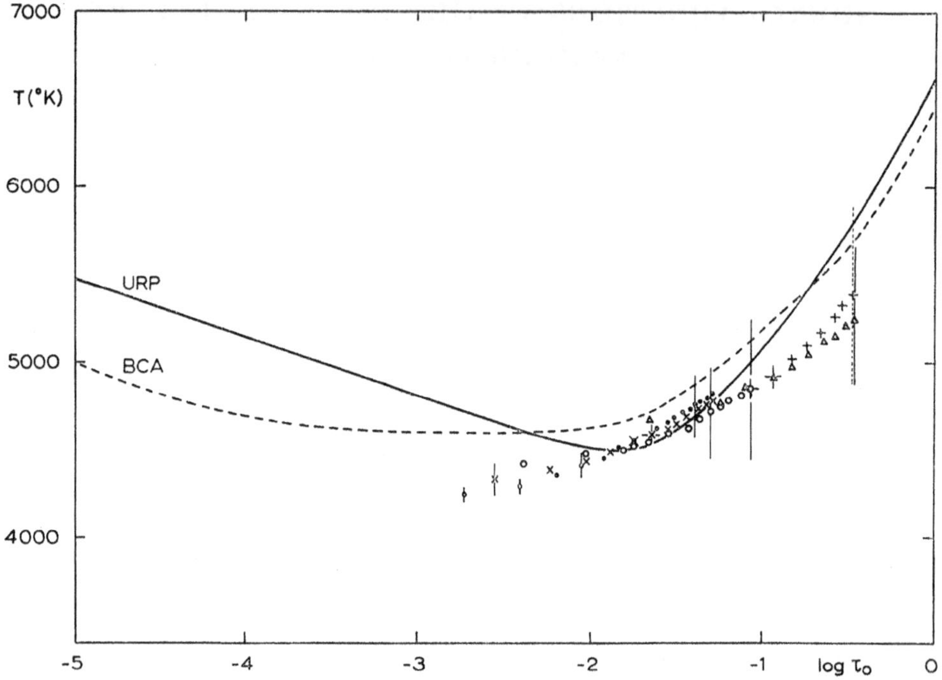

Fig. 1. Empirically derived source function from the centre-limb variation of the central intensities of lines from the Si I multiplet $4p^3D$–$4d^3F^\circ$. The solid line gives the temperature depth variation according to the URP 1964 model, the broken line represents the BCA 1967 model.

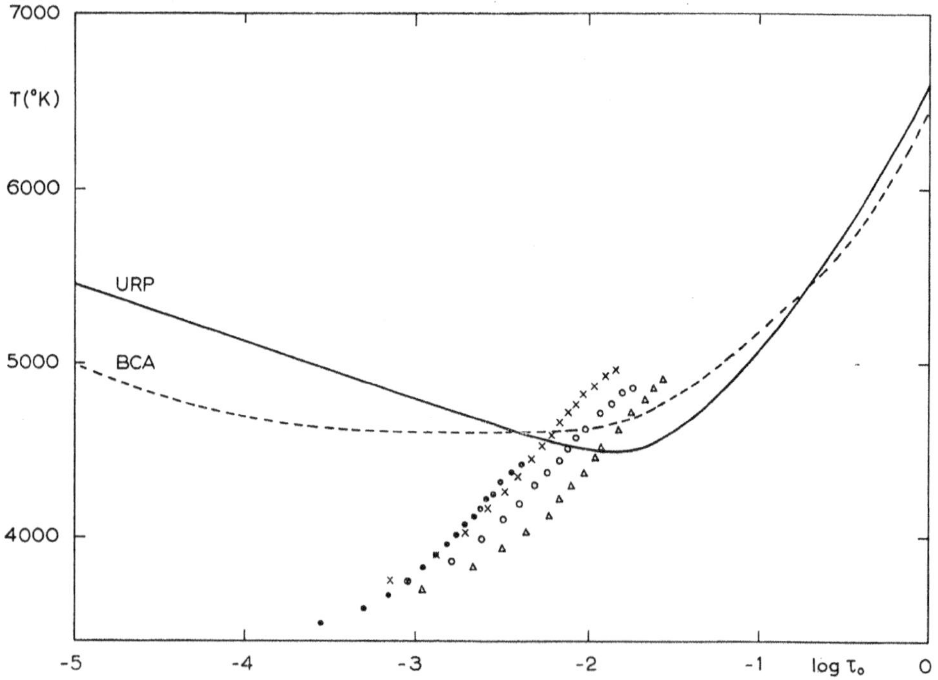

Fig. 2. The same as Figure 1 for the multiplet Si I; $4s^3P^0$–$4p^3P$.

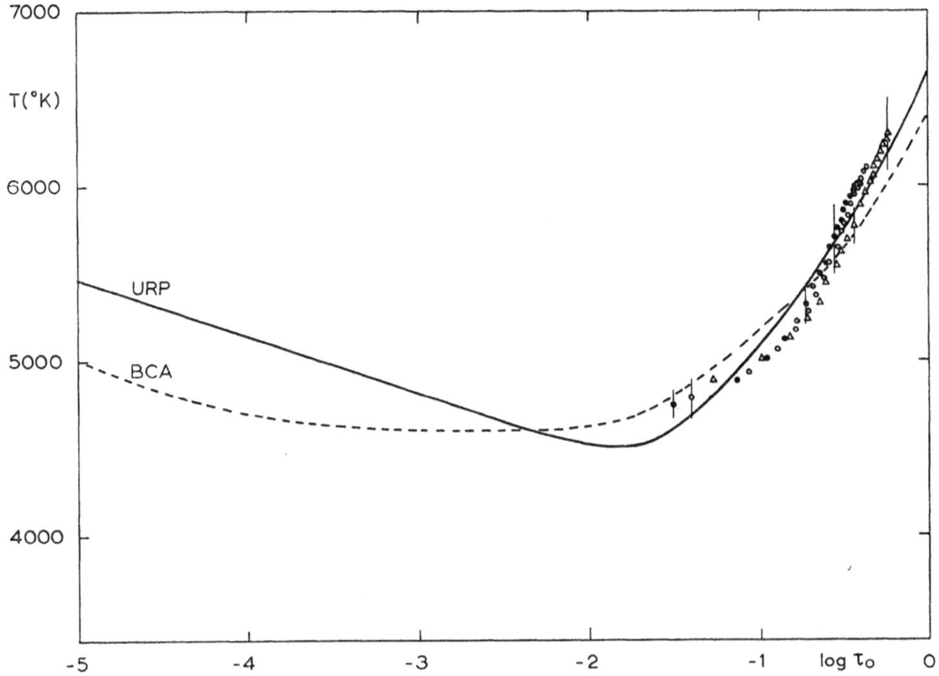

Fig. 3. The same as Figure 1 for the multiplet O I; 3^5S^0–3^5P.

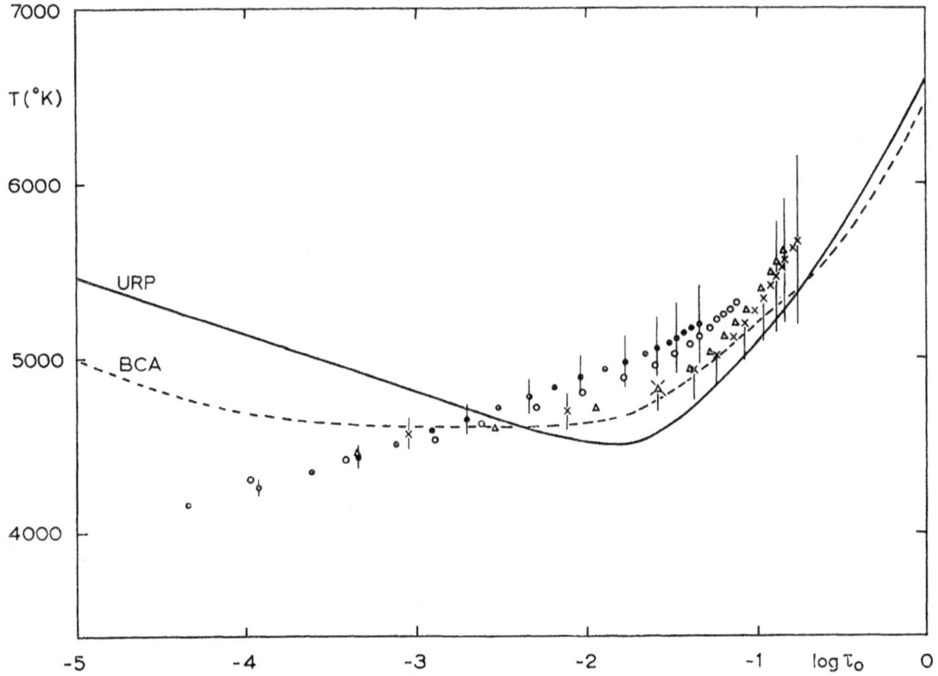

Fig. 4. The same as Figure 1 for the multiplet C I; $3s^3P^0$–$3p^3D$.

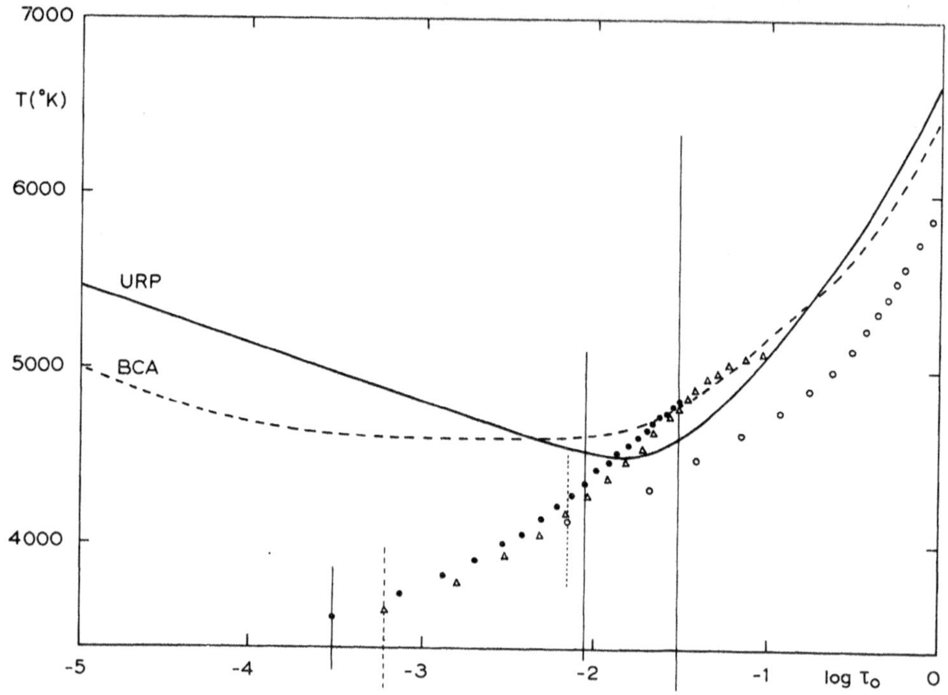

Fig. 5. The same as Figure 1 for the multiplet Sr II; 4^2D–5^2P^0.

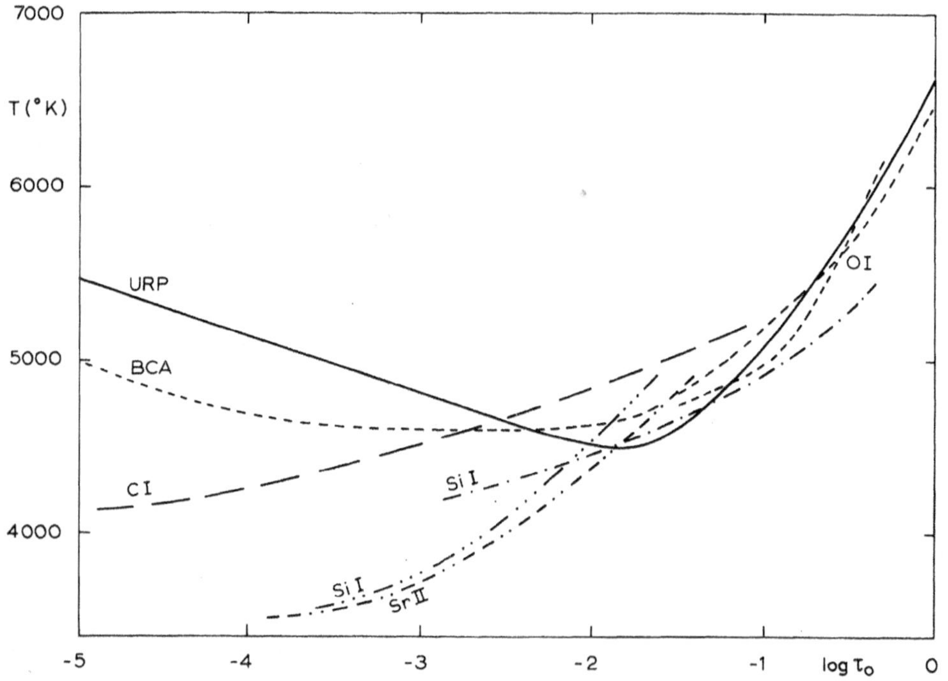

Fig. 6. Average excitation temperatures derived from centre-limb variations of the central intensities of infrared lines belonging to five different multiplets. These are compared with the kinetic temperatures for the URP and BCA models.

(1) The differences between $T(\tau_0)$ for the URP 1964 or the BCA 1967 models and the excitation temperatures for all multiplets are small or negligible for $\log\tau_0 \gtrsim -2$.

(2) There are important differences between the various excitation temperature functions found from the five multiplets for $\log\tau_0 \lesssim -2$. In our opinion this result proves that the hypothesis that the source function is identical to the Planck function cannot be correct in the uppermost photosphere and low chromosphere. Consequently this means that the derivation of the low chromospheric temperature from observations of central intensities of Fraunhofer lines on the basis of a L.T.E. assumption may lead to incorrect results.

References

De Jager, C. and Neven, L.: 1967a, *Solar Physics*, **1**, 27.

De Jager, C. and Neven, L.: 1967b, *Bull. Astron. Inst. Netherlands, Suppl. Series* **1** (no. 8), 325.

THE SOLAR H AND K LINES OF IONIZED CALCIUM

J. B. ZIRKER
University of Hawaii

(Received 7 June, 1967)

Abstract. Photographic observations of the mean H and K profiles are given for several positions across the solar disk ($0.1 < \mu < .6$). The core profiles of H and K are nearly identical at the same position on the sun.

Estimates of the mean Doppler width and the depth of formation of the K_2 emission peaks are given. A more elaborate analysis confirms that the K_2 peaks are formed in a relatively shallow layer ($\tau_0 \sim 3$) in which the Doppler width decreases rapidly inward. The relevance of these conclusions to current attempts to predict theoretically the H and K profiles is discussed.

1. Introduction

The H and K lines of ionized calcium are among the most sensitive indicators of chromospheric structure available to the astronomer. Their power in delineating the shape and vertical extent of solar atmospheric structures has been recognized since Hale's initial investigations. During the past few years, they have yielded important information on the structure of solar and stellar chromospheres. Thus, the concept of a highly inhomogeneous solar chromosphere, organized into columnar structures arranged along a 'coarse network', has emerged from careful studies of calcium spectroheliograms. A fascinating correspondence between the strength of the magnetic field and the intensity of the K-core emission has been established, principally by the Mt. Wilson investigators (BABCOCK and BABCOCK, 1955; LEIGHTON, 1959; HOWARD, 1959; SIMON and LEIGHTON, 1964). The oscillatory component of the chromospheric velocity fields has also been extracted from K-line observations (JENSEN and ORRALL, 1963).

In the hands of O. C. Wilson and his co-workers, the K line has become an important tool in the study of the atmospheres of G, K and M stars. Two important relations have been uncovered: the width of the K-emission core correlates with absolute magnitude (WILSON and BAPPU, 1957) and the brightness of the core declines with the age of the star (WILSON, 1963). These results give promise of new insights into the development of convection zones during the evolution of the cool stars.

If we are to extract maximum information from these solar and stellar K-line observations, we must have a thorough understanding of the origin of the K-emission core. An adequate theory for the dependence of the core profile upon such atmospheric parameters as temperature, density, turbulence, macroscopic motions, and magnetic field has still to be developed. Several explanations of the self-reversal of the core have been offered (GOLDBERG, 1964; SUEMOTO, 1963; JEFFERIES and THOMAS, 1960) but none has yet been generally accepted.

Nearly all the emphasis in current work is upon the prediction of solar H and K profiles, starting from some model atmosphere. The number of free parameters available in such a calculation is large, because the chromospheric model is very poorly known. Little, if any, attention has been given to the *analysis* of the observations, i.e. the attempt to extract relevant information about the chromosphere directly from the observations.

It is the purpose of this paper to present a new set of H and K observations and a preliminary analysis that sets limitations on any acceptable interpretation.

These observations were obtained photographically. They have the important virtues of extending close to the solar limb (to $\mu \simeq 0.1$) and of revealing the point-to-point variations of the profiles. Moreover the observations of H and K are precisely simultaneous, so that the effects of seeing are the same for each.

In the following section, the observations and their reduction are described. An analysis of the center-limb variation of the *mean* K profile follows. The results are discussed in the last section.

2. Observations

The data reported here are based on simultaneous, photographic observations of the H and K line profiles along a radius of the sun. The observations were made at the Sacramento Peak Observatory with the 12-inch coelostat and 13-m spectrograph. Table I gives some of the relevant parameters.

TABLE I

Observational Parameters

Date: July 10, 1964
Emulsion: IV-0 (70 mm)
Filter: Schott BG-1
Image diameter: 254 mm
Grating: 1200 line/mm, second order (dispersion = 3.54 mm/Å)
Slit width: 100 μ
Exposure time: 30 sec

During the exposures, the spectrograph slit lay along a solar radius in approximately a North–South direction and cut the solar limb near the South Pole. Figure 1 is a reproduction of the K-line spectrum. The solar limb is at the top of the figure; the chromosphere is visible in the line core as a bulge extending beyond the limb. The spectrum contains information on the variation of the K profile from the limb inward to a point corresponding to $\cos\mu = 0.7$. The spectra cover the intervals λ 3964–3988 and λ 3917–3938 at a dispersion of 3.54 mm/Å.

Considerable fine structure is visible in the K_2 red and K_2 violet emission peaks. Fine mottles as small as 2 sec of arc, and coarse mottles, of the order of 7 sec of arc, are clearly resolved. In addition, several arch-like features that connect the red and blue emission peaks are visible. White (1967) has given a convincing interpretation of these features in terms of columnar chromospheric structures.

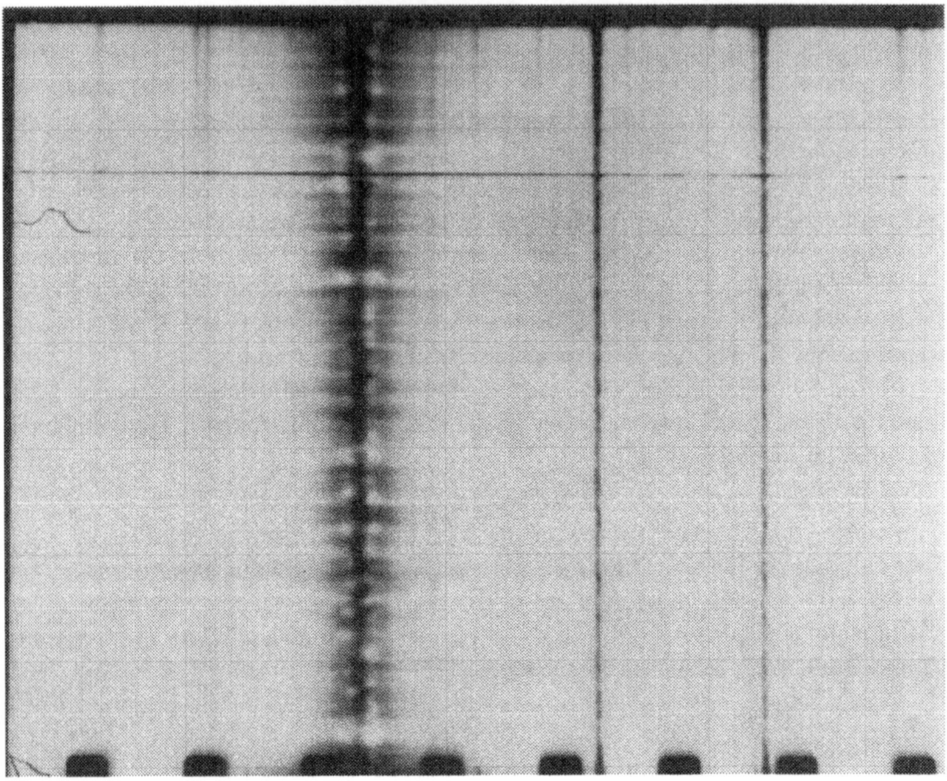

Fig. 1. Spectrogram of the core of the K line, made with a radial slit. The limb of the sun is at the top. Chromospheric emission appears beyond the limb at the line core.

In addition to the observations near the limb, a pair of calibration spectra (with exposure times of 30 and 15 sec) were obtained at the center of the disk with a nine-step wedge placed over the slit. Moreover, spectra of the disk center, covering the range $\lambda\lambda$ 3838–4020, were obtained the following day with a special long-frame camera back and with the step wedge in place. These latter spectra were used to determine the intensity, relative to the continuum, at several wavelengths near H and K.

As is well known, the true continuum is masked by the overlapping of many Fraunhofer lines in the spectrum below λ 4000 Å. In order to establish an absolute photometric scale in this spectral region, it would be necessary to compare the radiation of the sun at a given wavelength with a black-body source, and to make proper allowance for the transmission characteristics of the optical system.

Rather than undertake the extensive investigation required to establish an absolute photometric scale, I adopted the following procedure. The long-frame spectrum was traced on the Sacramento Peak microdensitometer and converted to a plot of relative intensity ($\log I$) vs. wavelength. In this conversion, a single characteristic curve was used (corresponding to λ 3954.3 Å). Previous tests had shown that the shape of the curve is independent of wavelength over the interval of interest. A straight-line envelope was then drawn to fit the highest intensity peaks (which occur at λ 4013 and λ 3909). This envelope was adopted as the continuum level. The intensity of any wavelength in the spectrum can now be related to the continuum intensity at that wavelength. In order to relate the intensities of the continuum at different wavelengths, I next assumed that the color temperature of the continuum below λ 4013 is 6500 °K. The variation of the continuum brightness is quite small ($\Delta \log I = .013$) over the wavelength interval from λ 4013 to λ 3909.

After carrying out this calibration procedure, HOUTGAST's (1965) preliminary absolute intensity scale came to my attention. In Table II, I compare the results of the above calibration procedure with those of Houtgast. The agreement is satisfactory. The K and H-line observations were traced on the SPO microphotometer in the direction normal to the dispersion (i.e. at a series of fixed wavelengths). Deflections on the tracings were converted into intensities in units of the continuum at the center of the disk at the wavelength of the center of K.

TABLE II

Comparison of Present Photometric Calibration with HOUTGAST's (1965)

	$\log I_\lambda / I_{4013.4}$	
λ	Houtgast	Zirker
4013.4	0.000	0.000
3.1	-.016	-.01
3999.9	-.002	-.01
93.4	-.013	-.02
90.8	-.021	-.03
88.1	-.025	-.03
77.3	-.131	-.12
60.4	-.179	-.17
57.4	-.102	-.09
54.2	-.081	-.05
50.7	-.081	-.05
45.7	-.130	-.10
40.5	-.281	-.26
26.8	-.277	-.25
23.8	-.162	-.12
14.8	-.057	-.02
11.5	-.057	-.01
09.2	-.057	-.01
01.3	-.072	-.02

These tracings at constant wavelength are, of course, limb-darkening curves and show the chromospheric fine-structure as fluctuations in intensity. Near the wavelengths of the K_2 and H_2 peaks, the fluctuations due to the fine structure are comparable with the overall darkening toward the limb. However, there is a striking variation of the *separation* of the H_2 and K_2 peaks toward the limb that is shared by all the fine-structure (see Figure 1). After some experimentation with the data, I decided to investigate this uniformity as a property of an *average* atmosphere, from which all the fine-structure has been smoothed.

The limb-darkening curves at the various wavelengths were therefore fitted, with a least-square procedure, by a quadratic form in $\cos\theta = \mu$;

$$I(\lambda, \mu) = a_\lambda + b_\lambda \mu + c_\lambda \mu^2 .$$

The position of the limb ($\cos\mu = 0.0$) was identified as the point of inflection of the limb-darkening curves at λ 3918 and λ 3986, at which wavelengths the intensity lies close to the continuum. From these smoothed limb-darkening curves, I constructed the mean H and K profiles illustrated in Figures 2 and 3 and tabulated in Table III.

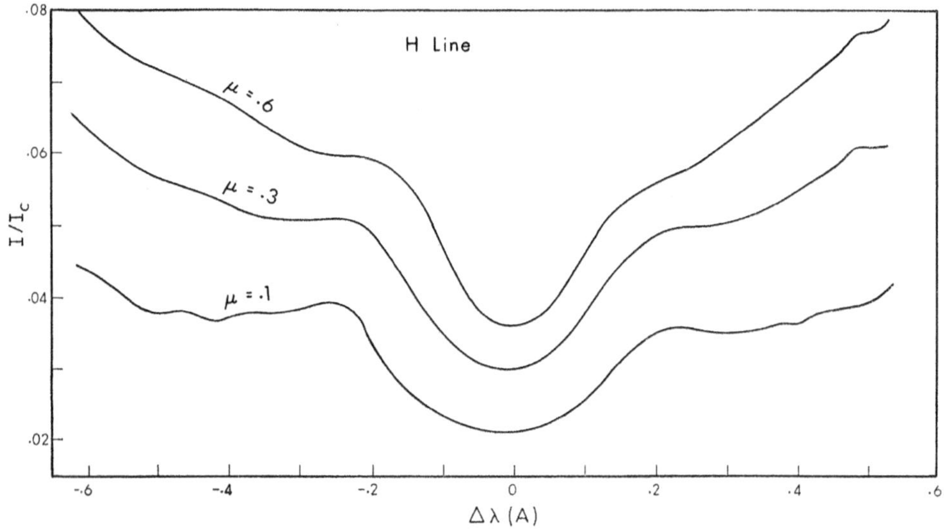

Fig. 2. Profiles of the H line across the solar disk. Intensities are in units of the continuum at H, at the center of the disk.

No correction has been applied to these data for light scattered by the grating into the line cores. The spectrograph was used in its normal single-pass mode during the observations. Previous comparisons between single and double pass observations (Evans and Waddell, 1962) suggested that the scattered light contribution to the present data might be as large as 1% of the continuum. It was therefore of considerable interest to compare the present observations with photoelectric, double-pass observations made with same spectrograph and grating by White and Suemoto (1967). The latter were kindly communicated to the author in advance of publication, and should be virtually uncontaminated with scattered light. Figure 4 compares the single-pass and double-pass limb-darkening curves at the cores of the H and K lines. The intensity units are the same for both curves, since White established the level of the continuum in essentially the same way that I did. The agreement is very good, which

TABLE III

H and K Line Profiles across the Solar Disk

H Line							K Line						
$\Delta\lambda$(Å)	μ=.6	μ=.5	μ=.4	μ=.3	μ=.2	μ=.1	$\Delta\lambda$(Å)	μ=.6	μ=.5	μ=.4	μ=.3	μ=.2	μ=.1
+.5	7.70	7.27	6.75	6.05	5.07	3.91	+.5	6.57	6.17	5.63	5.05	4.40	3.65
	7.50	7.11	6.58	5.90	4.96	3.86		6.40	6.02	5.50	4.95	4.35	3.66
	7.30	6.95	6.40	5.76	4.84	3.80		6.29	5.90	5.40	4.85	4.32	3.65
	7.09	6.75	6.25	5.65	4.72	3.77		6.14	5.77	5.30	4.76	4.28	3.65
+.4	6.85	6.57	6.07	5.50	4.61	3.65	+.4	5.95	5.64	5.20	4.71	4.20	3.60
	6.70	6.39	5.94	5.32	4.53	3.62		5.75	5.49	5.11	4.67	4.17	3.57
	6.50	6.20	6.80	5.20	4.44	3.55		5.57	5.36	5.07	4.70	4.20	3.60
	6.30	6.00	5.65	5.14	4.36	3.50		5.43	5.27	5.07	4.69	4.27	3.69
+.3	6.13	5.85	5.50	5.03	4.32	3.49	+.3	5.34	5.26	5.06	4.74	4.34	3.75
	5.94	5.73	5.43	4.97	4.30	3.50		5.27	5.27	5.08	4.80	4.39	3.79
	5.78	5.66	5.40	4.96	4.34	3.57		5.25	5.30	5.11	4.84	4.41	3.80
	5.69	5.60	5.34	4.94	4.36	3.50		5.25	5.30	5.15	4.85	4.39	3.77
+.2	5.57	5.50	5.23	4.83	4.25	3.46	+.2	5.28	5.25	5.10	4.75	4.26	3.64
	5.43	5.32	5.03	4.66	4.08	3.30		5.25	5.15	4.91	4.60	4.08	3.45
	5.25	5.07	4.80	4.45	3.75	3.20		5.15	5.00	4.70	4.35	3.81	3.23
	4.95	4.76	4.45	4.10	3.43	2.82		4.95	4.79	4.46	4.10	3.56	2.95
+.1	4.56	4.40	4.08	3.77	3.13	2.56	+.1	4.65	4.50	4.17	3.80	3.29	2.70
	4.20	4.00	3.70	3.45	2.89	2.38		4.39	4.20	3.93	3.55	3.07	2.50
	3.88	3.72	3.45	3.20	2.76	2.25		4.10	3.95	3.71	3.36	2.90	2.35
	3.65	3.54	3.25	3.05	2.68	2.17		3.94	3.80	3.56	3.25	2.80	2.26
.0	3.60	3.50	3.24	3.00	2.60	2.15	.0	3.87	3.75	3.50	3.23	2.80	2.25
	3.65	3.55	3.32	3.03	2.64	2.14		3.90	3.80	3.58	3.27	2.83	2.27
	3.88	3.70	3.45	3.10	2.70	2.15		4.05	3.96	3.72	3.34	2.90	2.33
	4.20	4.00	3.70	3.24	2.83	2.25		4.30	4.20	3.90	3.50	3.00	2.41
−.1	4.65	3.90	4.00	3.48	3.00	2.35	−.1	4.70	4.50	4.11	3.75	3.20	2.57
	5.18	4.75	4.38	3.83	3.27	2.53		5.10	4.86	4.45	4.05	3.46	2.75
	5.55	5.25	4.75	4.20	3.56	2.73		5.40	5.16	4.83	4.35	3.71	3.00
	5.79	5.55	5.25	4.54	3.85	2.98		5.65	5.40	5.07	4.65	4.01	3.80
−.2	5.90	5.70	5.40	4.85	4.23	3.35	−.2	5.75	5.54	5.27	4.85	4.31	3.65
	5.94	5.75	5.50	5.03	4.47	3.74		5.67	5.57	5.33	4.96	4.53	3.89
	5.95	5.80	5.50	5.07	4.55	3.87		5.55	5.50	5.31	5.02	4.60	4.08
	6.00	5.80	5.50	5.07	4.57	3.88		5.49	5.40	5.25	4.96	4.57	4.10
−.3	6.10	5.84	5.50	5.08	4.55	3.84	−.3	5.42	5.33	5.15	4.89	4.49	4.03
	6.24	5.90	5.55	5.05	4.50	3.80		5.44	5.27	5.10	4.78	4.38	3.94
	6.40	6.03	5.65	5.10	4.50	3.78		5.50	5.28	5.03	4.65	4.25	3.85
	6.55	6.20	5.77	5.18	4.50	3.75		5.59	5.31	5.00	4.58	4.15	3.75
−.4	6.70	6.35	5.90	5.27	4.55	3.70	−.4	5.70	5.38	5.05	4.60	4.10	3.65
	6.85	6.50	6.04	5.38	4.64	3.65		5.83	5.46	5.10	4.63	4.10	3.59
	6.97	6.63	6.13	5.50	4.70	3.77		6.00	5.60	5.20	4.70	4.15	3.53
	7.07	6.70	6.24	5.60	4.76	3.80		6.18	5.77	5.35	4.80	4.22	3.51
−.5	7.15	6.82	6.35	5.65	4.77	3.75	−.5	6.35	5.91	5.45	4.85	4.30	3.56
	7.28	6.95	6.47	5.77	4.90	3.84		6.54	6.05	5.58	4.95	4.39	3.65
	7.40	7.10	6.60	5.90	5.05	4.00		6.70	6.23	5.70	5.13	4.45	3.72
	7.57	7.25	6.79	6.10	5.21	4.19		6.85	6.40	5.84	5.23	4.55	3.75
−.6	7.82	7.48	7.00	6.81	5.39	4.32	−.6	6.95	6.48	6.00	5.30	4.57	3.25

suggests that the contamination by scattered light of the present observations is negligible. Also shown in Figure 4 are the observations of GOLDBERG *et al.* (1959).

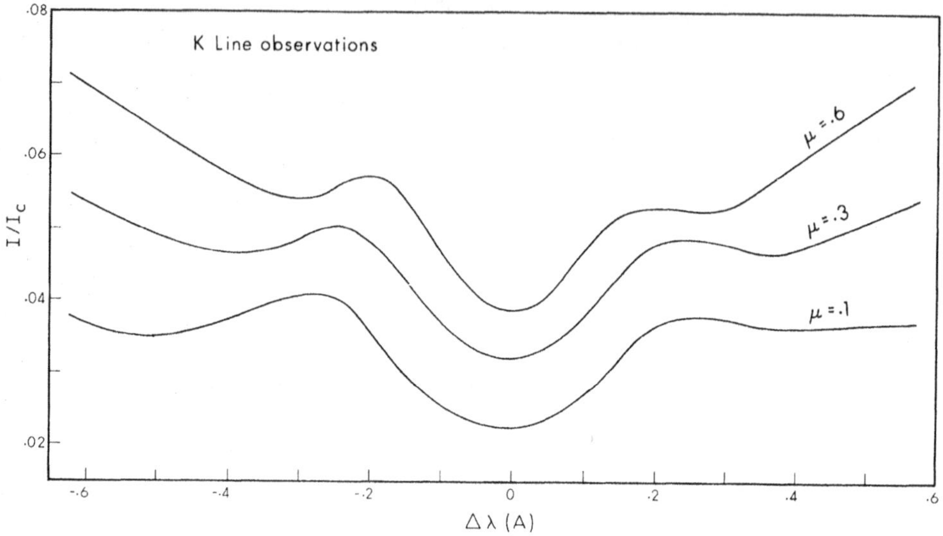

Fig. 3. Profiles of the K line. Intensities are in units of the continuum at K at the center of the disk.

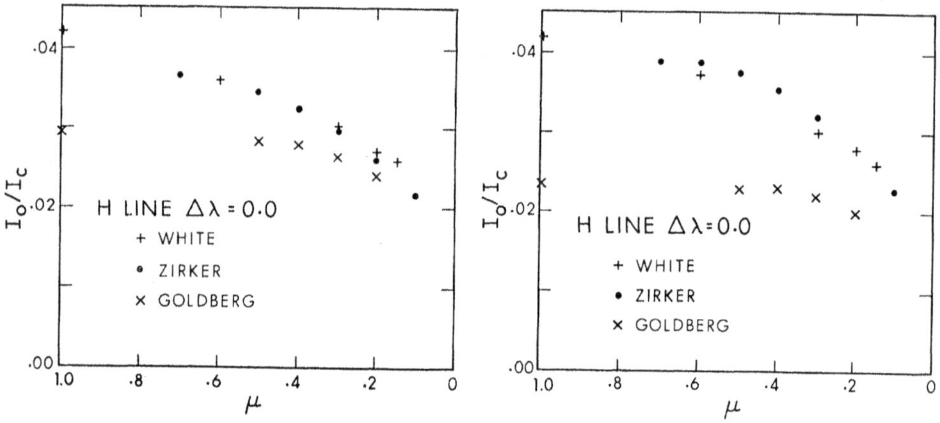

Fig. 4. The limb-darkening of the cores of H and K, as observed by three investigators.

These observations were evidently overcorrected for scattered light.

Since the slitwidth of 100 μ corresponds to only 0.0282 Å and no feature of the profiles is narrower than approximately 0.2 Å, no corrections for the apparatus function of the spectrograph were applied to the data.

3. Analysis

A. SOURCE FUNCTION INEQUALITY OF H AND K

It is instructive to apply WADDELL's (1962) test for source function equality at a fixed geometric depth to the H and K lines. Waddell showed that two lines in a multiplet, sharing a common ground level and having equal source functions at any geometric depth will have identical profiles at pairs of $\mu = \cos\theta$ in the ratio

$$\frac{\mu_1}{\mu_2} = \frac{g_2 f_2 \lambda_2}{g_1 f_1 \lambda_1}.$$

In the present application,

$$\frac{\mu_H}{\mu_K} = \frac{1}{2}.$$

Figure 5 compares the smoothed profiles of H at $\mu = 0.3$ and K at $\mu = 0.6$. The profiles are far from identical: K is displaced above H by about 1% of the continuum intensity in the core and its emission peaks (K_{2v} and K_{2r}) lie closer together than those of H. Similar disparities appear at other pairs of μ.

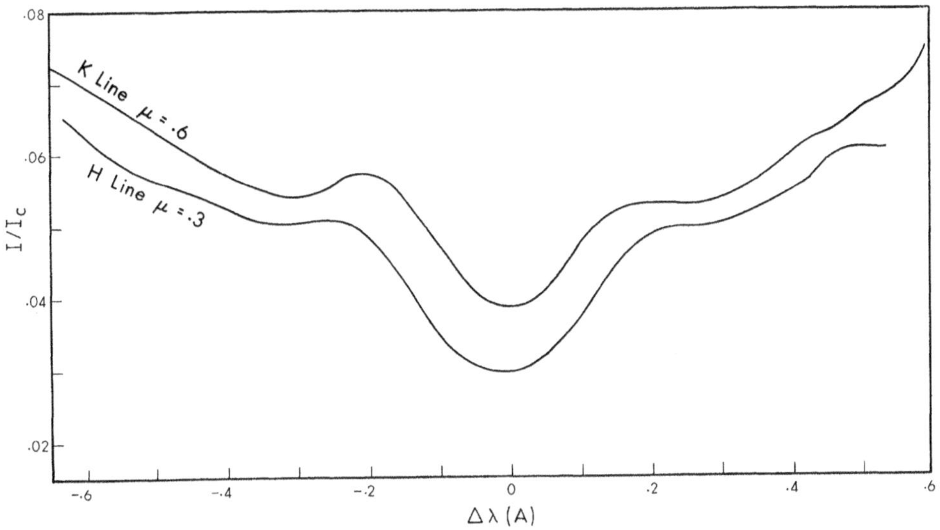

Fig. 5. Comparison of the profiles of $K(\mu = 0.6)$ and $H(\mu = 0.3)$. The lack of agreement suggests that the line-source functions are different at a given geometrical depth.

If, on the other hand, we compare H and K at the *same* position on the sun (e.g. $\mu = 0.6$) we find that the profiles are much more alike (see Figure 6). The discrepancy amounts to 0.25% of the continuum or less for displacements from the line centers less than 0.25 Å. Beyond the emission peaks, however, the profiles are clearly dissimilar.

This comparison suggests that H and K have nearly the same source function at the same *optical depth* (i.e., $S_H(\tau_H(\lambda)) = S_K(\tau_K(\lambda))$, unlike the NaD lines, which have the same source function at a given *geometrical* depth (Waddell, 1962)).

This result, while somewhat surprising, is quite insensitive to errors in the absolute

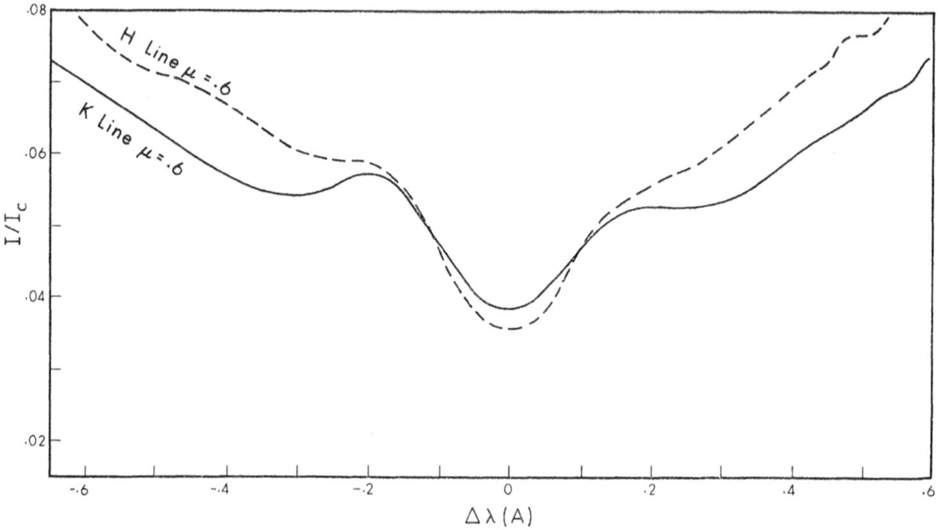

Fig. 6. Comparison of the profiles of $K(\mu = 0.6)$ and $H(\mu = 0.6)$. The agreement suggests that the line-source functions are identical at the same *optical* depths in the two lines.

photometric calibration. For example, in order to force the centers of the H and K profiles shown in Figure 5 to coincide, the level of the continuum at λ 3909 would have to be raised by 80%. While it is difficult to rule out completely such an error in my choice of the continuum level, the agreement with Houtgast's continuum makes such an error most improbable.

B. THE SOURCE FUNCTION OF THE K LINE

The bulk of this section is devoted to an attempt to extract from the K-line observations the run of the K source function with depth through a mean solar atmosphere. As a preface, it is useful to estimate the mean Doppler width of the K line absorption coefficient.

1. *Mean Doppler Width*

It is reasonable to assume that the absorption coefficient for the K line has a Doppler profile, at least out to some 2 or 3 Doppler widths from the line center.

We may estimate this width using the method of JEFFERIES and CURTIS (1965). We assume that the source function is independent of wavelength and depends only on geometric depth (or, equivalently, on the optical depth (τ_0) at the wavelength of the line center). We may determine the source function from the observations of the limb-darkening at any wavelength. Thus,

let

$$S(\tau_0) = a_\lambda + b_\lambda \tau_\lambda + \frac{c_\lambda}{2} \tau_\lambda^2$$

then

$$I(\lambda, \mu) = \int_0^\infty S\, e^{-\tau_\lambda/\mu}\, d\tau_\lambda/\mu = a_\lambda + b_\lambda\mu + c_\lambda\mu^2 .$$

Since the coefficients a_λ, b_λ, c_λ have been determined by the least-square fit to the limb-darkening observations at the wavelength λ, we may determine S in terms of τ_λ, the optical depth at this wavelength. Then, since S is a function of depth only, we may find corresponding values of $\tau_{\lambda 1}$ and $\tau_{\lambda 2}$, at which $S(\tau_{\lambda 1})=S(\tau_{\lambda 2})$. The slope of the relation between $\tau_{\lambda 1}$ and $\tau_{\lambda 2}$ is

$$\frac{d\tau_{\lambda 1}}{d\tau_{\lambda 2}} = \frac{\phi_{\lambda 1}}{\phi_{\lambda 2}} .$$

If we assume a Doppler profile for ϕ_λ, we may determine the Doppler width (Δ) as a function of depth for each wavelength pair. Table IV shows the Doppler widths obtained in this way.

TABLE IV

Doppler Widths

Wavelength pair	τ_0	$\Delta\lambda_D$(Å)
$\Delta\lambda=0.0$, $\Delta\lambda=.0565$	.39	.16
$\Delta\lambda=0.0$, $\Delta\lambda=.113$	.53	.175
$\Delta\lambda=0.0$, $\Delta\lambda=.170$	.71	0.200
Wavelength pair	$\tau_{.0565}$	$\Delta\lambda_D$(Å)
$\Delta\lambda=0.0565$, $\Delta\lambda=.113$	.47	.179
$\Delta\lambda=0.0565$, $\Delta\lambda=.170$	.62	.217

This method yields surprisingly large Doppler widths (of the order of 0.2 Å). The slight tendency for the width to increase inwards is probably not significant. Note that consistent widths are obtained from observations ranging from the line center out to the K_2 emission peak ($\Delta\lambda\approx 0.2$ Å). The method fails in the vicinity of the emission peaks.

These results suggest that the emission peak lies at about 1 or 1.5 Doppler widths from the line center, i.e., within the Doppler core of the absorption coefficient.

Additional preliminary information concerning the K line source function and an independent check on the mean Doppler width may be derived with a method devised by WHITE (1967). Assume that the K_2 emission peaks reflect the presence of a peak in the source function at some definite depth in the atmosphere (say $\tau_0=\tau_p$ in the optical depth at the line center). If the atmosphere is viewed at an angle θ to the normal, the emission peaks appear at a wavelength $\Delta\lambda_p$ (from the line center) such that the optical depth τ_λ down to the level $\tau_0=\tau_p$ has the value $\mu=\cos\theta$. This follows from the Barbier-Eddington relation, $I(\lambda, \mu)\simeq S(\tau_\lambda=\mu)$.

Thus, if the Doppler width is constant,

$$\tau_{\lambda p} = \tau_p \, e^{-(\Delta\lambda_p/\Delta)^2} = \mu$$

or

$$\ln \mu = \ln \tau_p - \left(\frac{\Delta\lambda_p}{\Delta}\right)^2 .$$

Figure 7 shows a plot of this relationship for the H and K lines. Observations of the profiles of individual fine-structures, at various positions on the disk, were used to construct this diagram. The slope of the line gives, in each case, a measure of the mean Doppler width (Δ). The intercept with the ln μ axis gives an estimate of $\ln \tau_p$.

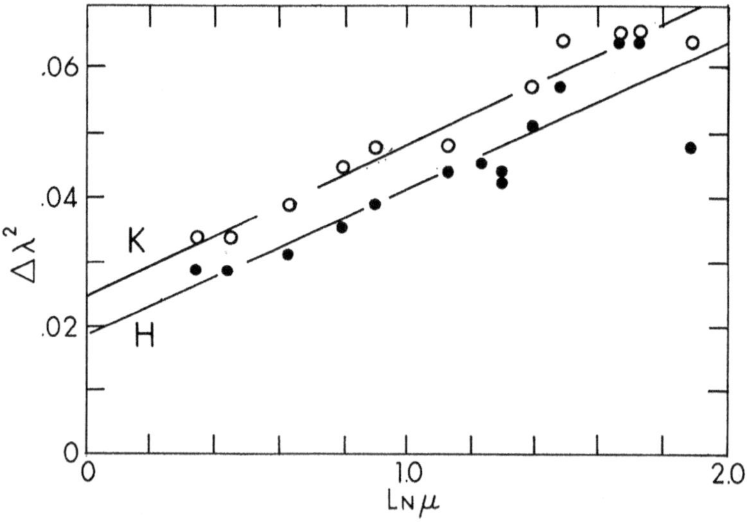

Fig. 7. Determination of the mean Doppler width and the depth of formation of the H_2 and K_2 emission peaks.

We find that $\Delta = 0.15$ Å,

$$\tau_p = 2.3 \text{ for the H line.}$$
$$\tau_p = 3.0 \text{ for the K line.}$$

This method gives a slightly smaller Doppler width than the preceding method, but the conclusion that the K_2 emission peak lies within the Doppler core is unmodified. The optical depth (τ_p) at which the peak in each source function occurs is astonishingly small in view of the great strength of the lines. ATHAY (1967) has suggested that White's method is extremely sensitive to small errors near $\mu = 1$, and yields gross underestimates of τ_p if the Doppler width varies rapidly in depth. As we shall see in the next section, this estimate of the optical depth of the source-function peak is confirmed by a more elaborate analysis.

2. *Depth Variation of the K Line Source Function and Doppler Width*

We have chosen to analyze the mean profiles of the K line in terms of a homogeneous, isotropic atmosphere. While the preceding analyses have yielded a useful estimate of the magnitude of the mean Doppler width in this atmosphere, we must recognize that the observations are not compatible with a constant (depth-independent) Doppler width.

We may show this directly. Assume that the Doppler width is constant. Then, the emergent intensity $I(\lambda, \mu)$ equals

$$I(\lambda, \mu) = \int_0^\infty S(\tau_0)\, e^{-\tau_0 \phi_\lambda/\mu} \left(\frac{\phi_\lambda}{\mu}\right) d\tau_0 .$$

Let $\ln \phi_\lambda/\mu = v$, $\ln \tau_0 = u$,

then

$$I(v) = \int_{-\infty}^{\infty} S(u) \exp\left[- e^{u+v} + u + v\right] du$$

$$I(v) = \int_{-\infty}^{\infty} S(u)\, \mathrm{K}(u + v)\, du .$$

If the Doppler width is constant, the emergent intensity is a function of ϕ_λ/μ, not of λ and μ separately. Thus, if the K_2 peak has an intensity I_p at $\mu = 1$, it must have the *same* value I_p at all $\mu < 1$. The wavelength at which the peak appears is fixed by the condition $v = \ln(\phi_\lambda/\mu) = \ln(\phi_{\lambda 1}/1.0) = \text{const.}$, where λ_1 is the wavelength of the peak at $\mu = 1.0$.

If we examine the mean K line profiles (Figure 3), however, we see that the intensity of the K_2 peak *decreases towards the limb*. Thus the Doppler width cannot be constant, but must vary with depth, angle, position, etc. If we attempt to analyze the profiles in terms of a homogeneous, isotropic atmosphere, we can only assume that the Doppler width is depth-dependent.

We have the following problem then: *given* a set of mean K profiles that correspond to different positions on the disk, *find* a depth-dependent source function and Doppler width that satisfy the observations.

After some preliminary experimentation with a perturbation method, the following iteration scheme was devised:

(a) Choose an initial depth-variation of the Doppler width. (In practice, I chose $\Delta = .15 = \text{const.}$)

(b) Use this Δ to extract the corresponding source function, $S(\tau_0)$, from observations at *one* position on the disk. (In practice, I chose the profile at $\mu = 0.6$, but supplemented it with the limb-darkening curve at the line-center.)

(c) Use this initial $S(\tau_0)$ to extract a new depth-variation of the Doppler width from the observations of the intensity of the K_2 peaks at various μ.

(d) Iterate, until the process converges. The final $S(\tau_0)$ and $\Delta(\tau_0)$ must predict the complete Doppler core at all other values of μ.

The details of the numerical procedures in each step are given in the Appendix.

The calculation converged after two iterations. Figure 8 illustrates the initial and second approximations to the source function and Doppler width.

Note that the source function has a maximum near $\ln \tau_0 = 1$, in accord with the prediction of White's method (see Section 3.B.1). The Doppler width *decreases* as we proceed inward into the atmosphere. One can show, from elementary considerations, that such a decrease is required to account for the decrease of the K_2 peak intensity toward the limb.

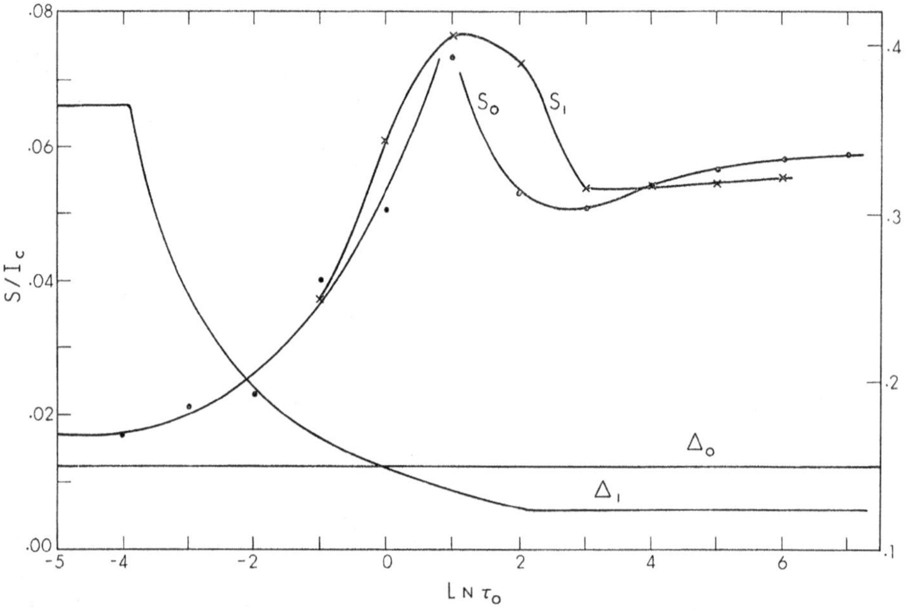

Fig. 8. Source function and Doppler width estimations.

Figure 9 shows profiles computed with the final source function (S_1) and Doppler width function (Δ_1) compared with the original observations. The representation of the data by these two functions is fairly good from the center to the emission peaks but is quite poor near K_1, especially near the limb. Evidently the iteration procedure described above has yielded an incorrect source function and Doppler width in the deeper layers ($\ln \tau \gtrsim 3$). This is hardly surprising, since no information regarding the shape of the absorption coefficient in these layers has been entered into the computation: the Doppler width was simply assumed constant ($=0.125$ Å). A more complete analysis, taking into account the variation of the Doppler width and the influence of the damping wings in the deep layers will be required to reproduce the observed K_1 intensities.

4. Discussion

The preceding analysis, while imperfect, has served to confirm that the K_2 emission peaks originate in relatively shallow layers ($\tau_0 \sim 3$), in a region in which the

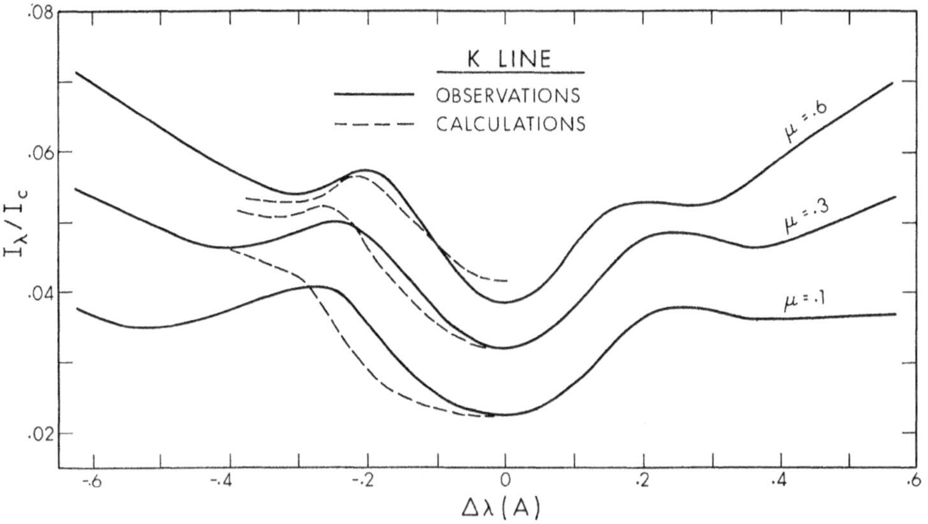

Fig. 9. Representation of the data by the source function and Doppler-width variations shown in Figure 8. The solid lines are the original observations.

Doppler width is large and increasing rapidly outward. It seems reasonable to identify this region with a steep chromospheric temperature rise.

Unfortunately, the magnitude of the Doppler width gives little direct information on the temperature. For example, at $\tau_K = 1$ the inferred Doppler width (0.15 Å) corresponds to a Doppler velocity of 11.4 km/sec. If the atomic motions are purely thermal, a temperature of 3×10^5 °K is indicated; but if non-thermal motions exist (equal to the velocity of sound) a temperature of 4500 °K is indicated.

Thus the most valuable clue on the chromospheric model arising from the preceding analysis is the low opacity of the chromosphere down to the region where K_2 is formed. Current attempts to predict theoretically the K profile begin with models that imply an opacity of 10^2 to 10^4 for this upper portion of the chromosphere (see a.o. Athay and Skumanich's paper in the present *Proceedings of the Bilderberg Conference; Solar Physics* **3**, 1968, No. 1). It seems clear that models with a steep temperature rise at an optical depth of $\tau_K \approx 3$ should be investigated.

We may estimate the amount of material down to this depth as follows. The optical depth is given by

$$\tau_p = \alpha_0 \bar{\chi}_{II} \left(\frac{N_{\text{Ca II}}}{N_e} \right) \int N_e \, dh = 8.3 \times 10^{-18} \, \bar{\chi}_{II} \int N_e \, dh .$$

Most of the contribution to τ_p takes place at such temperatures that $\bar{\chi}_{II}$ (the average fraction of calcium atoms that are doubly ionized) is approximately unity.

Thus if we require $\tau_p \approx 3$, then $\int N_e \, dh \approx 3.6 \times 10^{17}$. For example, if $N_e \approx 10^{10}$, the thickness of the calcium chromosphere can only be 360 km.

Whether such an optically thin chromosphere can, in fact, produce the required peak in the source function is a question that only calculations can settle. I urge the theoretical groups now investigating K profiles to attack this problem.

Appendix: Numerical Procedures Used in the Iteration Scheme

A. INVERSION OF THE EMERGENT INTENSITY INTEGRAL

Throughout the analysis, the convolution form for the emergent intensity

$$I(\lambda, \mu) = \int_{-\infty}^{\infty} S(\ln \tau_0)\, K(\ln \tau_0, \lambda, \mu)\, \mathrm{d} \ln \tau_0$$

was found most useful. In the first step of determining $S(\tau_0)$ with $\Delta = .15$ Å, the kernel (K) has the form

$$K(z) = \exp\left[- e^z + z \right],$$

with $\int_{-\infty}^{\infty} K\, \mathrm{d}z = 1$, $z = u + v$, in the notation given above.

A simple matrix inversion scheme was found to work satisfactorily. The emergent intensity was expressed as

$$I_i = I(v_i) = \sum K_{ij} S_j$$

where S_j is the source function at 10 equally spaced depths in the $\ln \tau_0$ scale ($-3 < \ln \tau_0 < 6$). Observations were chosen at values of $v_i = \ln(\phi_\lambda/\mu)_i = -u_i$ that place the maximum of the kernel (i.e. the contribution function) at the specified values of $u_j = \ln \tau_0$. Thus, three observations at $\Delta\lambda = 0$ ($\ln \mu = -3, -2, -1$) and seven observations at $\mu = 0.6$ ($\ln v = 0, -1, -2, \ldots, -6$) were chosen. I also assumed that the source function remains constant (and equal to its value at the end-point) outside the specific range $-3 < \ln \tau_0 < 6$.

Essentially the same procedure was used with a depth-dependent Doppler width. In this situation, the kernel is a function of λ and μ separately:

$$K(\lambda, \mu, \tau_0) = \frac{\tau_0 \phi(\lambda, \tau_0)}{\mu} e^{-\tau_\lambda/\mu}$$

where $\tau_\lambda(\tau_0) = \int_0^{\tau_0} \phi_\lambda\, \mathrm{d}\tau_0$, and $\phi_\lambda = e^{-[\Delta\lambda/\Delta(\tau_0)]^2}$.

As before, observations along the $\mu = 0.6$ profile were chosen at wavelengths $\Delta\lambda$ such that the kernel maximum fell at the specified values of $\mu = \ln \tau_0$. The equations specifying the source function are even worse – conditioned than with a constant Doppler width. As a result the source function could be determined simultaneously only at five optical depths: $\ln \tau = -3, -1, +1, 3, 5$. However, by using another set of points on the $\mu = 0.6$ profile, the source function at the intermediate depths (-2, 0, 2, 4, 6) were determined. The two solutions fit moderately well.

B. DETERMINATION OF THE VARIATION OF THE DOPPLER WIDTH

Knowing the source function $S(\tau_0)$ at some stage of the calculation, we wish to derive the corresponding Doppler width, $\Delta(\tau_0)$, that reproduces the observed K_2 intensities at each position (μ) on the disk.

At a fixed position (μ), the K_2 peak appears at some wavelength $\Delta\lambda_p$ in the emergent profile. Assume that the transformation between τ_λ and τ_0 is given by $\ln\tau_\lambda = \ln\mu + r_\lambda(\ln\tau_0 - \ln\tau_p)$ near the depth τ_p (measured at line center) where the peak in the source function exists. Here we assume that $\tau_\lambda(\tau_p) \approx \mu$, as required by the Barbier-Eddington relation and that r_λ is independent of τ_0. We determine r_λ, a free parameter, by forcing agreement between the observed K_2 intensity and the predicted emergent K_2 intensity:

$$I(\lambda(\mu), \mu) = \int S(\tau_0)\, K(\tau_0, r_\lambda, \lambda)\, \mathrm{d}\ln\tau_0 .$$

Thus, for each μ we find

$$r_\lambda = \frac{\mathrm{d}\ln\tau_\lambda}{\mathrm{d}\ln\tau_0} = \frac{\tau_0\, e^{-(\Delta\lambda/\Delta(\tau_0))^2}}{\tau_\lambda(\tau_0)},$$

and then

$$\Delta(\tau_0) = \Delta\lambda_p[\ln\tau_0 - (\ln\mu + r_{\lambda_p}(\ln\tau_0 - \ln\tau_p)) - \ln r_{\lambda_p}]^{-1/2} .$$

In practice, I limited the variation of $\Delta(\tau_0)$ to the range $-4 < \ln\tau_0 < 2$ and assumed Δ=constant (equal to the value at the end-point of the range) outside the range. A check on the method is the consistency of the $\Delta(\tau_0)$ derived from observations of $\Delta\lambda_p$ at different μ.

Acknowledgements

It is a pleasure to acknowledge numerous fruitful conversations with J. T. Jefferies and O. R. White on the subject of this paper. I also wish to thank H. Mauter of the Sacramento Peak staff for his help in obtaining the observations.

The research was supported by the National Aeronautics and Space Administration under grant #NRG 12-001-011.

References

Athay, R. G.: 1967, private communication.

Babcock, H. W. and Babcock, H. D.: 1955, *Astrophys. J.* **121**, 349.

Evans, J. W. and Waddell, J. H.: 1962, *Applied Optics* **1**, 111.

Goldberg, L.: 1964, *Astrophys. J.* **140**, 384.

Goldberg, L., Mohler, O. C., and Mueller, E. A.: 1959, *Astrophys. J.* **129**, 119.

Houtgast, J.: 1965, *Koninkl. Nederl. Akad. van Wetenschappen, Proc.* **B, 68**, #5.

Howard, R.: 1959, *Astrophys. J.* **130**, 193.

Jefferies, J. T. and Curtis, G. W.: 1965, *Proc. Second Smithsonian Conferences on Stellar Atmospheres*; *Smithsonian Ap. Obs. Report* #174.

Jefferies, J. T. and Thomas, R. N.: 1960, *Astrophys. J.* **131**, 695.

Jensen, E. and Orrall, F. Q.: 1963, *Astrophys. J.* **138**, 252.

Leighton, R. B.: 1959, *Astrophys. J.* **130**, 366.

SIMON, G. and LEIGHTON, R. B.: 1964, *Astrophys. J.* **140,** 1120.
SUEMOTO, Z.: 1963, *Proc. Jap. Acad.* **39,** 463.
WADDELL, J. H.: 1962, *Astrophys. J.* **136,** 231.
WHITE, O. R.: 1967, Private Communication.
WHITE, O. R. and SUEMOTO, Z.: 1968, *Solar Physics* (in press).
WILSON, O. C.: 1963, *Astrophys. J.* **138,** 832.
WILSON, O. C. and BAPPU, M. K. V.: 1957, *Astrophys. J.* **125,** 661.

EMISSION CORES IN H AND K LINES

I: *The Optically Thick Chromosphere*

R. GRANT ATHAY and A. SKUMANICH

High Altitude Observatory, Boulder, Colo., U. S. A.

(Received 1 June, 1967)

Abstract. Profiles of the H and K lines of Mg II and the K line of Ca II are computed using a two-level atom for five model atmospheres distinguished from each other mainly by the location of the temperature minimum. In the five models the temperature minimum and the chromospheric temperature are adjusted to give best agreement between computed and observed profiles. The parameters ε and r_0 are prescribed as functions of τ from a density model of the atmosphere. By comparing computed and observed profiles of the K_3, K_2 and inner K_1 components of the lines we determine both the approximate depth variation of $\Delta\lambda_D$ and the best of the temperature models. We find that the Doppler width increases rapidly with height in the chromosphere beginning from a value of 1.6 km/sec at $\tau_c \approx 10^{-2}$. This latter value corresponds closely to the thermal velocity of Mg atoms in the upper photosphere. The preferred temperature model is one for which the temperature minimum occurs near $\tau_c(\lambda\,2800) \approx 10^{-4}$–$10^{-5}$ with a value $T_{\min} \lesssim 4200°$ and which has a temperature near 7000° at $\tau_c = 10^{-6}$ where K_2 is formed. The intensity in K_3 is determined largely by $\mathrm{d}\Delta\lambda_D/\mathrm{d}\tau$ in the K_2 and K_3 regions.

1. Introduction

The faint self-reversed emission cores in the solar H and K lines of Ca II have long been a source of curiosity and speculation. Relatively few attempts have been made, however, to provide an explicit quantitative theory to explain these phenomena. The discoveries of similar, but more pronounced, emission cores in the H and K lines of Mg II and of self-reversal in the Lyman-α and Lyman-β lines in the solar spectrum, as well as the discovery of the Wilson-Bappu effect in stellar spectra, have prompted a more concerted effort to provide such a theory. Qualitative and semi-quantitative theories are no longer capable of furnishing adequate guidance to the understanding of self-reversal and its applications in observational astronomy.

In essence, those who have worked on the problem of self-reversal can be divided into three schools of thought typified by the following broad divisions: (1) those who would explain the reversals in terms of a mechanistic theory, such as a change from coherent to non-coherent scattering (cf. MIYAMOTO, 1957); (2) those who would explain them in terms of a Schuster-Schwarzschild type layer producing an emission core that is simply added to an independently formed monotonic absorption profile (cf. WILSON and BAPPU, 1957); and (3) those who would explain the profile in terms of a Milne-Eddington type model in which the profile is a direct mapping of the source function assumed to be a continuous function of height (cf. JEFFERIES and THOMAS, 1960). Most likely there are elements of truth in all three approaches, and, in point of fact, many of the apparent differences are superficial rather than fundamental.

Perhaps the most important distinctions between the three schools of thought are in the underlying assumptions. For example, group (2) assumes, either implicitly or explicitly, that there are in fact two distinct profiles formed in two distinct parts of the atmosphere. Usually, the profile is assumed to be made up of a normal absorption profile and a normal emission profile which show a composite self-reversal only because of the manner in which they add together. It is also usually the case that the emission core is assumed to be formed in discrete optically thin clouds overlying a more or less homogeneous medium. By distinction, group (3) usually asserts that the self-reversal results from the chromospheric temperature rise, essentially independently of any inhomogeneities in the chromosphere, and that the emission features are formed in layers where the opacity is moderately large.

No proposed explanation of the profiles has been both sufficiently accurate and sufficiently quantitative to serve as a basis on which to build a more complete theory. Our purpose in this paper is to provide such a basis. We treat the problem in a manner that appears to us to be the most direct, and we introduce just enough complexity to be consistent with the inherent difficulties of the problem and somewhat consistent with reality.

We begin by adopting a model atom with just two energy levels. Hence, we ignore all interlocking transitions, even between H and K themselves. We assume non-coherent scattering throughout the line. The model atmospheres used for computing profiles are spherically symmetric in all their properties, but we allow all of the parameters of the atmosphere influencing the line profiles to vary with depth. The basic method of calculating a profile follows that initiated by JEFFERIES and THOMAS (1960).

Our purpose, quite deliberately, is to introduce as few complications as possible in the model atom, but to make the model atmosphere as realistic as possible. We hope, in so doing, to provide a quantitative starting basis for improved models and improved calculations. We do not attempt to account for the more detailed features of the profiles, such as asymmetry in the emission peaks. On the other hand, as we shall show in the following discussion, we do obtain a first-order representation of the K_2 emission features and the inner portions of K_1 in both Mg II and Ca II with a realistic model chromosphere.

Parenthetically, we note that the resultant model, even though approached from a Milne-Eddington point of view, bears certain resemblances to a Schuster-Schwarzschild model. The role of the chromosphere in producing the self-reversed emission core is described primarily in terms of its temperature, density, and opacity at line center. Continuum opacity in the chromosphere enters only secondarily. Also, we note that similar attempts at fitting the Ca II profiles have been made by other authors (cf. JEFFERIES and THOMAS, 1960; LINSKY, 1966). Differences between our results and theirs arise from their neglect of one or more essential aspects of the model atmosphere and not from differences in the fundamental approach to the problem.

The ensuing calculations illustrate the effect of various atmospheric parameters on the amplitude and position of the emission peaks. Consequently, they offer a guide to the observed changes in the emission cores from point to point on the solar disk and they suggest the type of changes that may occur in different stellar atmospheres to produce the Wilson-Bappu effect, we discuss this latter application in a separate paper.

2. Method of Computing Profiles and Observed Profiles

In keeping with our attempt to minimize the complexity of computing a profile for the H or K lines, we restrict our discussion to the two energy levels involved in the line formation. This means that we treat H and K independently and ignore any coupling between the upper levels of the two transitions. Since, in reality, these levels are strongly coupled, we are ignoring an important feature of the problem. However, the coupling between the upper levels will affect primarily the relative intensities of H and K rather than the absolute intensities. By ignoring the coupling we tend to maximize any differences between H and K that may occur. We also ignore interlocking with other levels, such as the 3d levels of Ca II. The consequences of this are not entirely clear, but LINSKY (1966) reports that ignoring the 3d levels changes the computed intensities by no more than 1% of the continuum. Since the residual intensities in H and K of Ca II are of the order of 2 to 5% of the continuum intensity, this is not an insignificant error. We regard this and the coupling between excited p levels as refinements of the problem to be added at a later date.

By restricting the problem to two energy levels and ignoring the interaction of these levels with the continuum, we may write the frequency independent part of the line source function, S, in the form (JEFFERIES and THOMAS, 1960):

$$S = \frac{\int J_\nu \Phi_\nu \, \mathrm{d}\nu + \varepsilon B}{1 + \varepsilon} \tag{1}$$

where J_ν = mean intensity, Φ_ν = normalized absorption profile, $\varepsilon = C_{21}/A_{21}$, and B = Planck function.

This equation ignores stimulated emissions, which are not important in this particular problem. Equation (1) is combined with the transfer equation, using a method described in detail elsewhere (ATHAY and SKUMANICH, 1967, hereafter referred to as Paper I), to obtain

$$S = B + \frac{1}{\varepsilon + \delta} \frac{2}{\sqrt{\pi}} \int_0^\infty \frac{\phi_y}{\phi_y + r_0} \frac{\mathrm{d}H_y}{\mathrm{d}\tau_0} \, \mathrm{d}y . \tag{2}$$

The notation used is ϕ_y = the Voight absorption profile normalized to unity at line center, $y = \Delta\lambda/(\Delta\lambda_D(\tau_c = 1)) = \Delta\lambda/\Delta\lambda_D^1$, $r_0 = \mathrm{d}\tau_c/\mathrm{d}\tau_0$, $\delta = 2r_0 \int_0^\infty \phi_y/(\phi_y + r_0)\, \mathrm{d}y$, and H_y = the monochromatic flux at y, which depends upon S.

The mathematical procedure for computing a profile is briefly as follows: B, ε, r_0 and ϕ_y are all prescribed as functions of τ_0 or τ_c, whichever is more convenient. Equation (2) is then solved for S. S and B are combined to give the total source function

$$S_{t,y} = \frac{\phi_y}{\phi_y + r_0} S + \frac{r_0}{\phi_y + r_0} B , \tag{3}$$

which is used to compute I_y and H_y at $\tau=0$. Details of the method of solution are given in paper I and need not be repeated here.

The observational data that we wish to represent are shown in Figures 1 and 2. Figure 1 contains H and K profiles of Ca II in the undisturbed chromosphere (GOLDBERG, MOHLER, and MÜLLER, 1959) and in a plage (ZIRKER, 1965). Both the plage and

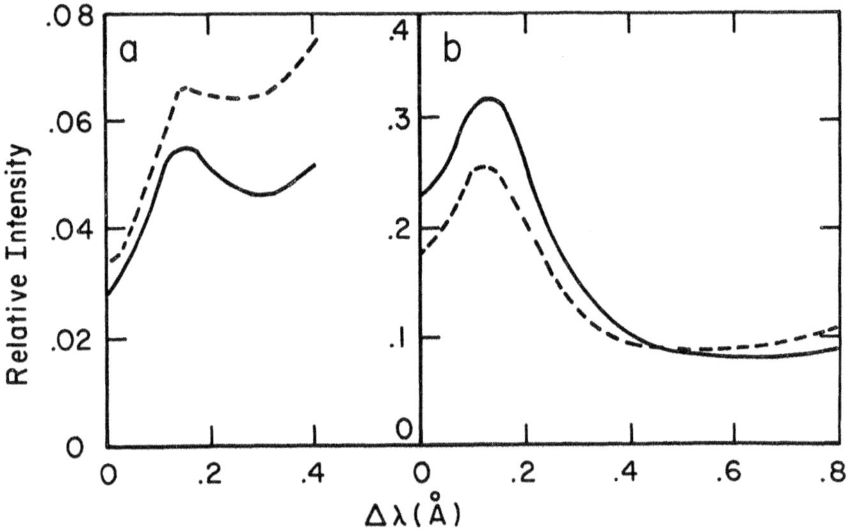

Fig. 1. Observed Ca II H and K profiles: (a) Center of quiet sun (GOLDBERG *et al.*, 1959), (b) plage (ZIRKER, 1965).

undisturbed chromosphere profiles vary considerably from point to point on the disk. We have picked those shown as being somewhat typical. We will, in fact, compute a representative set of active region profiles that more or less span the limits of the observed variations in profiles. Figure 2 shows the Mg II profiles averaged over the central third of the solar disk (PURCELL *et al.*, 1962).* We expect the Mg II profiles to vary considerably from point to point on the disk and in active regions. The Mg II profile shown, therefore, should be regarded as a composite of many rather diverse profiles and not as a profile truly representative of any particular region of the sun. On the date of the Mg II observations, 21 August 1961, the central regions of the solar disk contained several plages and these may have contributed substantially to the observed profiles. Also, corrections for scattered light have been made which are large at points near the saddle in the profile centered at about 0.6 Å from line center. Thus, these regions of the profiles are relatively inaccurate compared to the remainder of the profiles.

The spectral resolution with which the Mg II profiles were observed, ~0.03 Å, is

* Relative intensities within the profile were kindly supplied by Dr. K. Widing of the Naval Research Laboratory, and intensities relative to the local continuum were obtained from the absolute intensity measurements of WILSON *et al.* (1954) and from an assumed continuum flux of 60 ergs cm^{-2} sec^{-1} $Å^{-1}$.

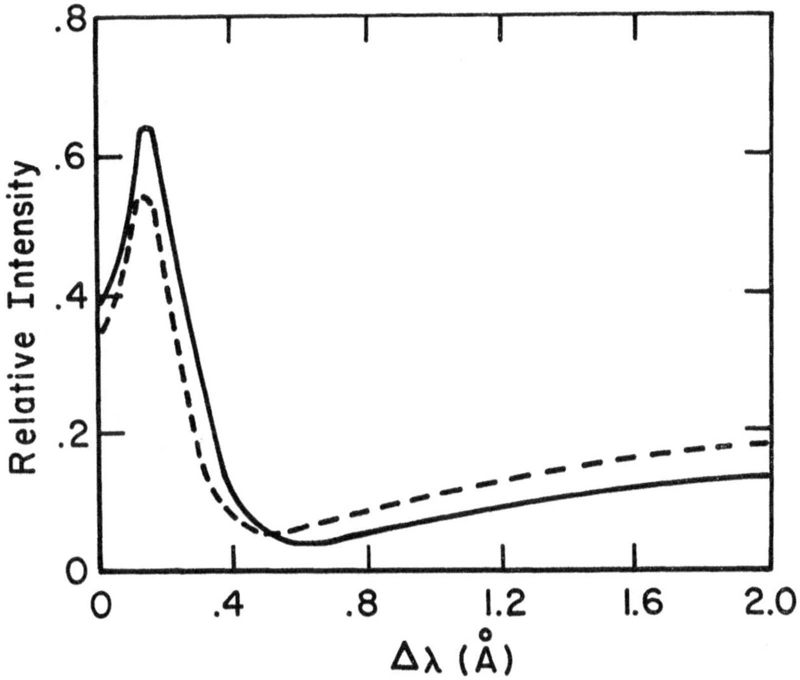

Fig. 2. Observed Mg II H and K profiles averaged over the central third of the solar disk on August 21, 1961, when several plages and sunspots were present (PURCELL *et al.*, 1962).

considerably narrower than either the K_3 core or the K_2 peaks midway between the K_{12} minimum and K_2 maximum. Hence, we do not expect any pronounced distortion of the profiles resulting from instrumental effects, but we do expect significant distortion from the poor spatial resolution. We have arbitrarily adopted a mean K_2 peak intensity of 60–65% for fitting purposes.

It is a point of considerable interest, as pointed out by GOLDBERG (1965), that the H and K relative intensities are reversed in Mg II as compared to Ca II on the undisturbed disk. In the plage reported by Zirker the relative intensities of H and K of Ca II are of the same sense as for the average disk for Mg II and are reversed from the Ca II profiles for the normal disk regions as reported by GOLDBERG *et al.* (1959). The Ca II profiles given by Goldberg *et al.* are measured relative to the continuum at λ 4000 rather than at the wavelengths of the lines. The relatively large separation of H and K for Ca II ($\sim$33 Å) provides a distinct possibility that the continuum may change appreciably between the two lines. This, plus the presence of numerous other absorption lines in this spectral region suggests that one should regard the published results with considerable caution.

We note at this point that if the source functions were identical in H and K the H profile would lie above the K profile in the K_3 core, intersect the K profile near the K_2 maximum, fall below the K profile on the declining part of K_2 farthest from line center, and again rise above the K profile in the K_1 region. This point was noted by GOLDBERG (1965), who also noted that this never seems to happen in the solar profiles. The assumptions made in this paper lead necessarily to an inequality between the source functions for the H and K lines in the K_3 and K_2 regions because of the difference in r_0, hence in optical thickness of the chromosphere for the two lines. The inequality is such that the H profiles always falls below the K profile in K_3 and K_2. Deeper in the atmosphere where the K_1 profile is formed the source functions both approach B, and therefore become equal. This equality brings the H profile back above the K profile in the K_1 region. In all cases our computed profiles agree with the types of profiles shown in Figures 1b and 2 with respect to the relative intensities of H and K. As already noted, the neglect of collisional coupling between the upper levels of the H and K transitions, which would tend to equalize the source functions if included, tends instead to maximize the difference between the source functions. Zirker's results show that the H profile falls below the K profile by about 25% throughout K_3 and K_2. In Mg II the observed profiles differ in intensity by about 20% at the K_2 peaks. The computed Mg II profiles differ by somewhat more than this amount when we reduce the chromospheric opacity by a factor 2 for the H line. This is indicative of the fact that we have overemphasized the differences between K and H by ignoring the coupling between the excited levels.

As might be expected, there is considerable freedom in the parameters that give satisfactory agreement with the observed profiles. Also, rather different sets of parameters that approximate the Mg II K profile tend also to give an acceptable fit to the Ca II K profile. Thus, with the crude data available one does not really gain much in the way of added discrimination by using the two sets of profiles. Furthermore, a set of parameters that fits the K profile of Mg II fits the H profile about equally well. The same is not true in Ca II because of the unexplained behavior of the H profile relative to K. We have, in effect then, just one profile to fit and from which we must determine all of the free parameters of the calculations. Because of the greater amplitude of the K_2 peak in Mg II, we deal primarily with the Mg II K profile.

3. Choice of Parameters

Of the four quantities B, ε, r_0 and ϕ_y only three are truly independent. In principle, at least, r_0 is determinable from the other three, assuming that abundances are known. However, even the three independent quantities permit far too many degrees of freedom if they are each allowed to vary arbitrarily with depth. It is necessary by some device to restrict the adjustable parameters to a number that is consistent with the information contained in the observations. We choose five points in the profile to describe its shape: the K_3 minimum, the K_2 peak, the K_{12} minimum, a point midway between K_2 and K_{12} and a point in K_1 at $\Delta\lambda = 2$ Å. To specify these five points we need seven quantities, intensities at the five points and the values of $\Delta\lambda$ at K_2 and K_{12}. Hence, we cannot allow more than seven adjustable parameters in the model atmosphere.

In order to make an intelligent choice of parameters it is necessary to understand their effect upon the nature of the profile, which we now briefly review. JEFFERIES and THOMAS (1960) showed that when $B(\tau)$ goes through a minimum near the top of the photosphere followed by a chromospheric temperature rise S tends to parallel B through the region of temperature minimum and to go through a maximum in the vicinity of initial temperature rise. This maximum in S supposedly produces the K_2 maximum in I_y at a particular value of y. The amplitude of K_2 depends mainly upon four factors: (a) the amount by which the temperature rises, (b) the coupling efficiency between S and B, (c) the optical thickness of the chromosphere, and (d) the shape and depth variation of ϕ_y. The coupling efficiency between S and B is measured by ε, which is directly proportional to electron density and depends only slightly upon temperature. For large densities, ε is relatively large and efficient coupling takes place. Hence, the amplitude of K_2 is strongly affected by $B(\tau)$ and $\varepsilon(\tau)$, both of which occur explicitly in Equation (1). The effect of ε represents a direct coupling between the radiating atoms and the free electrons, i.e., the thermal energy of the plasma.

An added source of coupling between S and B, hence an added effect on the K_2 amplitude, arises through the optical thickness (at line center) of the chromosphere, τ_0^*. This coupling can be thought of in terms of the mean number of scatterings experienced by a photon before it escapes the atmosphere. Photons generated at depths where the line opacity is large but the continuum opacity is small may escape the atmosphere either by diffusion in τ space or by diffusion in frequency space. The thermalization length, i.e., the number of scatterings required before the photons are destroyed by superelastic collisions, is given by $\mathbf{a}\varepsilon^{-2}$ for $a \geqslant \varepsilon$, Hence, when $\tau_0^* < a\varepsilon^{-2}$ most of the photons escape, and S is less than B. If diffusion in frequency is ineffective, i.e., $a < \varepsilon$, then most of the photons escape by diffusion in τ space and S/B is controlled mainly by ε. However, if diffusion in frequency is effective $(a > \varepsilon)$ so that photons escape primarily by this means, then S/B becomes a function of the parameter controlling the diffusion in frequency, i.e., $\mathbf{a}$.

To escape by diffusion in frequency photons must reach a value of y such that $\phi_y \leqslant (\tau_0^*)^{-1}$. The effectiveness of this diffusion is controlled by the shape of ϕ_y while the distance they diffuse is controlled by the shape of ϕ_y and τ_0^*. Hence, S/B becomes a function of r_0 since $\tau_0^* = f(r_0)$, as well as $\mathbf{a}$ and ε. Diffusion is most effective for a line with strong wings (large a) or for large r_0 (small line opacity). Thus, a large r_0 or broad wings on ϕ_y depress K_2 and small r_0 or narrow ϕ_y enhances K_2.

There is also a coupling between the diffusion in τ and y spaces and this introduces an effect on S/B from gradients in ϕ_y and r_0 with τ. For the moment, we ignore the effect of gradients in r_0 on S/B and consider the effect of gradients in ϕ_y on both S/B and K_2(max).

The effects on the K_2 amplitude of B, ε, r_0 and $\phi_y(a)$ so far discussed are direct effects in the sense that they determine $S(\tau)$, and they depend only on the local values of these parameters. There is a further effect of ϕ_y on K_2 of a less direct and more subtle nature. Both the value of S and the mapping of $S_{t,y}(\tau)$ into I_y by the processes of radiative transfer depend upon changes in ϕ_y with depth. If, for example, $d\phi_y/d\tau$

is negative through the regions of the atmosphere where S is near its maximum, two effects are present. S_{max} increases somewhat tending to enhance $I(K_2)$ but I spreads out in wavelength tending to weaken K_2 and enhance K_3 and K_{12}. Here we note only that the two effects exert opposing influences on $I(K_2)$ and we defer further discussion to the following section.

The influences of ε and B on the K_3 intensity are somewhat similar to their influences on K_2. However, εB is essentially determined by the K_2 intensity. Thus the only freedom we have with K_3 is in the choice of r_0 which controls the optical thickness of the chromosphere and in $\phi_y(\tau)$ which controls the greyness of the chromosphere. The optical thickness of the mean chromosphere is known within reasonable limits and we do not feel justified in varying r_0 enough to fit the K_3 profiles by adjusting this parameter alone. Thus, we use $\phi_y(\tau)$ to regulate K_3. We do, however, consider the influence of varying r_0 in Section 5, where we discuss the profiles in active regions.

Thus far we have noted five effects on the K_2 amplitude: the absolute values of B, ε and r_0 in the chromosphere, the bandwidth of ϕ_y, which is determined by the Voigt parameter **a**, and $d\phi_y/d\tau_0$. Hence, we must use at least two parameters for ϕ_y and if we use one each for ε and r_0 we are left with only two parameters to describe $B(\tau)$. Hopefully, we would like to conclude something about the location and value of the temperature minimum and a value of T for at least one point in the chromosphere. This requires a minimum of three parameters in $B(\tau)$. Since it is possible to estimate values of ε, r_0 and **a** from a density model of the atmosphere without knowing $B(\tau)$, it is tempting to resort to this and to use all of the six adjustable parameters in $B(\tau)$ and $d\phi_y/d\tau$. However, the wisdom in this depends upon how strongly ε, r_0 and **a** influence the final profile.

The principal influence of ε on the line profile is through the coupling efficiency between S and B. On the other hand, r_0 and **a** affect both the opacity of the chromosphere, hence the approach of S to B, and the mapping of $S_{t,y}$ into I_y. Thus, r_0 and **a** have a considerable stronger influence on the shape of the profile than does ε. Furthermore, ε is perhaps more accurately known than r_0. The sensitivity of the profiles to **a** and r_0 requires that we retain at least one free parameter for each of these quantities, and we need at least two parameters for $d\phi_y/d\tau$. However, we adopt fixed values of ε in order to permit three degrees of freedom in $B(\tau)$.

Values of ε based on densities given by THOMAS and ATHAY (1961) are given in Table I. These were computed for collisional cross-sections for Ca II and we assume the same cross-sections, hence the same ε's for Mg II. The results in Table I may be approximated by

$$\varepsilon = 10^{-2}\,\tau_c^{1/3} + 2\times 10^{-4}\,. \tag{4}$$

For ϕ_y we adopt

$$\phi_y = e^{-y^2 D^{-2}} + \frac{aD}{\sqrt{\pi}}\,y^{-2} \tag{5}$$

TABLE I

Estimated Values of ε and r_0

Approx. *Ht* (km)	$\tau_c(\lambda\ 3900)$	$\frac{d\tau_c}{dh}$ (cm^{-1})	Ca II $\frac{d\tau_0}{dh}$ (cm^{-1})	Ca II τ_0	Ca II r_0	Mg II $\frac{d\tau_0}{dh}$ (cm^{-1})	Mg II τ_0	Mg II r_0	ε
−300	1	10^{-7}	1	10^{7}	10^{-7}	2×10	2×10^{8}	10^{-8}	10^{-2}
100	10^{-3}	10^{-10}	10^{-2}	10^{5}	10^{-8}	2×10^{-1}	2×10^{6}	10^{-9}	10^{-3}
500	10^{-5}	10^{-12}	10^{-4}	3×10^{3}	10^{-8}	2×10^{-3}	6×10^{4}	10^{-9}	5×10^{-4}
1250	10^{-8}	10^{-15}	10^{-6}	10^{2}	10^{-9}	2×10^{-5}	2×10^{3}	10^{-10}	3×10^{-4}

where

$$D = \frac{\Delta\lambda_{\mathrm{D}}(\tau_c)}{\Delta\lambda_{\mathrm{D}}(1)} = 1 + \zeta\, e^{-(\delta\tau_c)^Z}. \tag{6}$$

D gives the depth dependence of $\Delta\lambda_{\mathrm{D}}$ and we choose values of ζ, δ and Z to give a selected magnitude, location and gradient, respectively, for the change in $\Delta\lambda_{\mathrm{D}}$. The second term in Equation (5) contains D rather than D^2 because the Voigt parameter is proportional to $\Delta\lambda_{\mathrm{D}}^{-1}$. This term is included in ϕ_y only for $y > 1$.

We have chosen a form for the broadening term, D, that has the approximate τ_c dependence of the velocity curve derived for the low chromosphere by UNNO (1959). However, we investigate the effects on the line profiles of varying each of the parameters in Equation (6). The most important effects of ζ are in determining the location of K_2 in wavelength and the relative separations of K_2 and the minimum in K_1. The parameters δ and Z influence rather markedly the intensity of K_2 and more especially the relative intensities of K_2 and K_3.

Because we have used three parameters in specifying D, whereas we have allowed room in the analysis for only two parameters, we cannot evaluate the three parameters uniquely. We therefore must regard the combination of δ and Z as essentially a single parameter.

The remaining two parameters r_0 and $B(\tau)$ are perhaps equally uncertain. If we use τ_c as the basic depth parameter, r_0 depends upon $d\tau_0/dh$ in the line and the scale height of τ_c. The absorption coefficient at line center depends upon total density, the relative abundance of the ion and the broadening velocity. So far as is known both Ca II and Mg II are the dominant ions of their species throughout the region of line formation. Hence, for our purposes the major uncertainties in $d\tau_0$ are in the total densities and relative abundances of Ca and Mg. The scale height of τ_c is also uncertain, particularly in the upper photosphere and low chromosphere. As a result of these combined uncertainties in both $d\tau_c/dh$ and $d\tau_0/dh$ we can make only rather crude estimates of r_0. The order of magnitude estimates given in Table I are obtained from total densities given by THOMAS and ATHAY (1961), relative abundances given by ALLEN (1955) and an assumed scale height for τ_c of 100 km. We emphasize that the estimated r_0 could perhaps be in error by over a factor of 3, particularly in the regions where $\tau_c > 10^{-3}$. Values of r_0 for Mg II differ from those for Ca II because of differences in $d\tau_c/dh$ and in the relative abundances of Ca II and Mg II. Mg is about a factor of 20 more abundant than Ca and for H^- absorption κ_c increases by about a factor of 3 from λ 3900 to λ 2800. It may be that additional sources of continuum opacity are present at λ 2800, but we do not consider this possibility.

The values of r_0 most important for our purposes are those in the optical depth range $\tau_c \approx 10^{-10}$ ($\tau_0 \approx 1$) to $\tau_c \approx 10^{-2}$, i.e., in the atmospheric regions producing K_3, K_2 and the inner K_1 profiles. When $r_0 < 10^{-10}$ we encounter oscillations in our solutions for $S(\tau)$ indicative of numerical instability. Thus, we have been forced to limit r_0 to values of 10^{-10} or greater. This is near the estimated value of r_0 at $\tau_0 = 1$ for Ca II but about an order of magnitude greater than the estimated r_0 for Mg II at $\tau_0 = 1$.

Absolute values of r_0 affect both the K_2 and K_1 component rather strongly. However, in K_2 a given change of r_0 may be effectively cancelled by a change in $B(\tau)$, whereas this is not the case in K_1. The location of the K_{12} minimum in y is strongly affected by r_0 and ϕ_y and only secondarily by $B(\tau)$. Hence in fitting the profile we are ultimately forced to adopt values of r_0 and ϕ_y that properly locate the K_{12} minimum.

We assume for r_0 a τ dependence and leave a single adjustable constant. The estimated values of r_0 in Table I vary approximately as $\tau_0^{1/3}$. We therefore adopt

$$r_0 = r_{0,1}\, \tau_0^{1/3} + 1 \times 10^{-10}, \tag{7}$$

which has the proper form for $r_0 > 10^{-10}$. Equation (7) seriously overestimates r_0 for MgII for regions with $\tau_0 < 10^2$. The resultant effect on I_y is very small, however, because the total optical thickness of the chromosphere is much larger than 10^2. The values of r_0 in Table I suggest $r_{0,1} \approx 4 \times 10^{-11}$ for MgII and 2×10^{-10} for CaII. We adopt these as initial values, but we shall investigate the effect on the profile of varying $r_{0,1}$.

In choosing among the wide variety of possible forms for $B(\tau)$ we again follow the same scheme adopted for D, viz., we assume the form of the τ_c dependence but leave adjustable constants to fix the amplitudes. Two sets of $B(\tau)$ curves for λ 2800 are selected, one with a relatively narrow minimum near $\tau_c = 10^{-2}$ and one that permits a broader minimum with the chromospheric rise in $B(\tau)$ beginning at an adjustable value of τ_c. The first set, designated by subscript 1, is represented by the equation

$$B_1 = 1 + 10\tau_c - \alpha_1\, e^{-30\tau_c} + A_1\, e^{-(10^7\tau_c)^{.17}}, \tag{8}$$

and the second set by

$$B_n = 1 + 10\tau_c - \alpha_n\, e^{-(300\tau_c)^{1/2}} + A_n\, e^{-(C_n\tau_c)^{x_n}}. \tag{9}$$

The coefficient of τ_c in the linear term approximates the limb darkening coefficient at λ 2800. The value of $B_{\min}$ is controlled by α_n. The two constants C_n and x_n are used jointly to determine the value of τ_c at $B_{\min}$ and therefore act effectively as a single adjustable constant. A_n determines the amplitude of the chromospheric rise in B.

Equation (9) is used to investigate three sets of temperature models, designated by subscripts $n=2$, 3 and 4, with minima occurring near $\tau_c = 10^{-3}$, 10^{-4}, and 10^{-5}, respectively. The corresponding values of C and x_n are $C_2 = 10^7$, $x_2 = .21$; $C_3 = 10^7$, $x_3 = .28$; and $C_4 = 10^8$, $x_4 = .28$. The values of x_n are dependent, of course, on our choice of C_n. However, the choice of C_n is not critical so long as $C_n \gtrsim 10^5$. The reason being that the observed K_2 peak intensity fixes $B(\tau)$ at $\tau_c \approx 10^{-6}$ within rather narrow limits. Thus, the only remaining freedom in $B(\tau)$ in the low chromosphere is in the value and location of $B_{\min}$, and, hence, the gradient in $B(\tau)$ between the minimum and $\tau_c \approx 10^{-6}$.

For the four models mentioned above we adopt somewhat arbitrarily $\zeta = 3$, $\delta = 10^4$ and $Z = \frac{1}{2}$. This gives a step-like rise in $\Delta\lambda_D$ with an amplitude of 4:1 and with the region of increasing $\Delta\lambda_D$ extending from about $\tau_c = 10^{-3}$ to 10^{-5}. In these models $\Delta\lambda_D$ is constant in the layers where K_2 and K_3 are formed and the values of I_2 and I_3 are not strongly influenced by D. From these models we select what appears to be the most favorable form for $B(\tau)$, which turns out to be model 3. We then construct model 5 by adopting the values of C_n and x_n for model 3 and changing the parameters in D to $\zeta=8$, $\delta=10^8$ and $Z=\frac{1}{4}$. This has the effect of increasing the amplitude of the rise in $\Delta\lambda_D$ and of both shifting and spreading out the regions of increasing $\Delta\lambda_D$ so that it extends from about $\tau_c=10^{-6}$ to 10^{-10}. K_2 and K_3 are now formed in layers where $\Delta\lambda_D$ is changing rapidly and this strongly influences the intensities of both K_2 and K_3 as well as their ratio. Parameters of the five models are summarized in Table II.

4. Results for Normal Regions

Rather than reproduce many series of calculations with different sets of parameters, we shall try to extract from the calculations the essential effects on the resulting profiles and to explain them wherever they are understood. This reflects our feeling that this set of calculations should serve only as a starting-point for improved calculations using improved data, a more realistic model atom and perhaps a more complex chromosphere.

The coupling between the chromospheric B and S that gives rise to the K_2 peak is conveniently represented, for fixed ϕ_y, by $\varepsilon\tau_0^*$, where, as before, τ_0^* is the opacity of the chromosphere. Since for constant r_0, τ_0^* is proportional to $r_0^{-1}\tau_c^*$, we have

$$\varepsilon\tau_0^* \propto \frac{\varepsilon\tau_c^*}{r_0}.$$

The values of ε and r_0 in Table I give $\varepsilon/r_0 \approx 10^6$ for Mg II and $\varepsilon/r_0 \approx 10^5$ for Ca II. Hence, the source functions are strongly coupled to the chromosphere for Mg II when $\tau_c^* \gtrsim 10^{-6}$ and for Ca II when $\tau_c^* \gtrsim 10^{-5}$. This difference between the coupling conditions for the two ions immediately provides a reason for the much stronger emission peaks in Mg II. Also, we note that because r_0 is larger by a factor of two for the H lines than for the K lines, we expect stronger coupling hence stronger K_2 peaks for the K lines than for the H lines.

When ϕ_y is allowed to vary, the choice of **a** and the location of the gradient in ϕ_y affects the K_2 amplitude markedly whereas ζ has only a secondary effect. Also, ε and A_n affect K_2 directly. For the combined effects we find the approximate proportionality

$$K_2(\max) \sim \frac{\varepsilon A_n}{(r_{0,1}^{1/2}\,\mathrm{Ca})^{1/2}} \tag{10}$$

or, since $\tau_0^* \sim C^{-1}$ and $\tau_0^* \sim r_0^{-\frac{1}{2}}$ we have

$$K_2(\max) \sim \varepsilon A_n \left(\frac{\tau_0^*}{a}\right)^{1/2},$$

TABLE II

Parameters for Models 1–5

Model	x_n	C_n	A_n	α_n	ζ	δ	Z	a	$r_{0,1}$	T_{ch}	$\bar{v}_{ch}$ (km/sec)
1	.17	10^7	51	1.20	3	10^4	$\frac{1}{2}$	.001	1×10^{-11}	7000	6.4
2	.21	10^7	53	1.55	3	10^4	$\frac{1}{2}$	.001	1×10^{-11}	7000	6.4
3	.28	10^7	100	1.20	3	10^4	$\frac{1}{2}$	.003	4×10^{-11}	7700	6.4
4	.28	10^8	250	1.05	3	10^4	$\frac{1}{2}$	.003	4×10^{-11}	9000	6.4
5	.28	10^7	320	1.25	8	10^8	$\frac{1}{4}$	.003	4×10^{-11}	9300	14.4

which is valid when the K_2 intensity is somewhat less than the continuum intensity.* For large K_2 amplitudes the ε dependence changes to a square-root law.

One effect on K_2 (max) of increasing ζ in these calculations is to increase S_{max} somewhat thereby tending to increase K_2 (max). A second effect is to broaden the K_2 (max) through the mapping of $S_{t,y}$ into I_y, which tends to cancel the enhancement of S_{max}.

The parameters **a**, $D(\tau)$, r_0, and C affect the general shape of the profile as well as the K_2 amplitude. Figures 3 and 4 illustrate these effects together with the effect of α. For $Z=\frac{1}{2}$ the principal control on the location of the K_2 maximum, which we label y_{max}, is through the value of ζ. To a rather good approximation

$$y_{max} = 2.5(1+\zeta), \tag{11}$$

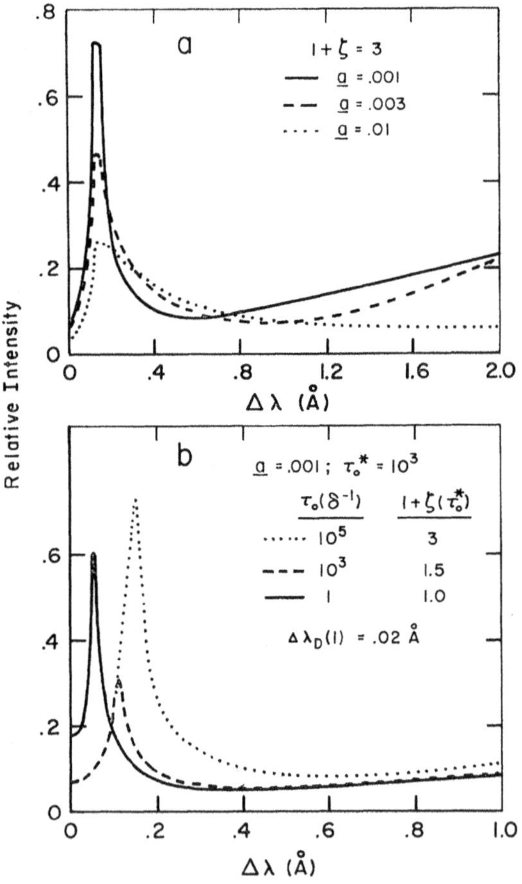

Fig. 3. Influence of **a** and δ on the Mg II K profile for model 3 with $A=150$, $\alpha=1.15$, $r_{0,1}=4\times10^{-11}$ and $\Delta\lambda_D(\tau_c=1)=.02$ Å.

* More accurately, this expression is valid when $\tau_0^* \gtrsim \mathbf{a}^{-1}$. If $\tau_0^* < \mathbf{a}^{-1}$ then $K_2\,(\max) \propto \varepsilon A_n \tau_0^*$ and is independent of **a** (cf. SKUMANICH, 1967).

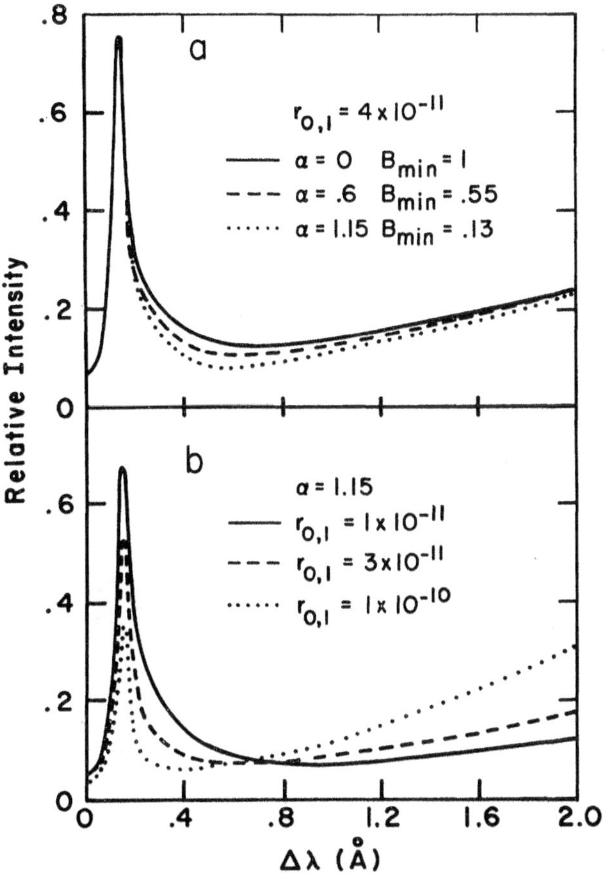

Fig. 4. Influence of α and $r_{0,1}$ on the MgII K profile, model 3 with $\mathbf{a} = .001$, $\zeta = 2$, and $\Delta\lambda_D$(chrom) = .06 Å.

i.e., y_{max} occurs near 2.5 doppler widths in the chromosphere. The observed K_2 maximum is at 0.15 Å. Hence for $Z = \frac{1}{2}$, $\Delta\lambda_D$ in the chromosphere is $\approx .06$ Å, corresponding to a doppler velocity of 6.4 km/sec. An upper limit on ζ is set by this value of $\Delta\lambda_D$ in the chromosphere and the thermal doppler width in the photosphere. For a temperature of 4000° the thermal doppler width for MgII is .015 Å. Since $\Delta\lambda_D^1$ cannot be less than this, ζ cannot exceed 3.

For $Z = \frac{1}{4}$ the situation is somewhat different because of the increased range in τ_c through which $D(\tau)$ increases. We have not performed sufficient calculations with this value of Z to establish a general rule for the location of y_{max}.

One of the principal difficulties in matching the observed profiles is to keep the separation between the K_2 maximum and the K_{12} minimum as small as it is observed to be. The location of the minimum is affected most strongly by **a** and less strongly by $r_{0,1}$ and ζ. The observed location of y_{min} is ≈ 0.6 Å for MgII, or $y_{min} \approx 4 y_{max}$. This favors the smaller values of **a** and $r_{0,1}$ and larger values of ζ. On the other hand, the value of I_y at $y \approx 10 y_{max}$ is essentially independent of ζ and shows opposite effects from changes in **a** and $r_{0,1}$. This allows us to select the parameters **a**, and $r_{0,1}$ somewhat independently. The location of y_{min} requires a value for **a** of about 10^{-3}, which we have adopted as a minimum value. The value of y_{min} further requires $\zeta > 2$ for models 1–4. Fortunately, this is close to the upper limit $\zeta \leqslant 3$ required by the location of y_{max}, so we adopt $\zeta = 3$. Finally, we determine a value for $r_{0,1}$ that gives a satisfactory value for I_y at $y \approx 10 y_{max}$ without seriously disrupting the value of y_{min}. In all cases we find that a value of $r_{0,1}$ near to the suggested value of 4×10^{-11} for Mg II is sufficiently good.

For each temperature model we adjust A_n and α_n to give the best intensities for the K_2 maximum and the K_{12} minimum. The maximum can be completely controlled so we obtain an unambiguous value for A_n once **a**, $D(\tau)$ and $r_{0,1}$ are specified. Unfortunately, the intensity of the K_{12} minimum cannot be fully controlled independently of the location. Thus, the values of **a**, ζ and $r_{0,1}$ selected to give the best 'shape' for the profile exert an influence on the intensity of the K_{12} minimum. We therefore consider this region of the profile in some detail.

For $\zeta = 2$–3, y_{min} falls near $y = 30$, and for $\mathbf{a} = .001$, $\phi_{30} = 10^{-6}$. Hence $\tau_{30} = 1$ corresponds to $\tau_0 \approx 10^6$, or, using the values of r_0 in Table I, to $\tau_c \approx 10^{-3}$. Since $(\varepsilon^2 \tau_0 / \mathbf{a}) \gg 1$ at this depth, we expect $S \approx B$, and since $\tau_c \ll 1$ we expect τ_{30} to be dominated by line opacity.

Because of backward irradiation by the chromosphere into the region of B_{min} and since the chromosphere contributes to the observed intensity to a point in the profile beyond the K_{12} minimum, the value of I_y at the minimum in the profile lies above the minimum value of B. The amount by which I_y (min) exceeds B_{min} depends, in part, upon **a**, ζ and $r_{0,1}$ as may be seen from inspection of Figures 3a, 3b, and 4b. If **a** or ζ are large or if $r_{0,1}$ is small, the 'wings' of the K_2 emission blend with K_1 and fill in the K_{12} minimum. This filling in of the K_{12} minimum restricts our knowledge of B_{min} to an upper limit, which we obtain by equating I_y (min) to B_{min}.

The observed K_{12} minima have reported intensities of about 5% for Ca II and 4% for Mg II. The corresponding radiation temperatures are 4200° for Ca II and 4250° for Mg II. Both profiles contain relatively large uncertainties but the very strong dependence of the intensity on T_{rad} makes these estimates of T_{rad} quite reliable. For example, if $T_{rad} = 4500°$, the K_{12} intensities increase to 8–9% for the two lines, which would seem to be beyond or near the limit of the probable errors in the profiles. We feel confident, therefore, that $T_{rad} < 4500°$. Computed intensities of the K_{12} minimum decrease as we decrease T_{min} down to values of about 4000° and thereafter remain essentially independent of T_{min}. Even at $T_{min} = 4000°$, the computed K_{12} intensities fall above the observed intensities. Hence we adopt values of T_{min} of about 4000° corresponding to $B_{min} \approx 0.2$. This determines the parameter α for each of the models. This adopted value of T_{min} represents a somewhat realistic value, but it does depend upon our choices of α, ζ and $r_{0,1}$. The values chosen for both **a** and ζ, within their allowed ranges, give the lowest K_{12} intensities and therefore tend to increase the reliability of the adopted T_{min}. Nevertheless, we can conclude only that T_{min} appears to be nearer 4000° than 4500°.

In summary, then, the four free parameters for each of the first four models are determined essentially from the following scheme: **a** from y_{min}, $r_{0,1}$ from I_y at $y \approx 30$, A_n from the K_2 amplitude and α from the requirement that $B_{min} \lesssim 0.2$. We then choose the particular model from the four that gives the best overall agreement with the Mg II and Ca II profiles.

Parameters for the first four models are given in Table II together with the associated values of the asymptotic chromospheric temperature, T_{ch}, and mean doppler velocity, $\bar{v}_{ch}$. (We delay comments on Model 5 until after our discussion of the first four models.) The temperature models are plotted in Figure 5. Note that each model gives $T \approx 6000°$ at $\tau_c \approx 10^{-6}$. This results from the adjusting of A_n to give the same K_2 intensity for each of the models. However, the value of T at $\tau_c \approx 10^{-6}$ depends upon the choice of ε, **a**, and $r_{0,1}$ and is therefore not as accurately determined as is implied by Figure 5.

Profiles of the Mg II K line for models 3, 4, and 5 are shown in Figures 6a and 6b, together with the observed profile to facilitate comparison. The computed profiles have been matched to the observed profiles at the K_2 peak and near $\Delta\lambda = 1$ Å. None of the

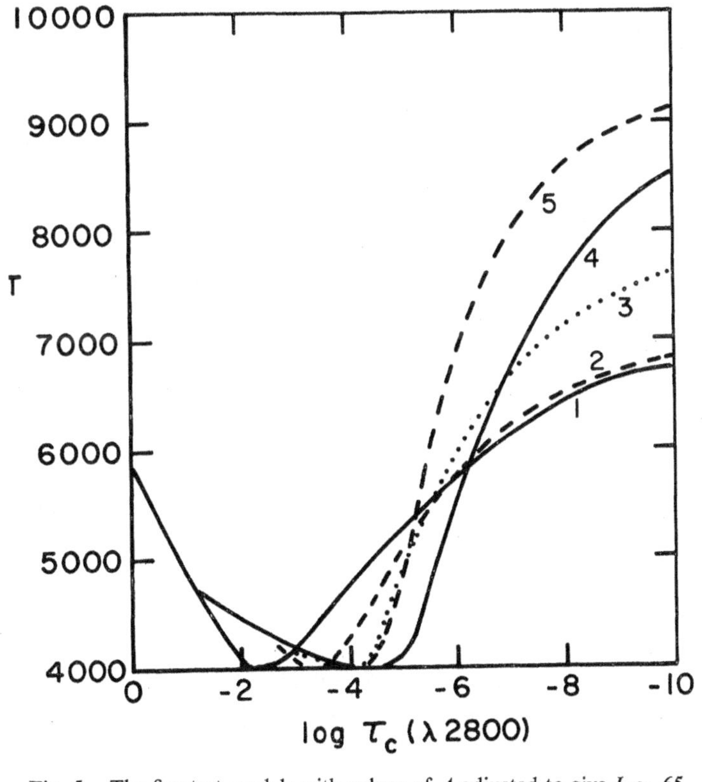

Fig. 5. The five test models with values of A adjusted to give $I_2 \approx .65$.

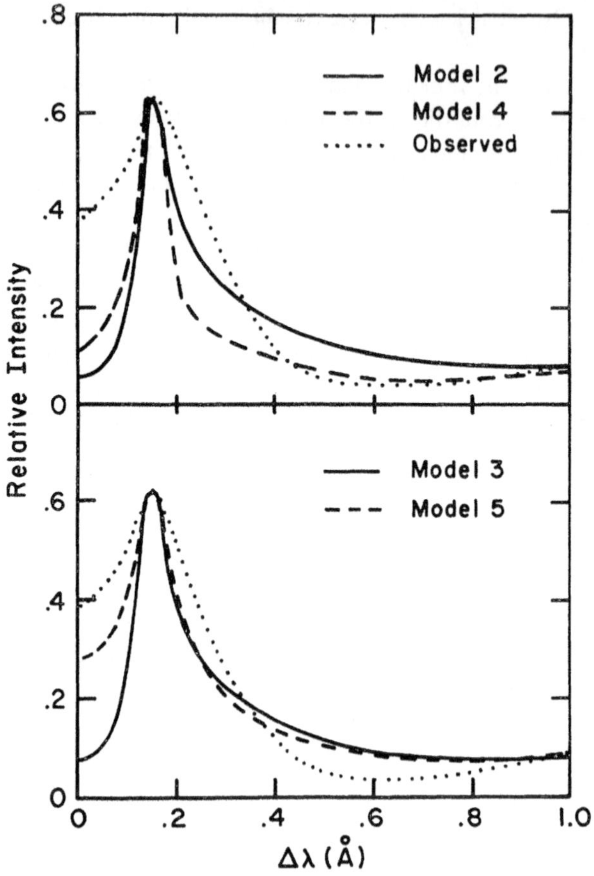

Fig. 6. Comparison of computed and observed Mg II K profiles for four models when the profiles are matched at the K_2 peak and near $\Delta\lambda = 1$ Å.

models gives a particularly close fit between the computed and observed profiles throughout the K_2 and K_1 regions. However, considering the relative crudeness of the parameters we have limited ourselves to and the uncertainties in the observed profile, we consider the results to be quite promising.

Model 1 gives clearly too broad a profile for K_2 and Model 4 gives clearly too narrow a profile. The best overall fit seems to be for Model 3. Figure 7 exhibits both the H and K profiles for this model, and Figure 8 shows the behavior of the two source functions. The H profile was computed with $r_{0,1}$ increased by a factor 2. It may be noted that the computed relative behavior of H and K agrees rather well with the observed behavior.

To compute Ca II profiles we use the temperature models shown in Figure 5 together with the same values of ζ and ε as were adopted for Mg II. We increase $r_{0,1}$ by a factor 10 to compensate approximately for the different line and continuum opacities for these lines and we leave **a** as an adjustable constant. The best agreement

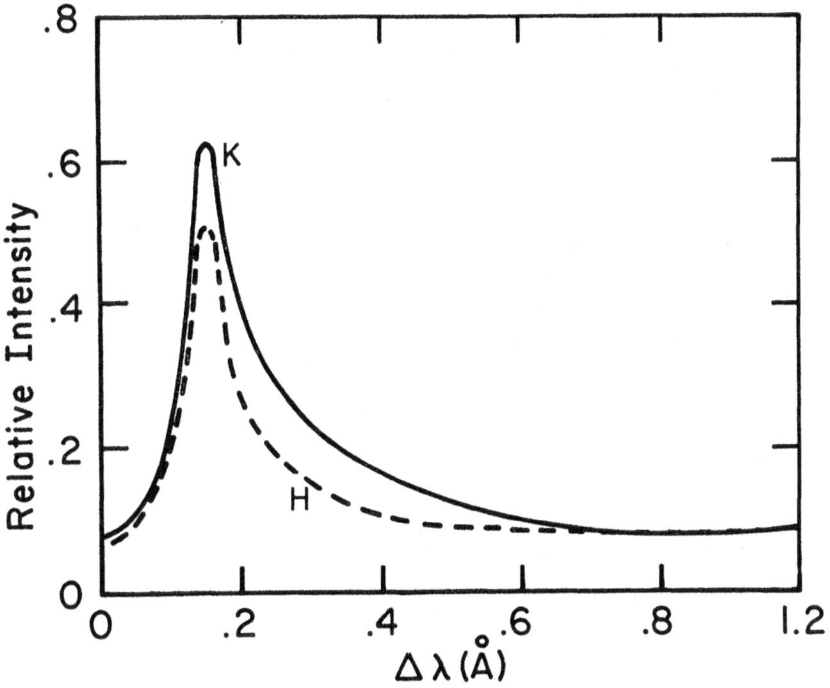

Fig. 7. The computed Mg II K and H profiles for model 3.

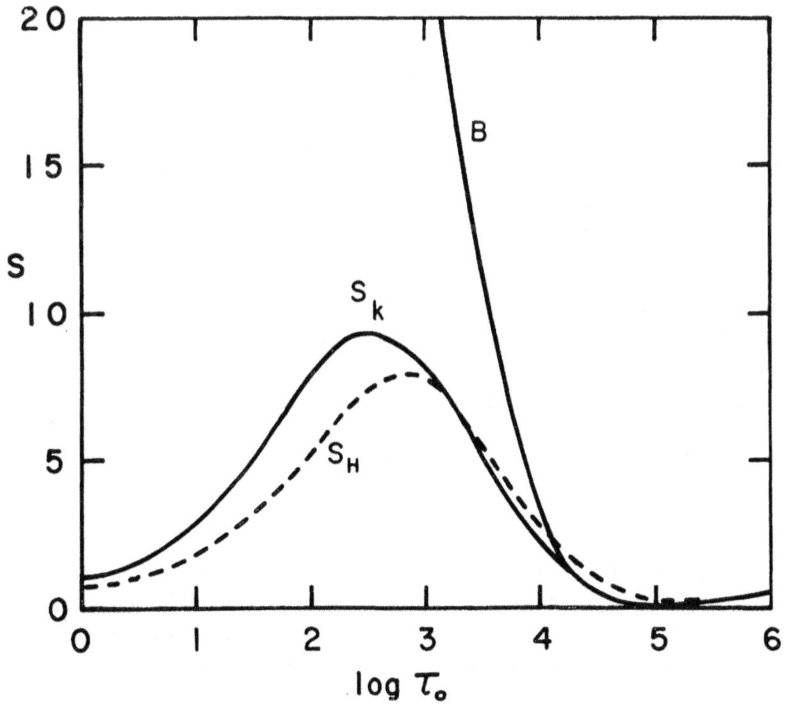

Fig. 8. Source functions for Mg II H and K lines for model 3 plotted against τ_0 for the K line.

between computed and observed profiles is for $\mathbf{a} \approx .003$. Figures 9a and 9b exhibit the four profiles. The best agreement here appears to be with Model 4, although a small adjustment of parameters, particularly $\mathbf{a}$ and $r_{0,1}$ would make the agreement for Model 3 equally good. We conclude therefore that the preferred model is number 3, which shows the temperature minimum near $\tau_c = 10^{-4}$.

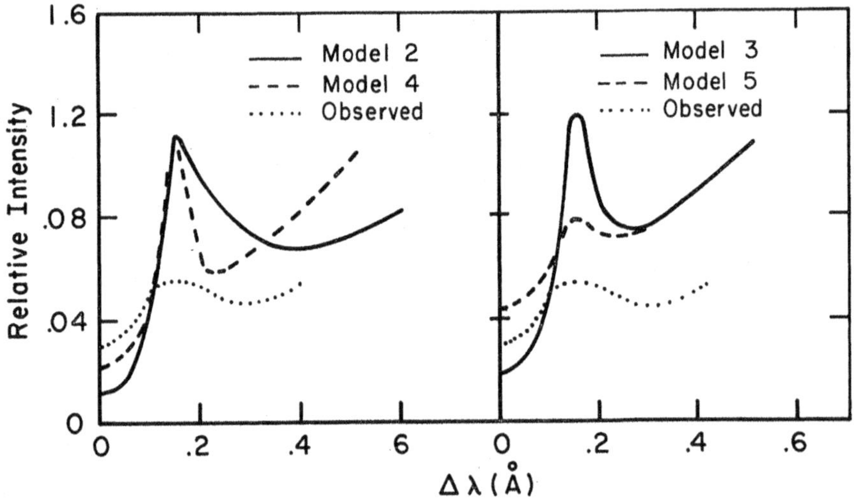

Fig. 9. Comparison of computed and observed Ca II K profiles for four models with $\mathbf{a} = 10^{-3}$.

For model 5 we select the values of C_n and x_n for model 3 but leave A_n and α_n free. We then adopt a new set of values for ζ, δ and Z and redetermine A_n and α_n as before. The values of δ and Z are selected to produce a gradient in $\Delta\lambda_D$ throughout the regions where K_3 and K_2 are produced, and the value of ζ is selected to give a proper ratio for I_2/I_3 in the Ca II K line. We use the Ca II line for this purpose because the profile is more accurately known near the core. The resultant values of A_n and α_n for $\zeta = 8$, $\delta = 10^8$ and $Z = \frac{1}{4}$ are given in Table II. The temperature model is plotted in Figure 5 and the resultant profiles are shown in Figures 6 and 9.

The primary effect of the different form for $D(\tau)$ in model 5 is to suppress I_2 and enhance I_3. I_3 increases because S increases near $\tau_0 = 1$ and I_2 decreases because of the mapping of $S_{t,y}$ into I_y. The decrease in I_2 is compensated for in model 5 by increasing A. I_3 for Mg II rises to about 0.3, which is in much better agreement with the observed I_3 of about 0.4. We have not felt justified in attempting a more refined fit because of the very strong possibility that the Mg II profiles are seriously modified by the plages that were present on the disk when the observations were made. We note from Figure 9b that the Ca II profile for model 5 has the correct ratio of I_2/I_3.

The fact that both the Mg II and Ca II profiles for model 5 agree reasonably well with the observed profiles, lends considerable confidence to our basic approach. It has been demonstrated (DUMONT, 1967) that when the $3d\,^2D$ levels of Ca II are included in the model atom the intensity in K_2 and K_3 are depressed below those computed using a simple two-level atom. Thus, the fact that our computed intensities in Ca II lie somewhat above the observed intensities is not particularly disturbing.

Obviously, improvements are needed. These include improved observations, an improved model atom with additional energy levels, an improved model atmosphere and perhaps improved collisional excitation rates for Ca II and Mg II. We have demonstrated only that there is approximate agreement between the observed profiles and the adopted model atmosphere. We emphasize again that some caution is needed in interpreting the results. The rather complex interactions of $\mathbf{a}$, r_0, ε and $D(\tau)$ on the computed profile necessarily results in a certain amount of ambiguity among these parameters. This ambiguity is undoubtedly reflected to some extent in the adopted temperature model. It is difficult to make a reliable estimate of the possible magnitude of these affects.

5. Active Regions

Ca II K profiles in active regions shows a variety of changes in the K_2 and K_3 regions from the profiles observed in quiet regions. One such change is shown in Figure 1b. The K_2 peak has intensified and moved somewhat closer to line center. K_3 has filled in and K_{12} has both filled in and moved further from line center. In some cases, the K_2 peaks intensify and move further from line center, and in others they move much closer to line center, even to the point where K_3 disappears and a single emission peak occurs.

As noted in the preceding discussion the K_2 amplitude is affected by all the parameters of the problem except α. It seems probable that in active regions all of the parameters change and one therefore cannot reach a general conclusion about the changes in any one of the parameters from the simple fact that K_2 is intensified. For example, an increase in K_2 may result from a rise in temperature in the chromosphere, a rise in density, an increase in the optical depth to the temperature minimum, or a reduction of ζ. On the other hand, the location of K_2 is affected mainly by the latter two factors. The influence of $D(\tau)$ on the intensity of K_2 and the shape of the profile has been discussed in the preceding. The effect of chromospheric opacity, which we characterize by C, also affects both the K_2 location and the K_3 intensity in a systematic way. It is of interest therefore to investigate the particular influence of the parameter C.

We show in Figure 10 several profiles computed for values of C ranging from 10^4–10^{12}. The parameters x_n, A_n, $\mathbf{a}$ and $r_{0,1}$ have the values 0.23, 80, .006 and 2×10^{-10} respectively, for each of the profiles. $A_n=80$ corresponds to $T_{\text{ch}}=12000°$ and increases K_2 by about a factor of 3 over its normal value. The parameter ε has been adjusted to give roughly the same K_2 intensity for the different profiles. The parameter ζ is set equal to 3 for $10^4 \leqslant C \leqslant 10^8$, and 0 for $10^8 \leqslant C \leqslant 10^{12}$. This latter change was necessary because of numerical instabilities encountered for combinations where both C and ζ were large. The profiles are plotted with $\Delta\lambda/\Delta\lambda_{\text{D}}$ (chromosphere) as abscissa so that the change in ζ is not reflected in the location of K_2.

The profile changes shown in Figure 10 are fairly typical of many active region

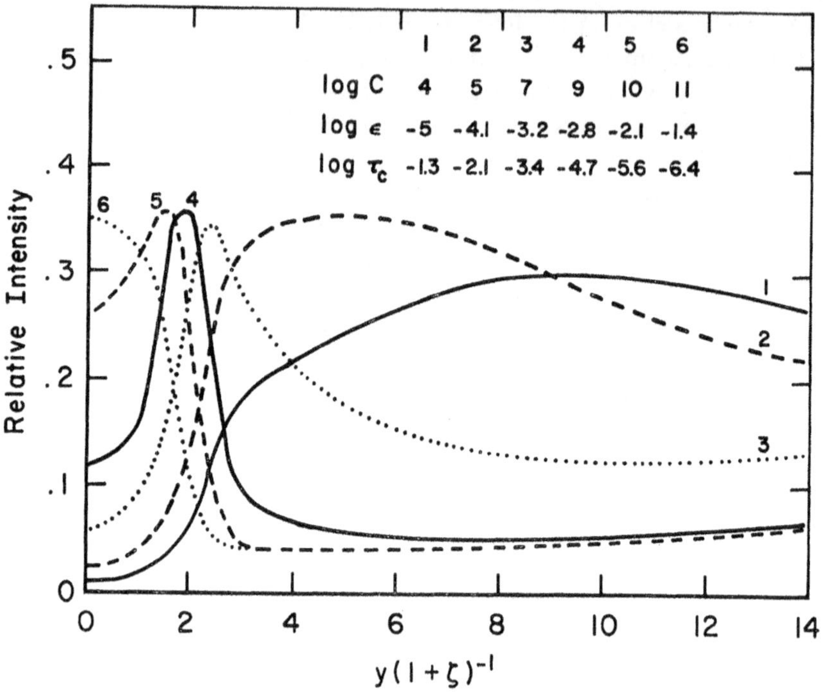

Fig. 10. Influence of C on the Ca II K profiles suggestive of the types of changes observed in active regions.

profiles. For small values of C the changes in K_3 are produced by changes in ε. However, for large values of C the filling in of K_3 is directly related to the decreased separation of the K_2 peaks produced by elevating the base of the chromosphere, i.e., reducing the chromospheric opacity. Similarly, wide separation of the K_2 peaks with no enhancement of K_3 results from increasing the chromospheric opacity by depressing its base. We emphasize again that effects of different $D(\tau)$ are undoubtedly present. However, the merging of the K_2 peaks to form a simple emission core is strongly suggestive of an increase in C. Since this behavior of the profile is found mainly over sunspots, it is not unreasonable to suppose that the base of the chromosphere is elevated. Widely separated K_2 peaks are associated with some flares and, again, it would not be unreasonable to suppose that the chromosphere begins somewhat lower in this case. We suggest therefore that a substantial part of the observed changes in the K profile from point to point on the solar disk results from changes in the location of the temperature minimum.

Acknowledgements

We are indebted to Dr. K. Widing for providing us with profiles of the Mg II lines corrected in an approximate way for scattered light.

References

ALLEN, C. W.: 1955, *Astrophysical Quantities*. Athlone, London.

ATHAY, R. G. and SKUMANICH, A.: 1967 (in press).

DUMONT, S.: 1967, Paper presented at Bilderberg Conference on the Photosphere and Chromosphere.

GOLDBERG, L.: 1965, in *Proceedings Second Harvard-Smithsonian Conference on Stellar Atmospheres.*

GOLDBERG, L., MOHLER, O. C., and MULLER, E.: 1959, *Astrophys. J.* **129,** 119.

JEFFERIES, J. T. and THOMAS, R. N.: 1960, *Astrophys. J.* **131,** 695.

LINSKY, J.: 1966, Paper presented at AAS Meeting.

MIYAMOTO, S.: 1957, *Pub. Astron. Soc. Japan* **9,** 146.

PURCELL, J. D., GARRETT, D. L., and TOUSEY, R.: 1962, *Space Res.* **3,** 781.

SKUMANICH, A.: 1967 (in press).

THOMAS, R. N. and ATHAY, R. G.: 1961, *Physics of the Chromosphere*. Interscience, New York.

UNNO, W.: 1959, *Astrophys. J.* **129,** 388.

WILSON, O. C. and BAPPU, M. K. U.: 1957, *Astrophys. J.* **125,** 661.

WILSON, N. L., TOUSEY, R., PURCELL, J. D., JOHNSON, F. S., and MOORE, C. E.: 1954, *Astrophys. J.* **119,** 590.

ZIRKER, J. B.: 1965, in *Proceedings Second Harvard-Smithsonian Conference on Stellar Atmospheres.*

CONTRIBUTION À L'ÉTUDE DU SPECTRE SOLAIRE DE L'HYDROGÈNE

Y. CUNY
Observatoire de Paris, Section d'Astrophysique (Meudon)

(Reçu le 18 septembre, 1967)

Résumé. Les raies et le rayonnement continu de l'hydrogène sont calculés, dans le cas solaire, en tenant compte des écarts à l'équilibre thermodynamique local.

La comparaison des résultats, interprétés par la théorie de la formation des raies de Jefferies et Thomas, avec les observations donne des renseignements sur le modèle de l'atmosphère solaire.

L'énergie émise par Ly α dépend très fortement de la densité électronique. Seuls les modèles à palier de température dans la haute chromosphère donnent un profil de Ly β à dépression centrale.

La raie Hα, qui est contrôlée essentiellement pour les photoionisations dépend surtout du modèle de photosphère; toutefois les modèles de chromosphère à fortes température et densité électroniques ont une influence sur le profil de Hα.

Première Partie

1. Introduction

Un modèle d'atmosphère solaire étant donné, il est possible de calculer le transfert du rayonnement dans les transitions d'un atome, ici l'hydrogène, sans l'hypothèse de l'équilibre thermodynamique local.

Pour traiter ce problème il faut, suivant l'étude théorique faite par Thomas et Jefferies résoudre simultanément les équations de l'équilibre statistique et les équations différentielles du transfert du rayonnement. Les équations de transfert du rayonnement sont intégrées à l'aide de la méthode de Feautrier (1964) qui est une application de la méthode générale d'intégration de Fox (1957). La méthode utilisée pour résoudre le problème pour un atome à plusieurs niveaux est décrite dans un article précédent (Cuny, 1967).

Le but de cette étude est de déterminer les caractéristiques d'un modèle donnant des résultats théoriques, pour le rayonnement émis par l'atome d'hydrogène, comparables aux observations. Dans cet article nous étudions les résultats obtenus avec un nombre restreint de modèles et essayons d'analyser les propriétés du modèle dont dépendent les intensités du rayonnement émergent. Nous avons analysé la formation du rayonnement continu de Lyman et des raies Ly α et Ly β d'une part, et de la raie Hα d'autre part. Les transitions de Lyman donnent des informations sur la haute chromosphère, les transitions de Balmer sur la basse chromosphère et la photosphère. Nous poursuivrons ultérieurement cette étude qui n'est que préliminaire.

Nous indiquerons d'abord les valeurs des paramètres atomiques utilisées.

2. Paramètres atomiques

Nous avons adopté pour les transitions liées-liées de l'hydrogène les probabilités de transition d'Einstein calculées par Menzel (1962) et pour les transitions liées-libres et libres-libres les facteurs de Gaunt de KARSAS et LATTER (1961). Dans ces calculs préliminaires, négligeant l'effet Stark, nous n'avons tenu compte que de l'amortissement radiatif et de l'amortissement du à l'hydrogène neutre. L'amortissement radiatif est prépondérant dans la chromosphère, négligeable dans la photosphère où pour les niveaux inférieurs l'amortissement est du essentiellement à l'hydrogène neutre: nous l'avons calculé à l'aide de l'expression donnée par GRIEM (1964).

Nous avons utilisé les sections de choc de GRYZINSKI (1965). Celles-ci donnent des termes de collision approximativement deux fois plus grands que ceux obtenus avec les sections de choc déduites des expériences de Fite et Brackmann (POTTASCH et THOMAS, 1959); nous avons donc effectué un certain nombre de calculs en divisant les sections de choc de Gryzinski par deux.

Pour calculer l'absorption continue de l'ion H^- qui est très importante dans la photosphère solaire nous avons utilisé la section de photoionisation de JOHN (1960).

L'étude faite dans cet article repose essentiellement sur l'analyse des résultats donnés par deux modèles.

3. MODÈLES D'ATMOSPHÈRES SOLAIRES

Les modèles sont définis par la température et la densité électroniques et la microturbulence.

Le modèle de photosphère est le modèle d'Utrecht 1964. La composition chimique est celle de VARDYA (1961), avec une abondance d'hélium égale à 0,125.

Les Figures 1 et 2 donnent les variations de la température et de la densité électroniques, et celle de la microturbulence en fonction de la profondeur géométrique.

Nous avons contrôlé que l'intensité continue calculée (avec l'hypothèse de l'équilibre thermodynamique local) et celle observée par LABS (1957) sont peu différentes aux longueurs d'onde qui nous intéressent.

Différents modèles de chromosphère ont été utilisés.

A. *Modèle HAO1*

Ce modèle est celui établi par ATHAY *et al.* (1954); il s'agit du modèle donné par l'Équation (2) de cet article.

Disons tout de suite que nous ne pensions pas à priori que ce modèle fût le meilleur existant. Ces calculs préliminaires ont pour but seulement de déterminer les différents facteurs du modèle qui influencent les profils.

Entre les modèles de photosphère et de chromosphère, nous avons établi un raccord assez arbitraire des courbes de température en fonction de l'altitude et calculé un modèle en équilibre hydrostatique.

A la base de la chromosphère, l'augmentation de la densité est due à l'ionisation de l'hydrogène.

B. *Modèle interspiculaire de Coates (1958)*

Ce modèle diffère du précédent dans la haute atmosphère lorsque l'altitude est supérieure à 2000 km. La température augmente très rapidement jusqu'à 3000 km,

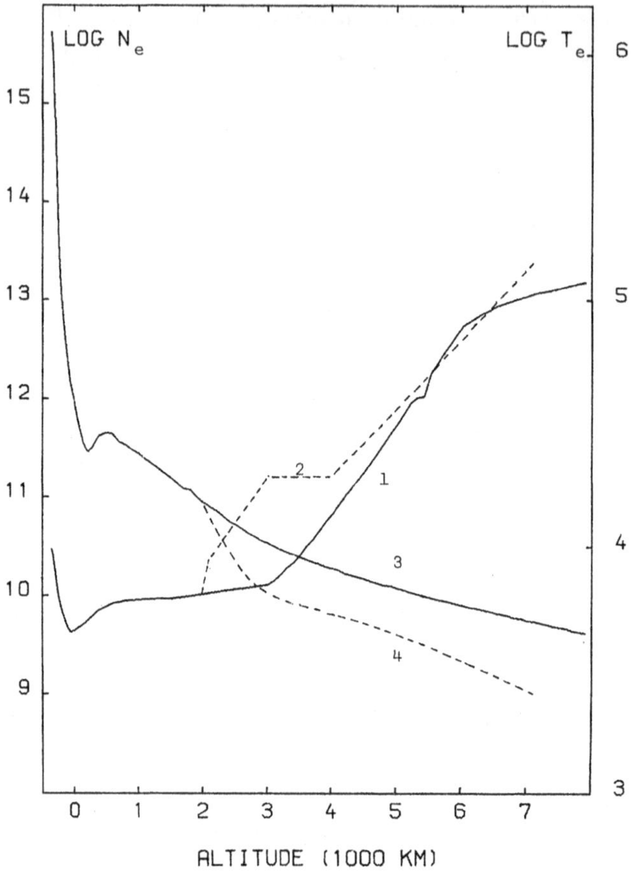

Fig. 1. Modèles de l'atmosphère solaire: (1) T_e = Modèle HAO1, (3) N_e = Modèle HAO1, (2) T_e = Modèle de Coates, et (4) N_e = Modèle de Coates.

est ensuite constante jusqu'à 4000 km (19000 °K), puis augmente à nouveau très rapidement.

A altitude égale, la densité électronique de ce modèle est inférieure à celle du précédent.

Nous signalerons ultérieurement les autres modèles utilisés.

Les raies et le rayonnement des différents continus de l'hydrogène ont des profondeurs de formation allant de la photosphère où l'hypothèse de l'équilibre thermodynamique local est valable à la très haute chromosphère où les écarts à l'équilibre thermodynamique sont très grands, ce qui conduit à la classification et aux remarques suivantes.

4. Classification des transitions de l'atome d'hydrogène

Nous pouvons classer les transitions de l'atome en trois groupes correspondant à des profondeurs de formation différentes.

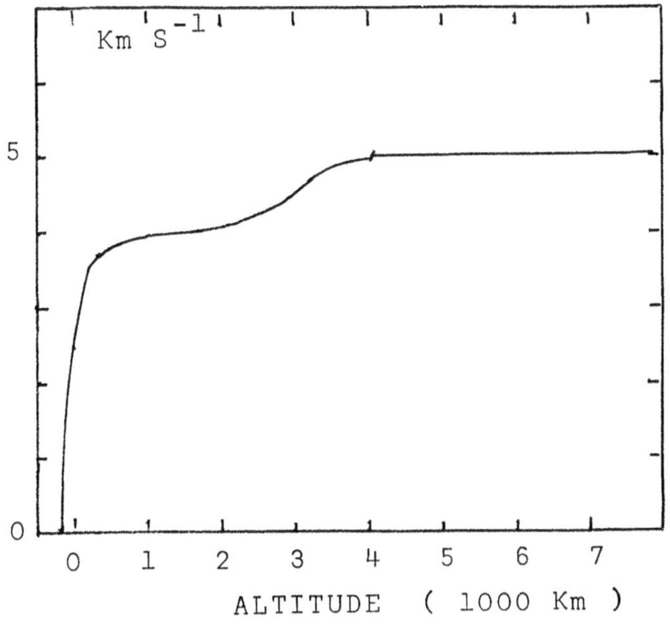

Fig. 2. Microturbulence.

A. *Transitions liées-libres, autres que le continu de Lyman*

Le rayonnement continu ionisant l'atome d'hydrogène à partir des différents niveaux d'énergie, à l'exception du niveau fondamental, est formé dans la photosphère pour laquelle l'hypothèse de l'équilibre thermodynamique local est valable. Pour un modèle de photosphère donné, les termes de photoionisation de ces niveaux sont calculés avec cette hypothèse.

Le rayonnement continu de Lyman est formé dans la chromosphère, hors des conditions de l'équilibre thermodynamique local.

B. *Transitions liées-liées, autres que celles de la série de Lyman*

Le corps des raies de la série de Balmer et de la série de Paschen est formé dans la chromosphère, à une altitude dépendant du modèle, mais en général inférieure à 2000 km, tandis que les ailes sont formées dans la photosphère.

C. *Les transitions de Lyman*

Le continu et les raies de Lyman sont formés dans la haute chromosphère à une altitude supérieure à 1000 km. A des altitudes inférieures, le bilan des processus radiatifs de ces transitions est nul.

Avant de donner et de discuter les résultats obtenus nous préciserons quelques points concernant d'une part la méthode utilisée pour traiter le problème hors de l'équilibre thermodynamique, d'autre part la précision des calculs numériques.

5. Méthode. Précision des calculs

A. Nous rappelerons d'abord l'expression de la fonction source d'une transition liée-liée (Thomas, 1957); lorsque la diffusion est complètement incohérente:

$$S_{jk} = \frac{2h\nu^3}{c^2} \frac{1}{\dfrac{b_j}{b_k} e^{\frac{h\nu}{kT}} - 1} . \tag{1}$$

Les notations sont celles utilisées dans l'article où nous faisons une étude de la méthode (Cuny, 1967). En particulier, nous rappellerons, lorsque ce sera nécessaire, les équations de cet article en faisant précéder leur numéro du chiffre I.

En introduisant l'expression des solutions b_k et b_j des équations d'équilibre statistique (Thomas, 1957), on obtient:

$$S_{jk} = \frac{J_{jk} + N_{jk}}{1 + D_{jk}}, \tag{2}$$

où

$$J_{jk} = \int_0^\infty J_\nu \phi_\nu \, \mathrm{d}\nu . \tag{3}$$

J_ν est l'intensité moyenne à la fréquence ν, ϕ_ν le profil d'absorption normalisé,

$$\int_0^\infty \phi_\nu \, \mathrm{d}\nu = 1 . \tag{4}$$

Pour la fonction source des transitions liées-libres on a des expressions analogues.

B. Calcul de J: formule de quadrature.

Le résultat de l'intégration numérique des équations de transfert dépend assez fortement de la formule de quadrature utilisée pour calculer le terme de rayonnement de Ly α.

Dans ce paragraphe nous supprimons les indices indiquant les niveaux définissant la transition.

Pour calculer J on partage l'intervalle d'intégration que l'on prend fini en une suite d'intervalles; dans chaque intervalle on admet que la fonction à intégrer peut être représentée par un polynôme: on peut donc utiliser les formules de quadrature de Gauss-Legendre ou de Tchebysheff.

1. Largeur utile: cas de Ly α. Lorsque les paramètres de la fonction source sont constants la largeur utile du profil est (I, 190):

$$x_{\mathrm{L}} = K \frac{1}{3} \frac{1}{(\pi 0{,}61)^{1/2}} \frac{a}{D}, \tag{5}$$

$$x = \frac{\Delta\nu}{\Delta\nu_{\mathrm{D}}} \tag{6}$$

a est la constante d'amortissement, D le paramètre du dénominateur de la fonction source.

Le facteur K doit être au moins de l'ordre de grandeur de 10. Avec K égal à 100 la précision est bonne.

Dans le cas de Ly α, à une profondeur optique de 10^5 les valeurs numériques sont les suivantes:

$$a = 0{,}4 \times 10^{-3}$$

$$D = 0{,}46 \times 10^{-4}.$$

On obtient donc, pour K égal à 10:

$$x_{\mathrm{L}} = 20.$$

A cette profondeur:

$$\Delta\lambda_{\mathrm{D}} = 0{,}04\ \text{Å},$$

ce qui donne pour la largeur utile:

$$\Delta\lambda_u = 0{,}8\ \text{Å} \qquad (K = 10).$$

Cette valeur est une valeur minimale. Un facteur K égal à 100 donnerait une valeur de $\Delta\lambda_u$ égale à 8 Å.

Nous avons exécuté des calculs avec pour Ly α différentes valeurs $\Delta\lambda_R$ de l'intervalle représenté par la formule de quadrature.

Nous avons limité la valeur de $\Delta\lambda_R$ afin de réduire le nombre de points de la formule de quadrature et par suite le temps de calcul.

2. Nombre de points de la formule de quadrature. Pour discuter l'influence de la formule de quadrature nous reprenons les résultats de l'étude du bilan des processus radiatifs faite dans l'article précédent:

$$\delta = 1 - \frac{J}{S}, \tag{7}$$

$$S = \frac{N}{D + \delta}. \tag{8}$$

La valeur numérique de δ est due d'une part à la dimension semi-infinie du milieu, d'autre part à la non-linéarité de la fonction source qui est très importante dans le cas de Lyα: il suffit pour s'en convaincre de se reporter à la Figure 7.

A une altitude donnée, soit τ_ν la profondeur optique de la transition à la fréquence ν:

$$\tau_\nu = \pi^{1/2}\, \tau\phi(x) \tag{9}$$

(lorsque la constante d'amortissement est petite la profondeur optique au centre de la raie est très peu différente de τ).

A la profondeur τ, nous avons montré que la contribution la plus importante à δ, due à la dimension semi-infinie du milieu est donnée par la fréquence définie par x_M:

$$\pi^{1/2}\, \tau\phi(x_M) = 0{,}61. \tag{10}$$

L'intervalle de fréquence contribuant à la valeur de δ est d'autant plus étroit que τ est petit.

Lorsque la fonction source varie lentement, ce qui est le cas lorsque N et D sont constants, une formule de quadrature à points relativement espacés donne une assez bonne précision.

La fonction source de Ly α présente une courbure variant très rapidement en fonction de la profondeur optique, que ce soit au voisinage du maximum ou dans les régions plus profondes où sont formées les ailes du profil. Les points de la formule de quadrature doivent donc être serrés. Nous avons vérifié que la valeur du maximum de la fonction source de Ly α dépend très fortement de la formule de quadrature choisie au voisinage de x_M défini par l'Équation (10) pour la profondeur optique τ du maximum. La valeur de δ à cette profondeur est due essentiellement à la non-linéarité de la fonction source.

Pour limiter le temps de calcul nous avons utilisé des formules à peu de points, répartis le plus judicieusement possible. L'erreur sur le maximum de la fonction source est inférieur à 20%. Lorsque la largeur $\Delta\lambda_R$ est trop petite l'intensité du rayonnement des ailes du profil est fortement surestimée; par contre la région centrale du profil dépend assez peu de l'étendue des ailes du profil utilisé.

3. Procédé itératif. Nous verrons plus loin que le continu de Lyman est formé à la même profondeur que les ailes de Ly α. Les équations de transfert du rayonnement de ces deux transitions sont donc couplées. Si l'on itère en traitant séparément le continu de Lyman et Ly α, en introduisant dans les équations de l'équilibre statistique le bilan des processus radiatifs de Ly α, on obtient à la deuxième itération une solution très éloignée de la solution physique, avec de fortes valeurs de δ négatives, conduisant à des populations et des intensités de rayonnement négatives. Il est donc nécessaire de résoudre simultanément les équations de transfert du rayonnement de Ly α et du continu de Lyman.

Les solutions des itérations pour Ly α et le continu de Lyman oscillent autour de la solution. J'ai arrêté les calculs à la 9ème itération: l'incertitude relative sur la solution est de l'ordre de grandeur de 10% (par défaut).

En substituant à la solution obtenue à la fin d'une itération une combinaison des solutions de cette itération et de la précédente il doit être possible d'accélérer très fortement la convergence: trois itérations devraient être suffisantes. Je modifierai ultérieurement le programme de calcul. Les autres transitions ne présentent pas de difficultés particulières.

Nous présenterons et discuterons les résultats obtenus pour la série de Lyman.

Deuxième partie: serie de Lyman. Résultats. Comparaison avec les observations

Dans les Tableaux I et II sont groupés d'une part des renseignements concernant les conditions imposées pour l'execution des calculs, d'autre part certains résultats.

Le numéro de référence indiqué sur la première ligne est reporté sur certains

TABLEAU I

Modèle HAO1

$N°$	1	2	3	4
N	2	2	2	3
$\Delta\lambda_R$	0,33	0,9	0,33	0,33
Col.	G	G	G/2	G
E	11,5	8,75	11,50	14
I	10,2	10	7,4	11,95
T_R(°K)	6975	6900	6940	6956

I est l'intensité maximale des profils d'émission.

TABLEAU II

Modèle de Coates

$N°$	1	2
N	2	3
$\Delta\lambda_R$	0,33	0,9
Col.	G/2	G
E	6,85	8,55
I	3,27	6,15
T_R	6730	6812

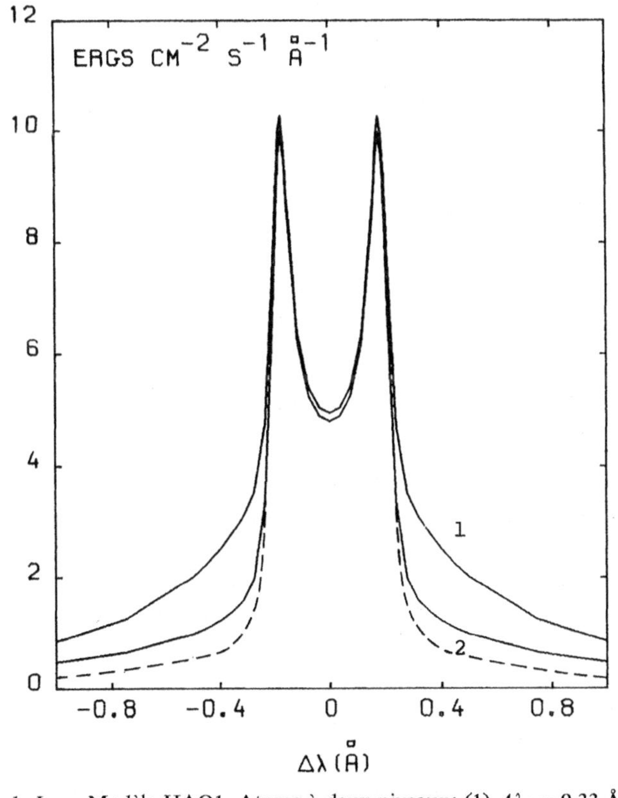

Fig. 3. Profils de Lyα. Modèle HAO1. Atome à deux niveaux: (1) $\Delta\lambda_R = 0{,}33$ Å, (2) $\Delta\lambda_R = 0{,}9$ Å, et- - - - $= \Delta\lambda_R = 2{,}5$ Å. (Profils intégrés sur le disque solaire.)

graphiques. N est le nombre de niveaux, $\Delta\lambda_R$ la largeur du profil de Ly α représentée par la formule de quadrature.

La ligne suivante indique si le calcul est exécuté avec les sections de choc de Gryzinski (G) ou celles-ci divisées par deux (G/2).

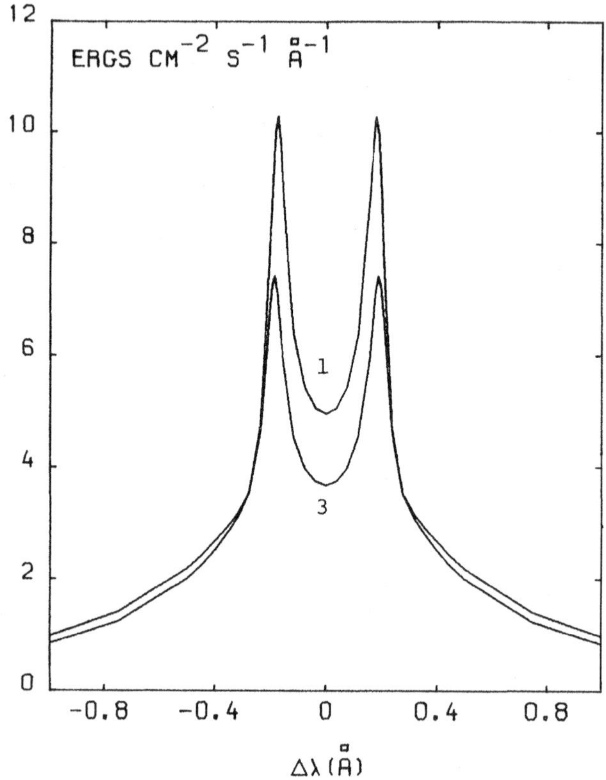

Fig. 4. Profils de Ly α. Influence des sections de choc. Modèle HAO1. Atomes à deux niveaux: (1) Sections de choc de Gryzinski, (2) Sections de choc de Gryzinski divisées par deux. (Profils intégrés sur le disque solaire.)

L'énergie E (ergs/s/cm^2) de Ly α est calculée pour une bande passante de 120 Å. L'intensité I des profils intégrés sur le disque solaire est donnée en ergs/s/cm^2/Å afin de rendre plus facile la comparaison avec les profils publiés par Tousey *et al.* (1964).

La température de rayonnement T_R du continu de Lyman est déterminée par l'équation:

$$\int_0^\infty J_\nu \alpha_\nu \, \mathrm{d}\nu = \int_0^\infty B(T_R) \alpha_\nu \, \mathrm{d}\nu . \tag{11}$$

Nous présenterons les résultats obtenus avec le modèle HAO1 et le modèle de Coates.

1. Modèle HAO1

Les profils de Ly α portés sur la Figure 3 montrent l'influence de $\Delta\lambda_R$. Nous avons porté sur ce graphique les ailes du profil de Ly α obtenues avec une valeur de $\Delta\lambda_R$ égale à 2,5 Å. Nous pouvons constater que la limitation de la valeur de $\Delta\lambda_R$ donne une valeur de l'intensité émise dans les ailes du profil beaucoup trop grande: l'erreur peut être d'un facteur 2.

La Figure 4 où sont reportés les profils 1 et 3 de Ly α montre l'influence des collisions Ce résultat sera discuté plus loin.

La Figure 5 montre que l'intensité du rayonnement de Ly α obtenue avec un atome à trois niveaux est légèrement supérieure à celle obtenue avec un atome à deux niveaux.

L'intensité du rayonnement de Ly α calculée avec le modèle HAO1 est supérieure à celle indiquée par Tousey *et al.* (1964).

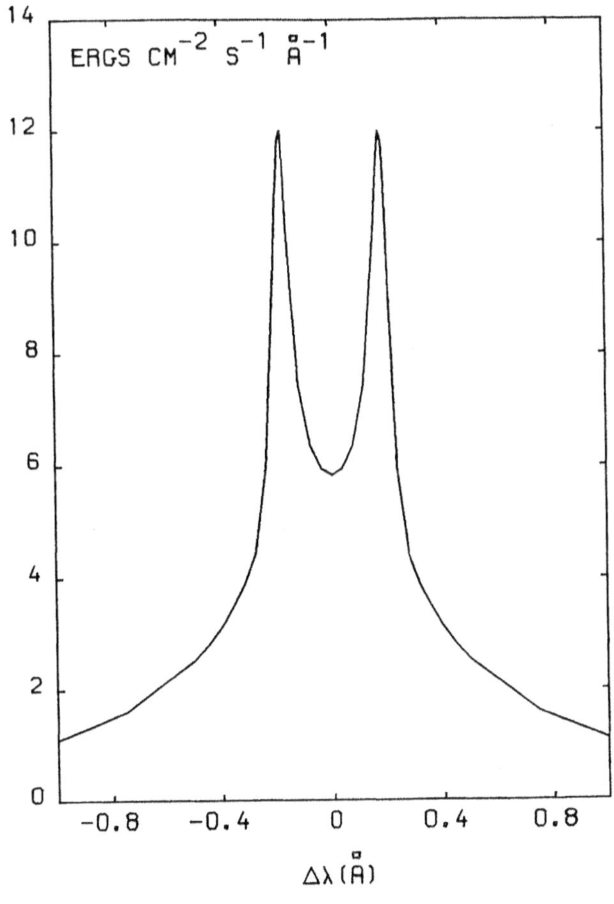

Fig. 5. Profil de Ly α. Modèle HAO1. Atome à trois niveaux. (Profil intégré sur le disque solaire.)

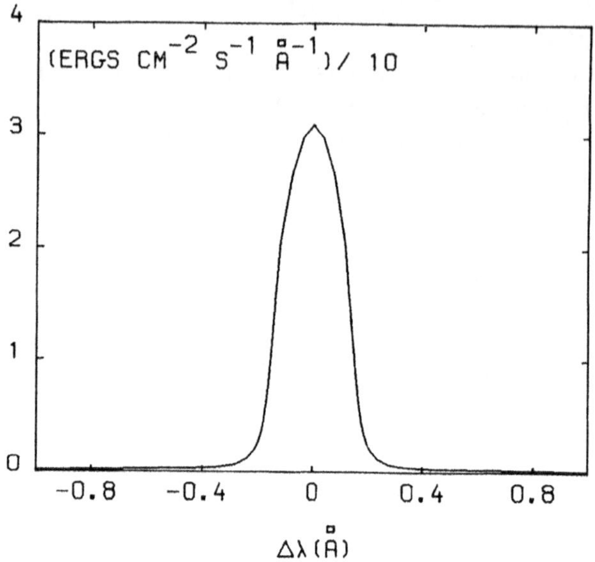

Fig. 6. Profil de Ly β. Modèle HAO1. Atome à trois niveaux. (Profil intégré sur le disque solaire.)

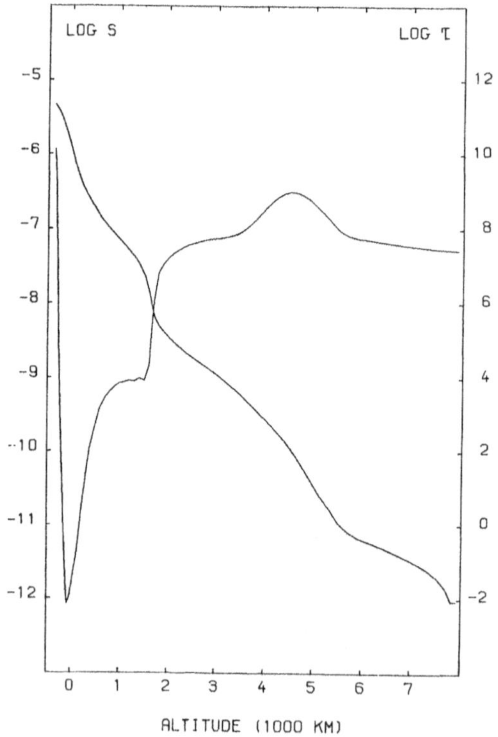

Fig. 7. Profondeur optique et fonction sourcé de Ly α. Modèle HAO1. Atome à trois niveaux.

Le profil de Ly β (Figure 6) ne présente pas la dépression centrale observée par TOUSEY *et al.* (1964).

La dépression centrale du profil de Ly α est due au maximum de la fonction source (Figure 7); la fonction source de Ly β (Figure 8) ne présente pas de phénomène analogue.

Les Figures 9 et 10 donnent dans le cas d'un atome à trois niveaux les coefficients d'écart à l'équilibre thermodynamique pour des valeurs de $\Delta\lambda_R$ de 0,33 Å et 2,5 Å.

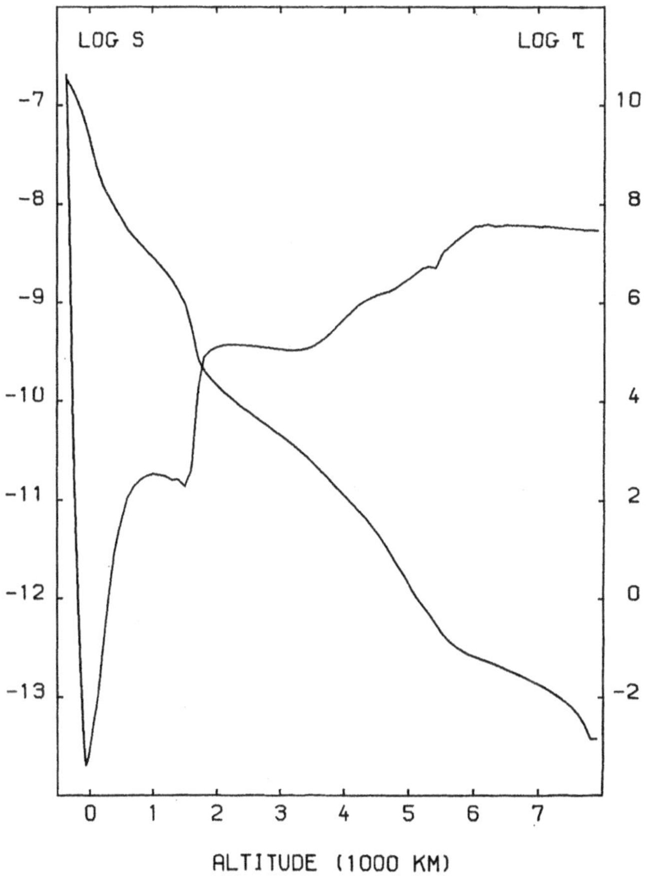

Fig. 8. Profondeur optique et fonction source de Ly β. – Modèle HAO1. Atome à trois niveaux.

2. MODÈLE DE COATES

Les Figures 11 et 12 montrent l'influence des collisions et du nombre de niveaux.

L'énergie émise, calculée avec le modèle de Coates, est peu différente de celle indiquée par Tousey.

Le profil de Ly β (Figures 13 et 14) présente un maximum central encadré par deux dépressions.

Les fonctions source de Ly α et Ly β présentent toutes les deux un maximum (Figures 15 et 16).

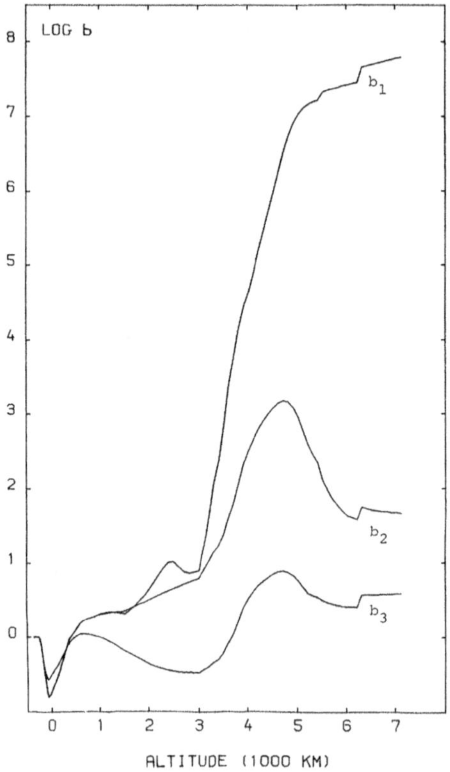

Fig. 9. Coefficients d'écart à l'équilibre thermodynamique. Modèle HAO1. $\Delta\lambda_R = 0{,}33$ Å.

A l'aide de la théorie de la formation des raies de Jefferies et Thomas, nous allons commenter les résultats obtenus.

Nous discuterons en particulier l'origine des dépressions centrales des profils, l'écartement des pics, le profil des ailes et la valeur de l'énergie.

Nous étudierons la raie Ly α, le continu de Lyman, puis Ly β.

3. Lyman α

L'étude de la formation des raies est basée sur l'expression de la fonction source donnée par Thomas et en particulier sur la signification et la valeur numérique des paramètres N et D.

Compte tenu de l'ordre de grandeur des différents termes, on a pour Ly α, pour

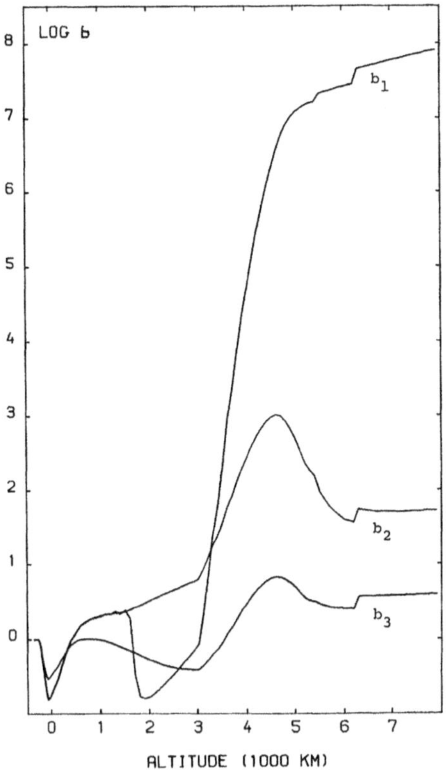

Fig. 10. Coefficients d'écart à l'équilibre thermodynamique. Modèle HAO1. $\Delta\lambda_R = 2{,}5$ Å

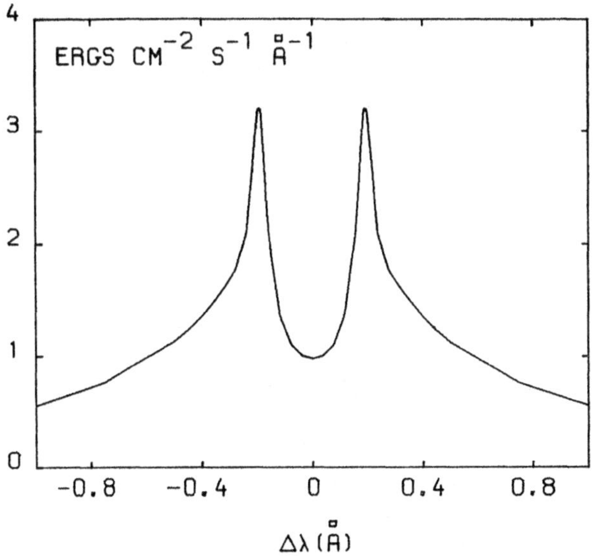

Fig. 11. Profil de Ly α. Modèle de Coates. Atome à deux niveaux. Sections de choc de Gryzinski divisées par deux. (Profil intégré sur le disque.)

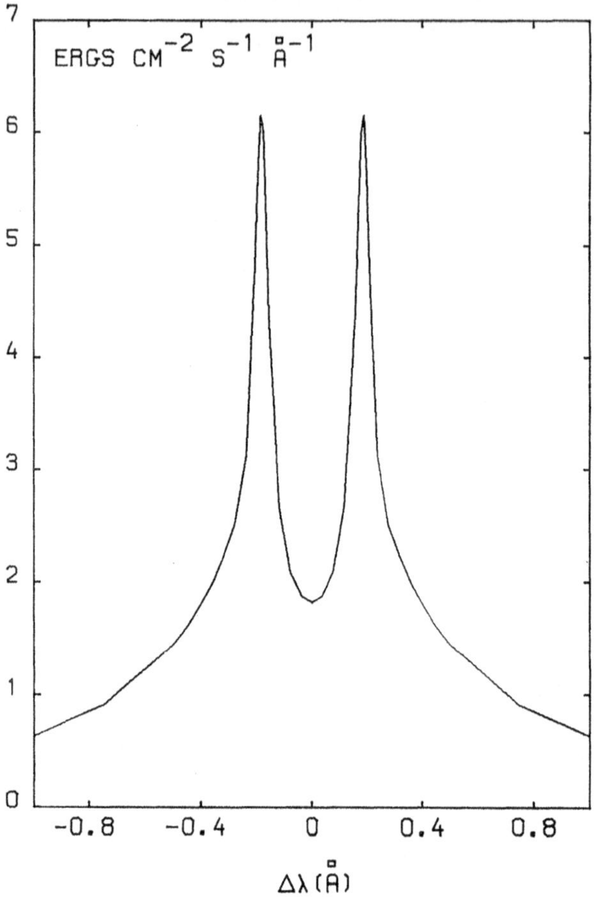

Fig. 12. Profil de Ly α. Modèle de Coates. Atome à trois niveaux. Sections de choc de Gryzinski. (Profil intégré sur le disque.)

un atome à deux niveaux (I, 329):

$$D_{12} = \frac{1}{A_{21}} J_2 \frac{R_1}{R_1 + R_2}, \tag{12}$$

$$J_2 = \int_{\nu 2}^{\infty} \alpha_\nu J_\nu \, d\nu . \tag{13}$$

Le paramètre D_{12} est déterminé par le terme de photoionisation J_2 du continu de Balmer.

Dans la chromosphère J_2 est constant:

$$J_2 = 0{,}300 \times 10^5 .$$

Le rapport des termes de recombinaison R_1 et R_2 du continu de Lyman et du continu de Balmer est peu différent de 2/5.

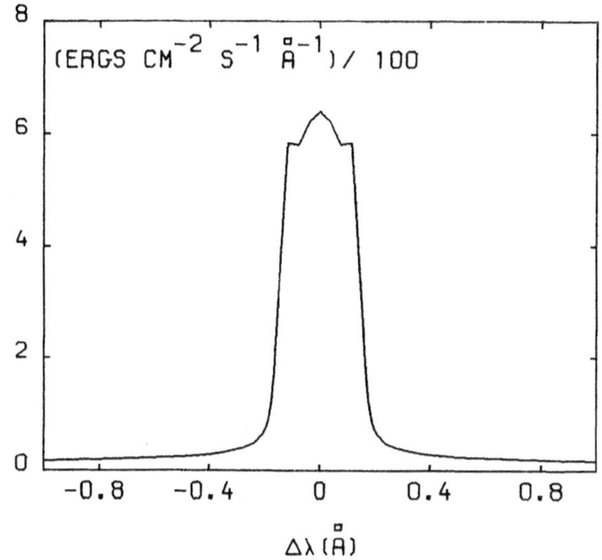

Fig. 13. Profil de Ly β. Modèle de Coates. Atome à trois niveaux. Sections de choc de Gryzinski. (Profil intégré sur le disque.)

On obtient donc:

$$D_{12} = 0{,}46 \times 10^{-4}.$$

Le numérateur N_{12} est déterminé par les excitations par collision C'_{12} et par les photoionisations à partir du niveau fondamental.

$$N_{12} = \frac{1}{B_{12}} \left(C'_{12} + J_1 \frac{1}{1 + \dfrac{R_1}{R_2}} \right), \tag{14}$$

$$\frac{1}{B_{12}} = 1{,}18 \times 10^{-10},$$

$$\frac{1}{\dfrac{R_1}{R_2} + 1} = \frac{1}{3}.$$

A. *Rôle des collisions et des photoionisations*

Nous allons discuter le rôle des collisions et des photoionisations.

Nous poserons:

$$N_{C12} = 1{,}18 \times 10^{-10}\, C'_{12},$$

$$N_{J1} = 1{,}18 \times 10^{-10}\, J_1 \frac{1}{1 + \dfrac{R_1}{R_2}}.$$

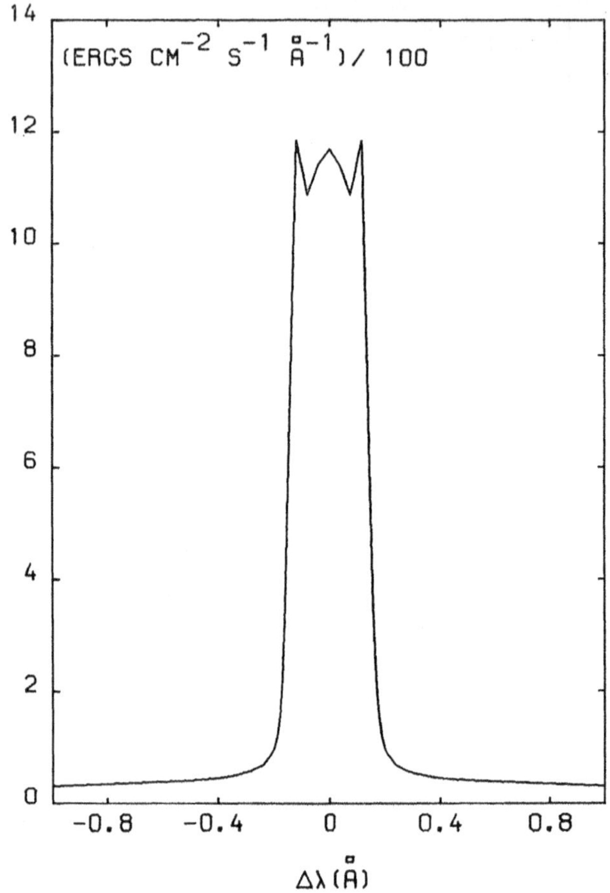

Fig. 14. Profil de Ly β. Modèle de Coates. Atome à trois niveaux. Section de choc de Gryzinski. Profil au centre du disque. (Nous utilisons la même unité d'énergie que pour le profil intégré sur le disque.)

Aux profondeurs où le continu de Lyman est transparent on a: $N_{J1} = 0{,}13 \times 10^{-11}$, si T_R est égal à 6900 °K.

Pour les deux modèles étudiés, HAO1 et de Coates, il y a égalité des termes de collision et de rayonnement à des profondeurs optiques au centre de Ly α respectivement égales à 2000 et 3300. A ces profondeurs le continu de Lyman cesse d'être transparent.

Dans les régions superficielles la fonction source de Ly α est déterminée par les collisions, aux grandes profondeurs par les photoionisations.

La grande valeur du terme de collision aux grandes altitudes est due à la température électronique élevée. On peut admettre que C'_{12} varie en fonction de la température comme:

$$A_c = T^{-1/2} e^{-X_{12}}.$$

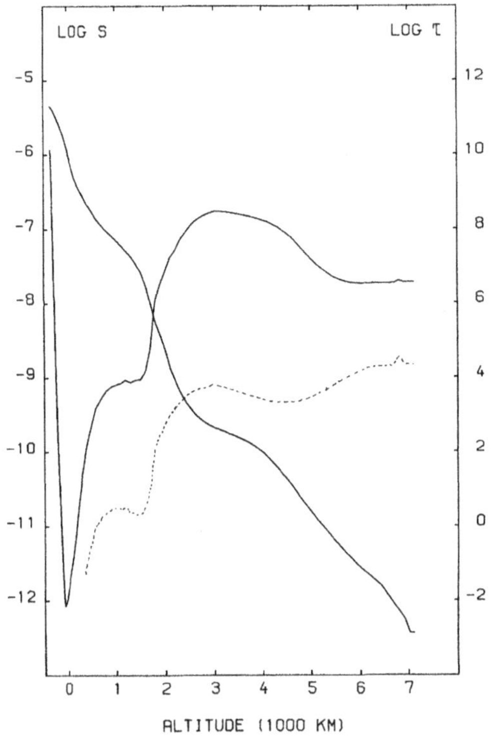

Fig. 15. Fonction source et profondeur optique de Ly α. Modèle de Coates. Atome à trois niveaux. En pointillés: fonction source de Ly β.

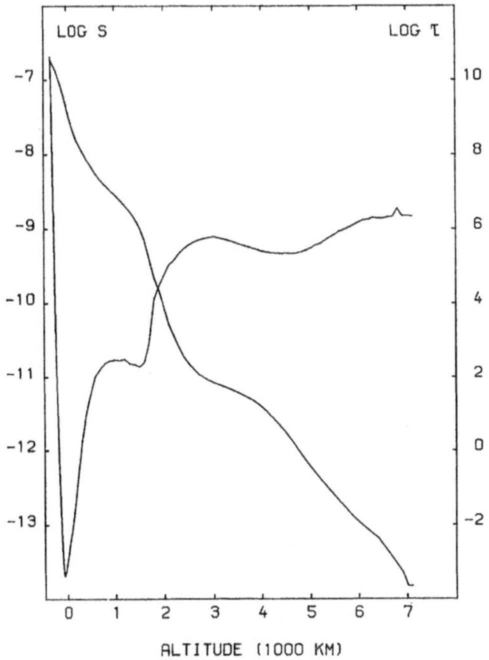

Fig. 16. Fonction source et profondeur optique de Ly β. Modèle de Coates. Atome à trois niveaux.

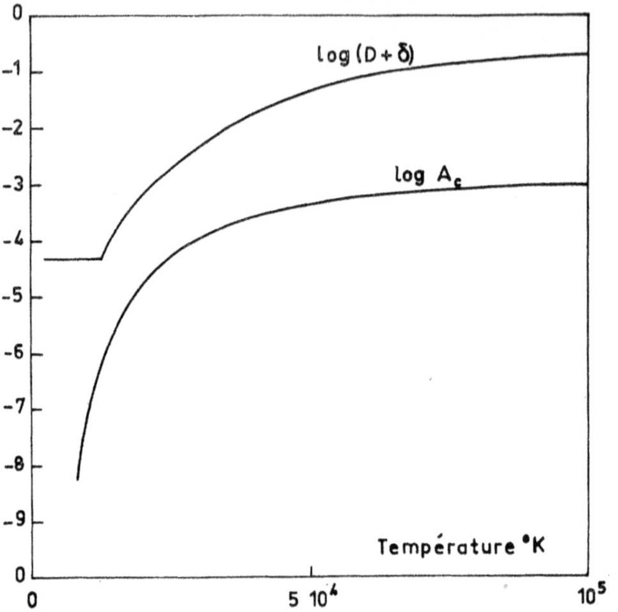

Fig. 17. $A_c(T_e)$. Influence de la température sur l'excitation par collisions du niveau 1 au niveau 2. $D+\delta$ = un exemple de variation de $D+\delta$.

La Figure 17 donne la variation de $\log A_c$ en fonction de la température:

$$\log A_c = -\tfrac{1}{2} \log T_e - 0{,}51 \left(\frac{10^5}{T_e}\right).$$

Entre 10^4 °K et 10^5 °K A_c varie d'un facteur 10^4. Dans un modèle de chromosphère la diminution de la densité électronique sur l'intervalle de hauteur correspondant est beaucoup plus faible.

On peut constater que la diminution du terme de collision devient très rapide aux températures inférieures à 20000 °K.

Le terme de photoionisation est très faible dans les régions superficielles. Lorsque le continu de Lyman cesse d'être transparent J_1 diminue, mais beaucoup moins rapidement que C'_{12}, l'effet de température étant compensé en partie par la diminution des écarts à l'équilibre thermodynamique.

A la profondeur où le terme de photoionisation devient prépondérant, le bilan des processus radiatifs de Lyman α n'est pas nul. Les résultats numériques montrent que le bilan des processus radiatifs de Ly α est nul, b_1 est alors égal à b_2, à une profondeur optique au centre de Ly α voisine de 10^7 avec les formules de quadratures utilisées.

La fonction source de Ly α variant très rapidement avec la profondeur, la profondeur limite τ_e(I, 172) calculée en supposant la fonction source constante n'est certainement pas valable. À titre d'indication, nous donnons cependant cette valeur:

$$\tau_e = \frac{1}{9} \frac{a}{D^2} \# 2 \times 10^4 .$$

Si l'on adopte un facteur K égal à 100 (I, 173) on obtient:

$$\tau_L \# 2 \times 10^8 .$$

L'altitude géométrique est égale à 1700 km.

Cette étude montre que les équations de transfert du continu de Lyman et de Lyman α sont couplées. Il est difficile d'admettre comme l'on fait MORTON et WIDING (1961) que la fonction source de Ly α dépende essentiellement des collisions. Ceci est vrai cependant dans la région où se forme le corps de la raie.

B. *Variation de la fonction source en fonction de la profondeur optique*

Le renversement du profil de Ly α, au centre de la raie, est du à la variation de la fonction source en fonction de la profondeur optique.

Rappelons l'expression de la fonction source:

$$S = \frac{N}{D + \delta},$$

où δ est le bilan des processus radiatifs par photon émis.

Dans les régions superficielles D_{12} est très inférieur à δ_{12}; $\delta_{12} \simeq 0{,}1$ ou 0,2, avec les modèles cités,

$$D_{12} = 0{,}46\ 10^{-4},$$

d'où:

$$S_{12} \simeq \frac{N_{12}}{\delta_{12}}.$$

La décroissance de δ_{12} étant plus rapide que celle de N_{12} la fonction source croît en fonction de la profondeur optique.

Nous avons porté sur la Figure 17 un exemple de valeur de log $(D_{12} + \delta_{12})$ afin de comparer sa variation à celle de N_{C12}.

Dans les régions très profondes:

$$\delta_{12} \ll D_{12},$$

$$S = \frac{N_{12}}{D_{12}}.$$

La fonction source décroît en fonction de la profondeur optique.

Nous ferons une remarque sur la position du maximum.

La température électronique à l'altitude où est formé le maximum de la fonction source est égale à 19000 °K dans le cas du modèle de Coates, 21000 dans le cas du modèle HAO1.

La position du maximum dépend d'une part de la variation de $\delta_{12} + D_{12}$, d'autre part de la variation de N_{C12}. La décroissance très rapide des excitations par collision au dessous de 20000 °K explique que la température électronique à l'altitude du maximum soit au moins de l'ordre de grandeur de 20000 °K.

Le maximum de la fonction source de Ly α est situé à une profondeur de température électronique d'autant plus élevée que le milieu est plus dense.

En plus des modèles cités nous avons utilisé le modèle interspiculaire de ATHAY et THOMAS (1961), modèle qui est très dense: le maximum de la fonction source de Ly α est situé à la profondeur de température électronique 28000 °K. L'émission de Ly α calculée avec ce modèle est beaucoup trop importante.

Nous ferons des essais systématiques ultérieurement afin de préciser la relation entre la profondeur optique τ_M du maximum de la fonction source et le modèle.

Les profondeurs optiques τ_M, au centre de la raie pour les modèles utilisés sont assez faibles:

Modèle	τ_M
HAO1	120
Coates	350
Athay Thomas	$\sim$100

C. *Écartement des pics*

La distance des pics dépend de la profondeur optique du maximum de la fonction source τ_M et de la largeur Doppler à cette profondeur.

On peut admettre que la position des pics x_0 est définie par le point du profil où la profondeur optique est égale à l'unité à la profondeur τ_M:

$$\pi^{1/2}\tau_M\phi(x_0) = 1\,,$$
$$\Delta\lambda_M = x_0 \Delta\lambda_D\,,$$

$\Delta\lambda_M$ est la distance des pics au centre de la raie, $\Delta\lambda_D$ la largeur Doppler à la profondeur τ_M (la valeur de x_0 ainsi définie est très peu différente de x_M).

Modèle HAO1:

$$\begin{aligned} \tau_M &\# 120 \\ V &= 19 \text{ km/s} \\ \Delta\lambda_D &= 0{,}077 \text{ Å} \\ a &= 0{,}23 \times 10^{-3} \\ \phi(x_0) &= 0{,}47 \times 10^{-2} \\ x_0 &\# 2{,}15 \\ \Delta\lambda_M &= 0{,}17 \text{ Å}\,. \end{aligned}$$

Le profil calculé donne une valeur de $\Delta\lambda_M$ égale à 0,18 Å.

Modèle de Coates:

$$\begin{aligned} \tau_M &\# 330 \\ V &= 18{,}26 \text{ km/s} \\ \Delta\lambda_D &= 0{,}074 \text{ Å} \\ a &\# 0{,}25 \times 10^{-3} \\ x_0 &\# 2{,}41 \\ x_M &\# 2{,}5 \\ \Delta\lambda_M &\# 0{,}185 \text{ Å} \end{aligned}$$

valeur qui est en accord avec celle donnée par le profil calculé.

Essayons de préciser la relation entre le modèle et la distance des pics.

Nous ferons d'abord la remarque suivante. Les valeurs de a et τ_M étant relativement faibles, les valeurs de x_0 sont égales à celles données par un profil purement Doppler.

Profil de Voigt		Profil Doppler
$a = 0,22 \times 10^{-3}$		
$\tau_M = 10^2$	$x_0 = 2,15$	$x_0 = 2,15$
$\tau_M = 10^3$	$x_0 = 2,64$	$x_0 = 2,63$

On obtient donc:

$$x_0 = 1,52 \left(\log \tau_M\right)^{1/2}$$

Cette relation suppose que la profondeur optique à la profondeur τ_M, à la fréquence définie par x_0, est égale à l'unité, et que la largeur Doppler est indépendante de la profondeur.

Il semble que la première hypothèse entraine au maximum une erreur d'environ 10% sur la valeur de x_0.

Pour les deux modèles étudiés la vitesse des particules varie assez peu entre τ égal à 10 et τ_M.

La distance des pics au centre de la raie est donnée par:

$$\Delta\lambda_M = x_0 \Delta\lambda_D$$

et dépend donc de la largeur Doppler à la profondeur τ_M. La largeur Doppler est essentiellement déterminée par la température électronique qui est élevée et voisine de 20000 °K.

Cette étude explique pourquoi la valeur $\Delta\lambda_M$ des profils observés par Tousey est toujours voisine de 0,2 Å.

D. *Étude de la dépression centrale du profil de Ly α*

Nous n'avons exécuté qu'un nombre limité de calculs et nous ne pouvons donner que quelques indications sur l'influence du modèle.

La dépression dépend de la variation de la fonction source entre τ égal à 1 et τ égal à τ_M. La fonction source jusqu'aux régions voisines du maximum est égale à N_{12}/δ_{12}. Il semble que la dépression est d'autant plus faible que la variation de N_{C12} entre τ_M et 1 est plus grande, c'est-à-dire que le gradient de température est plus élevé.

Les valeurs relatives des dépressions des profils de Ly α données par les modèles HAO1 et de Coates, qui sont respectivement égales à 0,5 et 0,3 semblent confirmer cette indication.

A la profondeur τ_M la température électronique est peu différente de 20000 °K pour les deux modèles; à la profondeur optique unité les températures sont égales à 50000 °K pour le modèle HAO1 et 40000 °K pour le modèle de Coates.

Modèle HAO1:

$H = 5500$ km	$Te = 50600$ °K
$H = 4500$ km	$Te = 20300$ °K

Modèle de Coates

$H = 5200$ km $Te = 40700\,°K$

$H = 3100$ km $Te = 19000\,°K$

Nous avons exécuté une série de calculs avec des modèles obtenus à partir du modèle HAO1 par modification du gradient de température au dessus de 2500 km: plus le gradient de température est faible, plus la dépression est grande. Les profils sont tracés sur la Figure 18. Le modèle donnant la plus grande dépression relative a une température constante égale à 30000 °K dans la haute chromosphère.

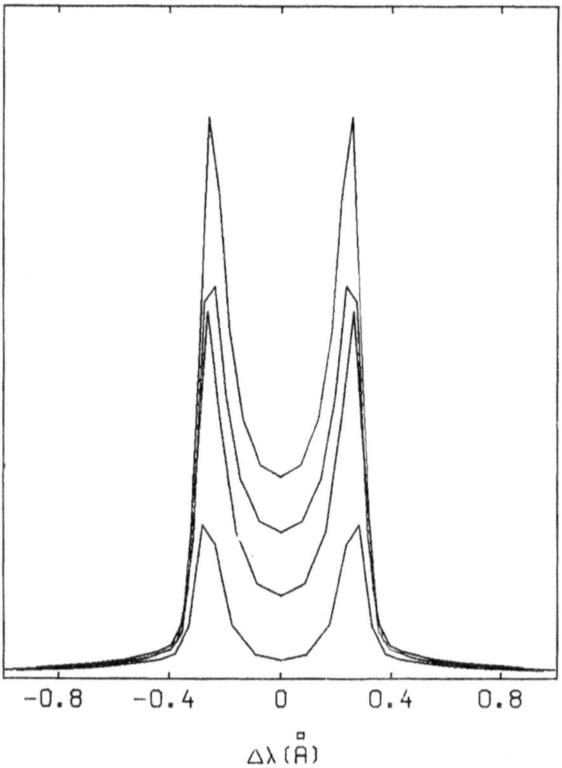

Fig. 18. Influence du gradient de température sur le profil de Ly α. La valeur relative de la dépression augmente lorsque le gradient de température diminue.

L'écartement des pics est du à la très grande microturbulence de ces modèles.
Nous expliquerons plus loin la variation de l'intensité.

E. *Étude des ailes du profil de Ly α. Continu de Lyman*

En présentant les résultats nous avons mis en évidence l'influence de $\Delta\lambda_R$ sur les ailes du profil de Ly α et sur le rayonnement du continu de Lyman (Tableaux I et II).

Nous ferons ultérieurement des calculs plus précis. Le continu de Lyman et les ailes du profil de Ly α doivent donner des renseignements sur le modèle à l'altitude à laquelle ils sont formés, qui est voisine de 2000 km.

F. *Puissance. Intensité du rayonnement*

La puissance reçue au sol calculée avec les différents modèles est supérieure à celle qui est mesurée, qui est peu différente de 6 ergs/cm^2 par seconde.

Nous avons déterminé dans le cas du modèle HAO1 la puissance reçue au sol en fonction de la bande passante considérée:

0,60 Å	25%
2 Å	50%
120 Å	100%

Une partie importante de la puissance reçue provient donc des ailes du profil que nous avons calculées avec une très mauvaise précision. Les calculs exécutés avec une trop petite valeur de $\Delta\lambda_R$ donnent une intensité de rayonnement dans les ailes surestimée.

Nous discuterons les résultats obtenus avec les différents modèles en partant de l'intensité du rayonnement émis dans la région centrale du profil et de façon plus précise en partant de l'intensité maximale du rayonnement.

Le Tableau I montre que l'intensité dépend des sections de choc, ce qui est justifié par la théorie de la formation de Ly α: la fonction source aux profondeurs où est formé le corps du profil est déterminée par les collisions.

L'intensité du rayonnement dépend également du modèle.

Le rapport des intensités maximales des profils donnés par les modèles HAO1 et de Coates est voisin de 2. Le rapport des termes de collision aux profondeurs τ_M des deux modèles est égal à 3,4, le rapport des densités électroniques est égal à 1,5. Il semble que le facteur densité électronique soit déterminant. La variation de l'intensité du rayonnement dans la série de calcul exécutée avec les modèles à divers gradients de température est due à la densité électronique; le maximum de la fonction source est toujours formé à une altitude de température électronique peu différente de 20000 °K: à l'altitude correspondante la densité électronique est d'autant plus faible que le gradient de température du modèle est lui-même plus faible.

4. Le continu de Lyman

Les paramètres de la fonction source (I, 78) du continu de Lyman sont dans le cas d'un atome à deux niveaux:

$$N' \simeq (C_{12} + P_{12}) \frac{P_2}{R_2 + C_{12} + R_{21}} \simeq P_{12} \frac{P_2}{R_{21}}, \tag{15}$$

$$D' \simeq R_2 \frac{C_{12} + R_{21}}{P_2 + C_{12} + R_{21}} \simeq R_2 . \tag{16}$$

Au numérateur le terme P_{12} est prépondérant.

Nous avons montré que les équations de transfert du rayonnement de Ly α et du continu de Lyman sont couplées. Si l'on traite séparément le continu de Lyman et

Ly α, on doit introduire le bilan des processus radiatifs de Ly α dans les équations d'équilibre statistique pour que le procédé itératif converge.

On obtient:

$$D' = R_2(C_{12} + \Delta_{21}) \frac{1}{P_2 + C_{12} + \Delta_{21}}. \tag{17}$$

Les solutions obtenues au cours des premières itérations sont assez éloignées de la solution physique du problème. A la deuxième itération, pour τ supérieur à τ_M, Δ_{21} est négatif: les intensités du rayonnement et les populations des niveaux sont négatives.

Nous avons du résoudre simultanément les équations de transfert du rayonnement de Ly α et du continu de Lyman en utilisant la double précision.

Nous avons signalé l'accord assez bon entre les résultats théoriques et les observations pour le rayonnement du continu de Lyman.

5. Ly β

A. *Fonction source. Dépression centrale du profil*

Nous allons étudier la raie Ly β et l'origine du renversement du profil au centre de la raie.

Les expressions de N_{13} et D_{13} sont obtenues à partir de celle de: b_3/b_1. Si on néglige les termes d'ionisation par collision, on obtient:

$$\frac{b_3}{b_1} = \frac{\begin{vmatrix} C_{12} + C_{13} + P_1 + P_{12} + P_{13} & -(C_{12} + R_{21}) & R_1 \\ -(C_{12} + P_{12}) & P_2 + C_{12} + C_{23} + P_{23} + R_{21} & R_2 \\ -(C_{13} + P_{13}) & -(C_{23} + P_{23}) & R_3 \end{vmatrix}}{\begin{vmatrix} C_{13} + R_{31} & -(C_{12} + R_{21}) & R_1 \\ C_{32} + R_{32} & P_2 + C_{12} + C_{23} + P_{23} + R_{21} & R_2 \\ -(P_3 + C_{13} + C_{32} + R_{32} + R_{31}) & -(C_{23} + P_{23}) & R_3 \end{vmatrix}}. \tag{18}$$

Soient m_1, m_2, m_3 les mineurs des éléments de la première colonne.

$$\frac{b_3}{b_1} = \frac{P_{13} + C_{13} + P_1 \dfrac{m_1}{m_1 - m_3} + (P_{12} + C_{12}) \dfrac{m_1 + m_3}{m_1 - m_3}}{R_{31} + C_{13} - P_3 \dfrac{m_3}{m_1 - m_3} - (R_{32} + C_{32}) \dfrac{m_3 + m_2}{m_1 - m_3}}. \tag{19}$$

$$m_1 = R_3 R_{21} + (P_2 + C_{12}) R_3 + (C_{23} + P_{23})(R_2 + R_3) \simeq R_3 R_{21}. \tag{20}$$

$$m_2 = -(C_{12} + P_{12}) R_3 + (C_{23} + P_{23}) R_1, \simeq -R_3 R_{21}. \tag{21}$$

$$m_3 = -(R_1 + R_2) R_{21} - R_1 (P_2 + C_{12} + C_{23} + P_{23}) - R_1 C_{12}, \simeq -(R_1 + R_2) R_{21}. \tag{22}$$

$$m_1 - m_3 \simeq R_{21}(R_1 + R_2 + R_3),$$

$$m_1 + m_2 = (C_{23} + P_{23})(R_1 + R_2 + R_3) + P_2 R_3,$$

$$m_2 + m_3 \simeq -R_{21}(R_1 + R_2 + R_3),$$

$$\frac{b_3}{b_1} = \frac{P_{13} + N'}{R_{31} + D'}.$$

Le terme le plus grand de D' est R_{32}, ce qui donne pour le paramètre du dénominateur de la fonction source:

$$D_{13} \# \frac{R_{32}}{R_{31}} = \frac{A_{32}}{A_{31}} = 0{,}794 . \tag{23}$$

Considérons maintenant le numérateur de la fonction source:

$$N_{13} = \frac{2h\nu^3}{c^2} e^{-x_1} e^{x_2} \frac{1}{R_{31}} \times$$
$$\times \left[C_{13} + (P_{12} + C_{12}) \left(\frac{C_{23} + P_{23}}{R_{21}} + \frac{P_2}{R_{21}} \frac{R_3}{R_1 + R_2 + R_3} \right) + P_1 \frac{R_3}{R_1 + R_2 + R_3} \right] . \tag{24}$$

L'étude systématique des valeurs numériques montre que les termes les plus importants sont le terme de collision C_{13} et le terme dépendant de P_{12}:

$$P_{12} \times P_{23}/R_{21} .$$

Nous poserons

$$N_{13} = N_{C13} + N_{J12} ,$$
$$N_{C13} = \frac{1}{B_{13}} C'_{13} , \tag{25}$$
$$\frac{1}{B_{13}} = 0{,}800 \times 10^{-9} ,$$
$$N_{J12} = \frac{1}{B_{13}} \frac{g_3}{g_1} \left(\frac{J}{\frac{2\,h\nu^3}{c^2}} \right)_{12} . \tag{26}$$

Etant donnée la grande valeur de D_{13} la fonction source dans les régions où la transition Ly β devient transparente est peu différente de:

$$N_{13}/(1 + D_{13}) .$$

En fait la valeur de δ_{13} à la surface est voisine de 0,7, ce qui donne:

$$S_{13} = \frac{N_{13}}{D_{13} + \delta_{13}} ,$$
$$= \frac{N_{13}}{1{,}5} .$$

Aux grandes profondeurs:

$$S_{13} = N_{13}/D_{13} ,$$
$$= N_{13}/0.8 .$$

D'après les résultats de l'intégration numérique δ_{13} est négligeable par rapport à D_{13} à une profondeur optique voisine de 30.

La fonction source de Ly β varie approximativement comme N_{13}.

Soit τ_M la profondeur optique au centre de Ly β à la profondeur où la fonction source de Ly α atteint sa valeur maximale. Pour tous les modèles utilisés à la profondeur τ_M le terme de rayonnement N_{J12} de Ly α est supérieur au terme de collision N_{C13}.

Dans le cas du modèle HAO1, l'augmentation rapide de N_{C13} due au gradient de température élevé compense la diminution de N_{J12}. La fonction source croît lorsque τ diminue.

Dans le cas du modèle de Coates la profondeur τ_M est située dans la région où la température électronique est constante. Lorsque τ décroît à partir de τ_M la fonction source commence par décroître; lorsque la température électronique croît à nouveau, les excitations par collisions et la fonction source croissent.

La forme du profil de Ly β dépend de la valeur de τ_M et de l'intervalle de profondeur où le terme de collision reste inférieur au terme de photoexcitation de Ly α, c'est-à-dire de la longueur du palier de température. Le modèle présentant une température constante dans la haute chromosphère donne un profil à grande dépression centrale (Figure 19).

Il semble que la présence d'un palier de température soit obligatoire pour que le profil de Ly β présente une dépression. La valeur de la dépression doit dépendre de la longueur de ce palier.

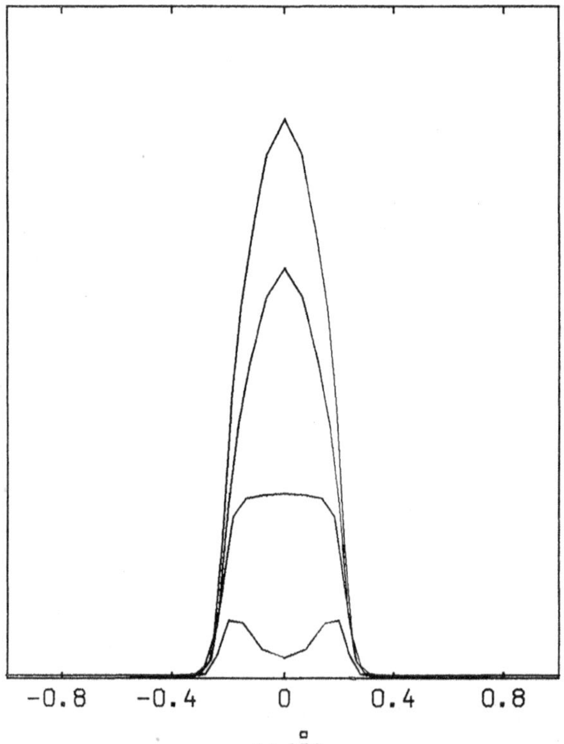

Fig. 19. Influence du gradient de température sur le profil de Ly β. Le modèle donnant une depression centrale du profil présente un palier de température.

Le maximum de la fonction source de Ly β étant formé à la même altitude que le maximum de la fonction source de Ly α, il doit y avoir une relation entre la distance des pics des profils des deux raies.

B. *Distance des pics du profil de Ly β*

Nous admettons que pour les deux raies, Ly α et Ly β, les profondeurs optiques aux longueurs d'onde définies par les pics et à la profondeur du maximum des fonctions source sont égales:

$$[\tau_M \phi(x_0)]_{\mathrm{Ly}\,\alpha} = [\tau_M \phi(x_0)]_{\mathrm{Ly}\,\beta}, \tag{28}$$

d'où

$$f_{13}\lambda_{13}\phi(x_{0\beta}) = f_{12}\lambda_{12}\phi(x_{0\alpha}), \tag{29}$$

relation qui suppose que les largeurs Doppler sont indépendantes de la profondeur.

Les profils d'absorption peuvent être considérés comme Doppler pour les valeurs de $x_{0\alpha}$ et $x_{0\beta}$:

$$x_{0\beta}^2 = x_{0\alpha}^2 - 1{,}826\,.$$

On peut calculer la distance des pics du profil de Ly β donnée par le modèle de Coates:

$$\begin{aligned} x_{0\alpha} &= 2{,}5 \\ x_{0\beta} &= 2{,}1 \\ \Delta\lambda_{\mathrm{D}} &= 0{,}062\ \text{Å} \\ \Delta\lambda_M &= 0{,}13\ \text{Å}\,, \end{aligned}$$

qui est égale à la valeur déduite du profil calculé.

Le rapport des distances des pics des profils de Ly α et Ly β obtenus avec le modèle de Coates est égal à 1,5, rapport qui est égal à celui indiqué par Tousey pour les profils observés.

Les distances des pics des profils de Ly α et Ly β observés au même point de disque solaire doivent donner une indication sur la température cinétique.

La relation (29) donne:

$$\Delta\lambda_{\mathrm{D}} = 0{,}74\left(\Delta\lambda_M^2(\mathrm{Ly}\,\alpha) - 1{,}4\,\Delta\lambda_M(\mathrm{Ly}\,\beta)\right)^{1/2} \tag{30}$$

où $\Delta\lambda_{\mathrm{D}}$ est la largeur Doppler de Ly α.

L'équation (30) n'est valable que si la largeur Doppler est indépendante de l'altitude.

Il serait intéressant d'avoir des profils de Ly α et Ly β observés au même point du disque solaire.

Troisième partie: Les raies Hα et Hβ. Résultats. Comparaison aux observations

Nous présentons ici les résultats des calculs exécutés avec le modèle HAO1 – De Jager.

1. Influence des transitions de Lyman

Dans le cas d'un atome à trois niveaux nous avons calculé les coefficients d'écart à l'équilibre thermodynamique en supposant nul le bilan des processus radiatifs des transitions de Lyman (Figure 20) afin de comparer les résultats à ceux obtenus sans cette hypothèse (Figure 10).

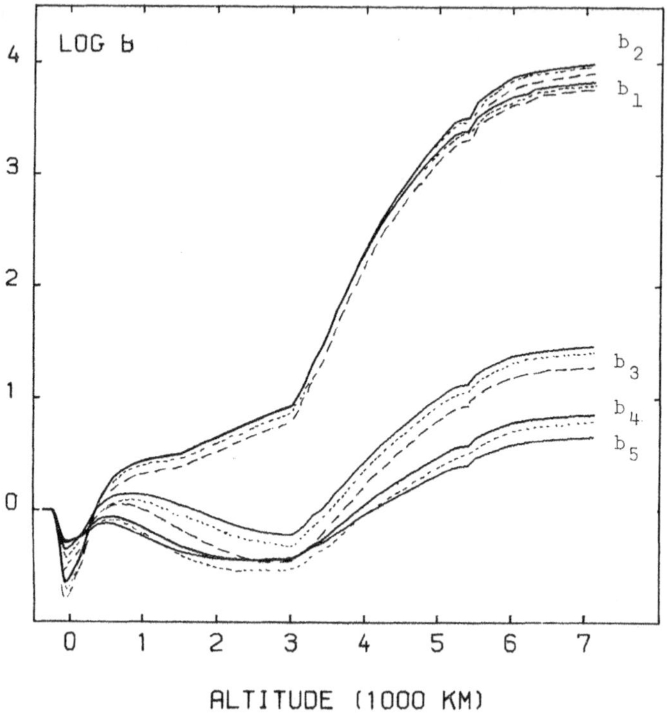

Fig. 20. Coefficients d'écart à l'équilibre thermodynamique. Influence du nombre de niveaux (avec bilan détaillé des processus radiatifs des transitions de Lyman). ——— = Atome à 5 niveaux, - - - - = Atome à 4 niveaux, et — — — = Atome à 3 niveaux.

Dans la basse chromosphère lorsque l'altitude est inférieure à 3000 km, les valeurs de b_2 et b_3 données par les deux calculs sont égales. On peut donc calculer, avec le modèle HAO1, les raies de la série de Balmer en admettant que le bilan des processus radiatifs des transitions de Lyman est nul.

Dans la haute chromosphère les fonctions source obtenues pour Hα (Figure 21) sont très différentes: la remontée de la fonction source de Hα lorsqu'on tient compte du transfert du rayonnement des transitions de Lyman, est due à l'action de Ly α et Ly β. Cette remontée est sans effet sur le profil de la raie Hα qui est transparente à cette altitude.

Tous les calculs concernant la série de Balmer ont été exécutés en supposant nul le bilan des processus radiatifs des transitions de Lyman, hypothèse qui est réaliste pour le modèle HAO1 mais qui ne l'est pas obligatoirement pour tous les modèles de chromosphère.

2. Influence du nombre de niveaux. profils de Hα et Hβ

Nous avons calculé les coefficients d'écart à l'équilibre thermodynamique pour un atome à 4 et 5 niveaux (Figure 20).

Dans la basse chromosphère les coefficients d'écart à l'équilibre thermodynamique croissent légèrement avec le nombre de niveaux.

A l'altitude où est formé le centre du profil de Hα, vers 2000 km, le rapport b_3/b_2 croît avec le nombre de niveaux: l'intensité résiduelle, au centre du profil de Hα, augmente donc également.

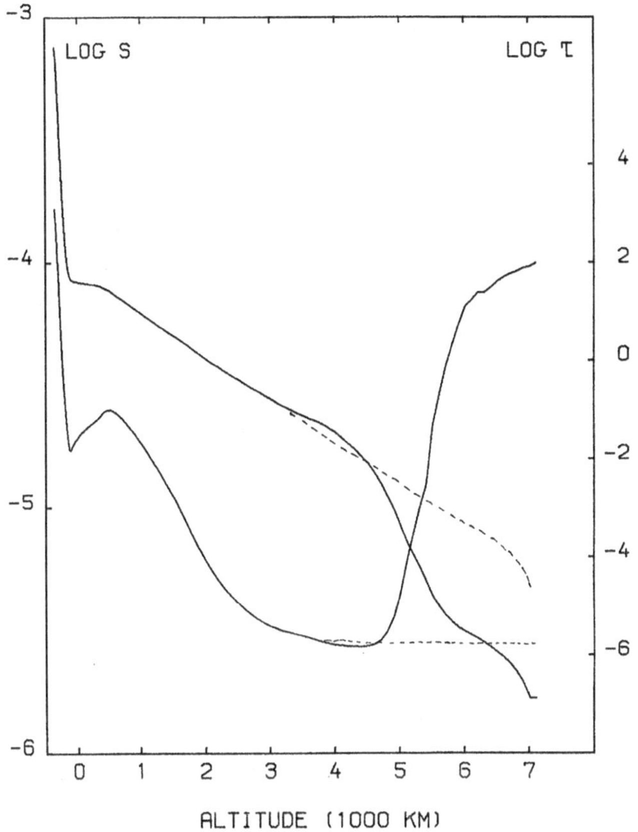

Fig. 21. Fonction source et profondeur optique de Hα Modèle HAO1 – De Jager. ——— = Résolution du problème du transfert du rayonnement pour toutes les transitions, et ------ = Bilan détaillé des processus radiatifs des transitions de Lyman.

Hα

Nombre de niveaux	Intensité résiduelle
3	0,121
4	0,135
5	0,145
Observations	
White (1963)	0,154
David (1961)	0,169

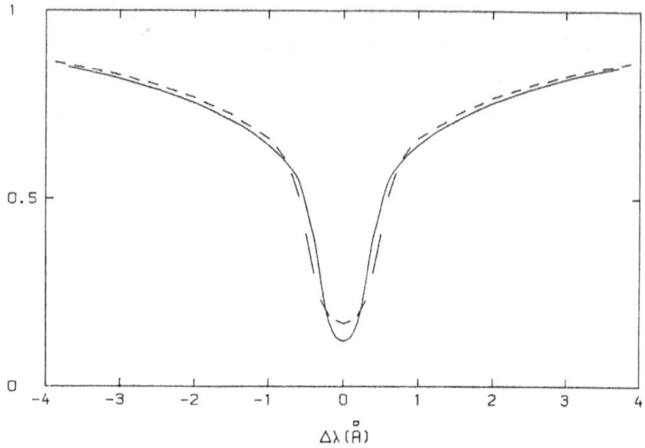

Fig. 22. Profil de Hα. ——— = Modèle HAO1 – De Jager. Atome à trois niveaux, et ----- = Profil observé par David. Microturbulence nulle.

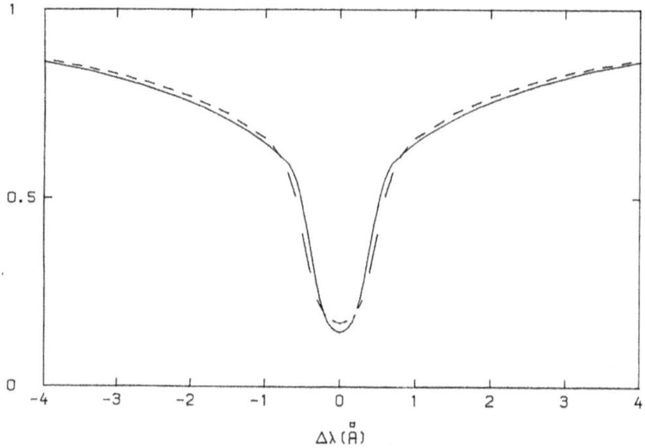

Fig. 23. Profil de Hα. ——— = Modèle HAO1 – De Jager. Atome à cinq niveaux, et ------ = Profil observé par David.

Le profil de Hα (Figures 22 et 23) est plus étroit que le profil observé.

Hβ

Nombre de niveaux	Intensité résiduelle
4	0,1174
5	0,120
Observations	
White (1963)	0,130
David (1961)	0,133

L'intensité du rayonnement dans les ailes de Hβ est trop grande: l'effet Stark ne peut être négligé (Figure 24).

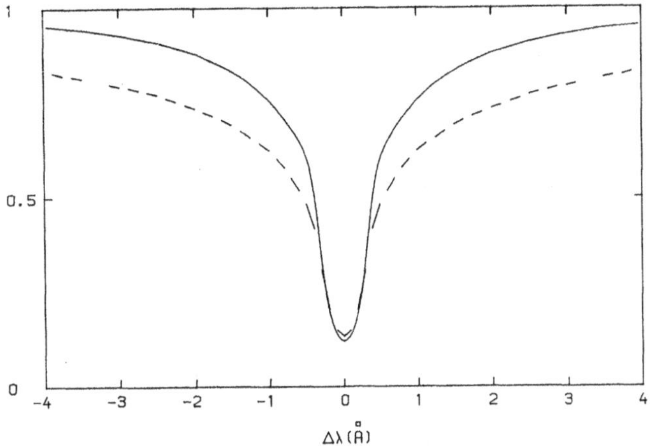

Fig. 24. Profil de $H\beta$. ——— = Modèle HAO1 – De Jager. Atome à cinq niveaux, et ------ = Profil observé par David.

Nous allons faire l'étude théorique de la fonction source de $H\alpha$ et expliquer en particulier la décroissance rapide de la fonction source dans la basse chromosphère.

3. Etude théorique de la fonction source de $H\alpha$. Discussion des résultats

A. *Etude théorique*

Les expressions des paramètres de la fonction source de $H\alpha$ sont déterminées en supposant nul le bilan des processus radiatifs des transitions de Lyman, dans le cas d'un atome à trois niveaux.

$$\frac{b_3}{b_2} = \frac{\begin{vmatrix} -C_{12} & C_1 & C_{12}+C_{13} \\ P_2+P_{23}+C_{23}+C_{12} & R_2 & -C_{12} \\ -(C_{23}+P_{23}) & R_3 & -C_{13} \end{vmatrix}}{\begin{vmatrix} C_{13} & C_1 & C_{12}+C_{13} \\ C_{23}+R_{23} & R_2 & -C_{12} \\ -(P_3+C_{13}+C_{23}+R_{32}) & R_3 & -C_{13} \end{vmatrix}}, \tag{31}$$

$$\frac{P_{23}+P_2\dfrac{m_2}{m_2+m_3}+C_{23}}{R_{23}+P_3\dfrac{m_3}{m_2+m_3}+C_{23}}.$$

De l'expression de b_3/b_2 on déduit celles des paramètres N_{23} et D_{23} de la fonction source.

$$N_{23} \# \frac{1}{B_{23}}\left[J_2\frac{R_3}{R_3+R_2}+C'_{23}\right] \tag{32}$$

$$D_{23} \# \frac{1}{A_{32}}\left[\frac{R_2}{R_3+R_2}J_3+C'_{32}\right]. \tag{33}$$

Dans la chromosphère entre la profondeur où commence à se former Hα et le minimum de température R_2/R_3 varie de 0,6 à 1,3. On peut adopter comme valeur moyenne de $R_3/(R_3+R_2)$ et $R_2/(R_3+R_2)$: 0,5.

Dans la chromosphère J_2 et J_3 sont constants.

$$
\begin{aligned}
J_2 &= 0{,}306 \times 10^5, \qquad \text{(Modèle De Jager)}\\
J_3 &= 0{,}200 \times 10^6,\\
N_{23} &\# 1{,}408 \times 10^{-9}(0{,}153 \times 10^5 + C'_{23}),\\
D_{23} &\# 2{,}28 \times 10^{-8}(0{,}100 \times 10^6 + C'_{32}).
\end{aligned} \tag{34}
$$

Nous donnons les profondeurs optiques au centre de Hα et les altitudes auxquelles les termes de collision représentent 10% et 50% des paramètres de la fonction source.

	H		τ
		N_{23}	
10%	260 km		35
50%	0		70
		D_{23}	
10%	2400 km		0,4
50%	85 km		39.

Le terme de collision du numérateur de la fonction source ne devient important que dans la très basse chromosphère; par contre le terme de collision du dénominateur n'est jamais négligeable; dans la basse chromosphère, son augmentation rapide, due elle même à l'augmentation rapide de la densité, explique la décroissance de la fonction source.

Dans le cas du modèle HAO1 les termes de collision sont inférieurs aux termes de photoionisation aux profondeurs où est fermé le centre du profil ($\tau = 1$, altitude: 2000 km); mais ils en sont assez proches. On peut prévoir qu'un modèle donnant des termes de collision plus grands, la densité ou la température électroniques étant plus élevées, donnera une intensité résiduelle au centre de Hα plus grande.

B. *Influence du modèle de chromosphère*

La Figure 25 donne le profil calculé avec le modèle interspiculaire de Athay et Thomas (1961).

Il est facile de contrôler que l'intensité résiduelle très grande est due aux collisions; la densité et la température électroniques de ce modèle sont supérieures à celles du modèle HAO1, à l'altitude où est formé le centre du profil.

HAO1	Athay-Thomas
$H = 2000$ km	$H = 1200$ km
$T = 6400\,°$K	$T_e = 7900\,°$K
$N_e = 0{,}891 \times 10^{11}$	$N_e = 0{,}219 \times 10^{12}$

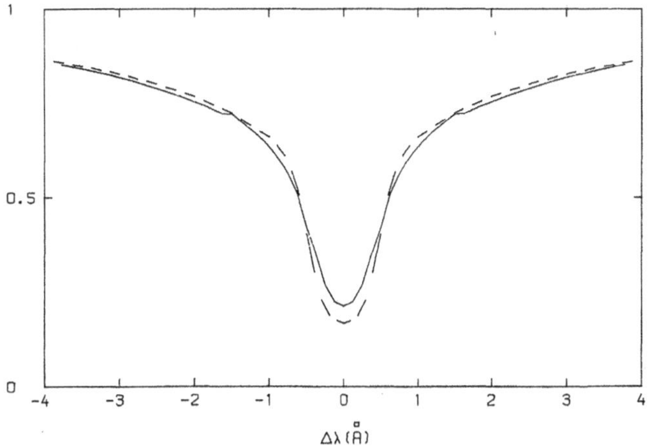

Fig. 25. Profil de Hα. ——— = Modèle interspiculaire de Athay et Thomas. Atome à trois niveaux, et ------ = Profil observé par David.

Le modèle de Athay et Thomas donne une largeur de profil qui se rapproche de celle du profil observé et qui est due à la température électronique élevée.

Le calcul exécuté avec le modèle spiculaire, qui donne une intensité résiduelle très importante, n'a guère de sens, les spicules n'occupant qu'un très faible pourcentage du volume de l'atmosphère à l'altitude où est formé le centre du profil de Hα.

Les profils calculés avec les modèles HAO1 et le modèle interspiculaire de Athay et Thomas montrent l'influence du modèle de la basse chromosphère sur le corps de la raie.

Nous ferons ultérieurement une étude systématique afin d'obtenir un meilleur accord avec les observations. Pour calculer correctement le profil de Hα, il serait nécessaire de tenir compte des hétérogénéités de l'atmosphère solaire.

C. *Influence du modèle de photosphère*

Les ailes du profil de Hα sont formées dans la photosphère, depuis une longueur d'onde assez proche du centre de la raie puisque la profondeur optique n'est que de quelques dizaines d'unités au minimum de température.

Dans la photosphère la fonction source de Hα diffère peu de la fonction de Planck. De plus on doit tenir compte de l'absorption continue et utiliser l'expression de la fonction source:

$$S = \frac{S_L + \frac{r}{\phi} S_c}{1 + \frac{r}{\phi}}.$$

Il en résulte qu'à une longueur d'onde très proche du centre de la raie l'hypothèse de l'équilibre thermodynamique local est valable.

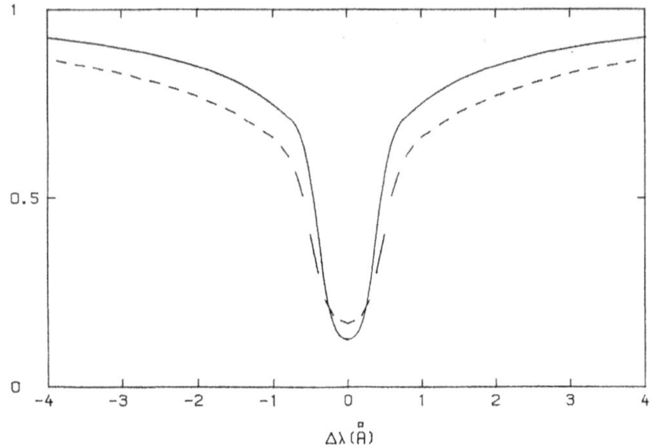

Fig. 26. Profil de Hα. ——— = Modèle de photosphère froid, donnant une intensité continue très inférieure à celle qui est observée, et ------ = Profil observé par David.

Le profil (Figure 26) calculé avec un modèle de photosphère très froid, donne une intensité relative du rayonnement dans les ailes beaucoup trop grande.

Conclusions

L'étude des profils théoriques de Ly α et Ly β calculés à partir d'un modèle en appliquant la théorie de la formation des raies de Jefferies et Thomas nous a permis de dégager certaines caractéristiques du modèle de la haute chromosphère.

La dépression centrale du profil de Ly β est due à un palier de la température électronique de la chromosphère. La valeur de la dépression, ainsi que celle du profil de Ly α dépendent de la longueur de ce palier.

La fonction source de Ly α, aux profondeurs où est formé le centre du profil, est déterminée par les collisions: l'énergie émise dépend donc de la température et de la densité électroniques. Pour les modèles qui donnent une valeur de l'énergie comparable à celle qui est mesurée, le maximum de la fonction source est formé au voisinage de l'altitude de température électronique 20000 °K. L'énergie dépend de la densité électronique à l'altitude correspondante. Le modèle de Coates, qui est moins dense que le modèle HAO1, donne une valeur de l'énergie proche de celle qui est mesurée.

Enfin l'étude de la formation des pics des profils de Ly α et Ly β permet d'expliquer leur distance et permet de comprendre pourquoi la distance des pics des profils observés de Ly α est toujours voisine de 0,2 Å.

Le profil de la raie Hα donne des renseignements sur la basse chromosphère. L'intensité résiduelle du centre du profil est essentiellement déterminée par les processus de photoionisation; cependant l'effet des collisions n'est pas négligeable. Les modèles donnant des termes de collision trop grands, la température ou la densité électronique étant trop élevées, donnent une intensité résiduelle trop grande.

Summary

The continuous and line spectrum of hydrogen emitted by the sun is computed, taking deviations of local thermal equilibrium into account. The theory of line formation as given by Jefferies and Thomas has been applied to the computations; a comparison of these results with the observations gives information on the solar atmospheric model.

The energy emitted by Lyman α has been computed for a two- and three-level atom: it appears to increase slightly with the number of levels. The energies, computed with the HAO model and the model of Coates, are larger than the observed values. The core of the Lyman α profile is determined by collisions; the peaks are formed at a height where the electron temperature is about 20000°K; the energy depends very strongly on the electron density of the model.

It is shown that the profile of Lyman β only shows a central self reversal if the model of the high chromosphere has a temperature plateau.

An interpretation of the observed distance of the peaks of the Lyman α and Lyman β line profiles is possible; it can also be shown why the distance of the Lyman α peaks to the line centre is always of the order of 0.2 Å.

The residual central intensity of Hα increases slightly with the number of atomic levels; the value computed with a five level atom, with the HAO chromospheric model, and the Utrecht photospheric model (1964) does not differ very much from the observed value: it is slightly smaller than the observed value and the computed profile is narrower than the observed profile.

In the case of the HAO model the source function of Hα is dominated by the photo-ionization terms; nevertheless, the collision terms are not much smaller than the photo-ionization terms; the residual central intensity of Hα computed with a chromospheric model similar to the interspicular model of Athay and Thomas but assuming a higher electron temperature and density is larger than the observed value.

Références

Athay, R. G. et Thomas, R. N.: 1961, *Physics of the Solar Chromosphere*. Interscience Publishers, New York.

Athay, R. G., Menzel, D., Pecker, J. G., et Thomas, R. N.: 1954, *Astrophys. J. Suppl.* **1,** 505.

Coates, R. J.: 1958, *Astrophys. J.* **128,** 83.

Cuny, Y.: 1967, *Ann. Astrophys.* **30,** 143.

David, K. H.: 1961, *Z. Astrophys.* **53,** 37.

Feautrier, P.: 1964, *Compt. Rend.* **258,** 3189.

Fox, L.: 1957, *The Numerical Solution of Two-point Boundary Problems in Ordinary Differential Equation*. Clarendon Press, Oxford.

Griem, H. R.: 1964, *Plasma Spectroscopy*. McGraw-Hill Book Company.

Gryzinski, M.: 1965, *Phys. Rev.*, 336.

John, T. L.: 1960, *Monthly Notices Roy. Astron. Soc.*, 41.

Karsas, W. J. et Latter, R.: 1961, *Astrophys. J. Suppl.* **55,** 167.

Labs, D.: 1957, *Z. Astrophys.* **44,** 37.

Menzel, D.: 1962, *Selected Papers on Physical Processes in Ionized Plasmas*. Dover Publ., New York.

Morton, D. C. et Widing, K. G.: 1961, *Astrophys. J.* **133**, 596.
Pottasch, S. R. et Thomas, R. N.: 1959, *Astrophys. J.* **130**, 941.
Thomas, R. N.: 1957, *Astrophys. J.* **125**, 260.
Tousey, R., Purcell, J. D., et Austin, W. E.: 1964, *Space Res.* **4**, 703.
Vardya, M. S.: 1961, *Astrophys. J.* **133**, 107.
White, C. R.: 1963, *Astrophys. J.* **137**, 1217.

GPSR Compliance

The European Union's (EU) General Product Safety Regulation (GPSR) is a set of rules that requires consumer products to be safe and our obligations to ensure this.

If you have any concerns about our products, you can contact us on ProductSafety@springernature.com

In case Publisher is established outside the EU, the EU authorized representative is:

Springer Nature Customer Service Center GmbH
Europaplatz 3
69115 Heidelberg, Germany

Batch number: 10761599

Printed by Printforce, the Netherlands